D1481133

PARTICLES IN NATURE

THE CHRONOLOGICAL DISCOVERY
OF THE NEW PHYSICS

No. 2616
$23.95

PARTICLES IN NATURE

THE CHRONOLOGICAL DISCOVERY OF THE NEW PHYSICS

JOHN H. MAULDIN

TAB BOOKS Inc.

Blue Ridge Summit, PA 17214

5306-9195

UNIV. OF CALIFORNIA
WITHDRAWN

FIRST EDITION
FIRST PRINTING

Copyright © 1986 by TAB BOOKS Inc.
Printed in the United States of America

Reproduction or publication of the content in any manner, without express
permission of the publisher, is prohibited. No liability is assumed with respect to
the use of the information herein.

Library of Congress Cataloging in Publication Data

Mauldin, John H.
Particles in nature.

Bibliography: p.
Includes index.
1. Particles (Nuclear physics)—Popular works.
2. Nuclear physics—Popular works.   I. Title.
QC793.26.M38   1986   539   85-27710
ISBN 0-8306-0416-2
ISBN 0-8306-0516-9 (pbk.)

Cover photograph courtesy of CERN.

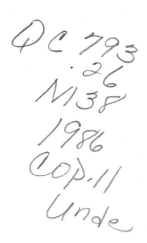
QC 793
.26
M138
1986
cop.11
Unde

Contents

Acknowledgments

APPRECIATION IS EXPRESSED TO THE FOLLOW-ing accelerator laboratories for providing current and helpful brochures, information, journals, tables, photographs, slides, and other materials (more than could be listed in the references): Deutsches Elektronen Synchrotron (DESY), Hamburg, West Germany; European Laboratory for Particle Physics (CERN), Geneva, Switzerland; Fermi National Accelerator Laboratory (Fermilab), Batavia, Illinois; Lawrence Berkeley Laboratory (LBL), Berkeley, California; and Stanford Linear Accelerator Center (SLAC), Stanford, California. Credit must also be given to the U.S. Department of Energy for supporting the American facilities and work discussed in the text and shown in the plates.

Because the number of major particle physics research centers is so large, I regret that it was not possible to contact very many of them or to report on more of their current work.

Besides the Nobel winners and other physicists named in this book, recognition should be given to the thousands of other physicists, mathematicians, accelerator engineers, technicians, programmers, and others around the world who have worked both independently and cooperatively to give us the present particle view of nature.

Appreciation is due, among others, to Paul Davies, author of *Superforce* and his publisher Simon and Schuster for permission to re-express approximately his excellent treatment of the difficult subject of gauge field theory and symmetry.

Appreciation is expressed to the editors and artists at *Scientific American* for publishing many excellent articles and diagrams pertinent to particle physics. In particular, the following diagrams inspired my figures: Feb 1964, p. 78, 81; Oct 1964, p. 39; May 1968, p. 23, 26, 27; Apr 1981, p. 63; May 1982, p. 71.

Finally, appreciation is due my wife Susan for her patient support and assistance, and to the several Nobel winners and other prominent particle physicists whom I was privileged to be in the classes of or to otherwise encounter during and after university studies, and to various friends in the field who shared their studies with me.

Color Plates 1 through 8 were painted by me.

Introduction

ONE WAY OF DESCRIBING PHENOMENA IN NA-
ture, indeed the entire universe, is called the
particle view. Describing nature by means of particles has a very long history, which parallels the
progress of science itself. The particle view is a
philosophical point of view as well as a very useful
theory. This book traces the history of this point
of view from ancient civilizations to the latest
speculations at the frontiers of physics and
astronomy. The emphasis is more on the progress
of ideas than on historical analysis.

Many issues and results in modern particle
physics can be understood without mathematics
because of the power of the particle view. The subject matter of this book is intended to be accessible to anyone having some secondary school
mathematics and science background. The discussion is mostly verbal, aided by diagrams and symbols, and does not rely on mathematics. The most
important mathematical relations used in the
physics of particles are displayed and explained
as important parts of our science-based culture.

Numerical estimates of physical quantities are
given frequently. However, the formal mathematical analysis needed by a physicist is not
given. No prior knowledge of physics is needed to
read this book, and all necessary principles are
presented.

If physics itself is to be defined, one approach
is to state that physics is the systematic investigation of the laws that govern the observed behavior
of particles. Along the way we would have to
define what a particle is and tell what kinds there
are. This is the task that begins in Chapter 1. In
some sense this book is a general physics book that
tries to show what physics is. Particles pervade virtually every aspect of the vast field of physics, and
we will find it necessary to explore many of those
areas. Understanding the most recent discoveries
in particle physics requires knowing about many
of the basic concepts and laws of physics. Chapters
2 through 5 provide a background in terms of particles, and many later chapters fill in background
as needed.

There are other physical viewpoints than one using particles. A major one is the wave model, which was developed later than the particle view and is perhaps less obvious. Almost everything that happens in nature could be described in terms of waves, although the explanations can be quite cumbersome. We shall need to discuss the wave view because some of the behavior of particles is virtually impossible to explain otherwise. Physics can no longer be done in terms of just waves or just particles but requires a certain mixture of the two views.

This book on physics for the general reader, using the particle point of view, has some goals beyond describing the findings and activities of physicists. It is hoped that the beauty and excitement of recent developments in particle physics will be apparent, and that they illustrate the power of science. Recent discoveries almost overshadow the scientific revolutions that occurred at the beginning of this century. Learning how this progress has occurred will not only be a study of intellectual progress and the ingenuity of human minds but will also be of practical importance.

In our increasingly technological society, progress and survival depend ever more on the applications of physics. Most advanced societies have chosen to fund work of increasing sophistication and expense in particle physics. This work promises to have spin-offs of practical value to every citizen comparable to the effects of the space program. Elected officials and ordinary citizens, many without formal study in science, must make ever more complex decisions on what pure and applied research should be funded. We are at a time of exciting promise, when major new but expensive experiments are finding results that may, again, change the human conception of the universe and might even give us new ways to solve our energy problems. Different fields of physical science are joining together, with physicists, astronomers, earth scientists, computer experts, and many others in many specialties finding their work increasingly involved in the establishment of new knowledge about particles. The largest and smallest features of the universe may be about to

be combined in one grand theory whose implications have hardly been explored.

The search for understanding has a long history, traced back to ancient Greece, Sumer, Egypt, China, prehistoric America, and several other civilizations that developed concerns beyond mere survival. This journey of the intellect is illustrated by means of interest in and discoveries about particles as building blocks of nature. Special emphasis is given to contemporary physicists who are constructing the latest revolution in science. It is impossible to give an account that remains up-to-date because there are so many physicists and others pursuing so many new avenues, and no one person can keep up with it all. This book is written at a time when it seems that a major new theory is being accepted, but many details remain to be worked out. Much of the new structure could collapse almost overnight if unusual new experimental discoveries are made.

The impression might be given inadvertently that physical science has progressed in a linear manner—that is, each student of nature builds on the work of known predecessors. This is far from the case both in earlier times and in the present. The history of science contains many branchings, and many dead ends. Scientists in different countries are more in contact now than in past centuries, but it still occurs that some discoveries are made independently when the time is ripe for them. The reader will encounter several cases where recognition is given to two or more people who did similar work without knowing of each other at the time.

Only since 1901 have Nobel Prizes been available for physicists, so some caution is needed in judging relative worth of scientific contributions over the centuries. Galileo in the 17th century, for example, deserved at least one Nobel Prize. As Nobel winners are described during the book, it should be apparent that the trend in Nobel Prize awards is parallel to the search for the ultimate building blocks of matter. The prizes have usually been awarded for progress in finding the deeper and deeper structures of matter. The Nobel Prize is not invariably given to the "best" contribution. Moreover, many physicists will be cited who have not yet received a prize, and many will never be so recognized. Nobel winners whose work is outside the scope of this book are not mentioned here.

Along this intellectual journey a number of philosophical themes are emphasized. First, there is the notion of particle itself, as a building block or elemental constituent. Second, studying the behavior of particles introduces the concept of determinancy—that starting with known particles doing known things, the future course of events can be completely predicted. The absence of determinancy—called *indeterminancy*—turns out to be of great importance. A third common theme is *duality*. There are many complementary or dual theories in physics, such as waves versus particles, indeterminancy versus determinancy, local versus global effects, and so on. A fourth theme is *symmetry*—how things appear the same from several different vantage points. Symmetry occurs over and over in physical theories and provides a way to obtain powerful conclusions from seemingly innocent questions. These and other connections between concepts derived from experimental physics and abstract philosophical issues should help the reader form a broader mental picture of the world and the ways of pondering it.

Finally, the nature of science itself will be illustrated by our journey through particle physics. The way scientific knowledge is established is closely parallel to the way early particle-based views have turned into revolutionary pictures of the whole universe. The division of science into experiment and theory and the way those components work together is well illustrated by the progress in particle physics. The development of the human intellect is impressive when it is seen that careful work with some mathematical symbols can predict a new particle whose existence is confirmed with a hundred million dollars of equipment.

To aid the reader who wants to make the most thorough possible study of particle physics (short of specializing in it in college and graduate school), study questions with answers are provided in the Appendix for each chapter. Those familiar with the basic concepts of physics might be able to pass over the early chapters, but it should be cautioned that a particle theme pervades all parts of the book. The physics of particles is pursued in a unified way that emphasizes understanding rather than mathematical description. The extremely large and small numbers that arise in this subject matter defy intuition and are best understood by accumulating a background in more ordinary physics. The book can be read even if the equations and the few sentences of explanation that follow them are found to be too abstract. The meaning of each equation is also given in words. New terms are shown in italics where they are introduced. Sometimes symbols are used in place of words. Occasionally the same symbol has different meanings in different chapters, and each is defined at the point used. Remarks intended for advanced readers are placed in brackets [ ]. Many recent and nontechnical references that support or extend the subject matter are listed by chapter and placed at the end of the book.

Chapter 1

The Particle View

T HE PARTICLE VIEW OF THE WORLD HAS ONE of the oldest histories in science and human thought. By following the developments that constitute the particle view, we shall encounter much basic science—its subject matter, methods, history, and philosophy. To get started in this discussion we shall need to *bootstrap* ourselves. This metaphor, which is becoming familiar from its use in various technological applications, refers to the fanciful notion that one could pull oneself up into the air by pulling on one's boots. Here, in order to begin to define some basic terms, some concepts pertaining to particles and matter must be mentioned long before they are discussed fully in the proper chapters. While this may sound illogical, it is the way much learning occurs. First a little is learned while much remains confusing. Then some of the confusing details are clarified and learned while new unfamiliar material is introduced. Bit by bit we forage in the field of knowledge until we have mastered a part of it. In this chapter we will acquire a preview of physics as it pertains to particles. The reader will need to

be able to recognize some chemical elements by name (but need not know their properties).

1.1 WHAT IS A PARTICLE?

A *particle* can be defined, for purposes of starting discussion, as a unit of matter. Matter, in turn, refers to the material stuff we see everywhere in our environment, not only the solids and liquids but also the air around us and the distant stars. Matter has weight and occupies space. The earliest humans knew instinctively that they were taking the particle view of nature when they threw rocks, fruits, and other objects either to satisfy curiosity or to accomplish some dimly-perceived purpose. Some parts of the natural world were seen to exist in the form of chunks of material.

This primordial particle theory must have gone further and included the notion of *identity*. When several objects appeared more or less identical, and behaved similarly when dropped or thrown—for example, fruits from a tree, these parts of nature were seen to have identity. Early

1

human perception was especially attuned to the similarities and differences of objects. Those early humans who visited beaches must have seen and handled what nature provides as a good working example of particles: grains of sand of nearly identical properties in great abundance, and very small. Smallness is necessarily associated with particles for several good reasons, as we shall see. Even big objects—coconuts or boulders, for example—have always been observed to act as particles under certain conditions. A thrown coconut follows very much the same path as a grain of sand thrown with similar speed (Fig. 1-1). A boulder falls in the same way as a pebble, although it remained until the 17th century to establish this for certain.

Better than arbitrarily defining a particle is to use what is called an *operational definition*, a philosophical procedure often used in physics. If something has the behavior and qualities of a particle, then it must *be* a particle for all practical purposes. We can use criteria such as identity, size, discreteness, and indestructibility to establish what is a particle and what is not. A tree is not much of a particle until a hurricane hurls it through the air to follow a path similar to what raindrops follow. Water itself is not a particle, but when it is broken into nearly identical small drops, it behaves like particles. A planet is not a particle

to those living on it, but it is a particle as far as its sun is concerned.

Eventually we will be concerned with establishing what are *elementary* particles. The search for basic or elementary building blocks of nature has occurred since ancient Greece, or earlier. One kind of building block we shall encounter often is the *atom*, from the Greek transliteration "atomos" and therefore a concept discussed 2500 years ago. For purposes of discussion we will need not only to know that atoms are the basic units of all common materials but also that they come in a limited number of distinct kinds, called chemical elements. These useful facts were established quite recently in the history of science and will be explained more fully in later chapters. The history of natural philosophy, which eventually became the science of physics, is partly a history of successive refinements of what are the units or atoms of nature—in essence, what constitutes fundamental particles.

The property of being *discrete*, attributed to particles to denote their individuality as units, is also known as the state of being *discontinuous*. If an apparently solid piece of matter is actually composed of numerous small particles, then the supposed *continuity* of matter is a fiction, contradicted by the *discontinuity* of its particles. Viewing matter as being continuous (the same regardless of how small a piece one examines), or as being discontinuous (showing units or particles when examined at a sufficiently small scale) is an example of a duality. These two different views are mutually exclusive: one may believe one or the other, but not both together. Philosophers in early Greece, in pondering whether apparently continuous materials must be made of tiny units or atoms had to suppose the indestructibility of the atoms. Else these constituents of matter in turn could be further broken down. The Greeks apparently avoided considering a never-ending series of particles made of particles made of particles, and yet this could turn out to be the structure of nature, according to 20th century discoveries.

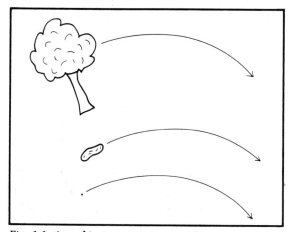

Fig. 1-1. Any object can act as a particle.

1.2 PRE-SCIENTIFIC PARTICLES AND ELEMENTS

In this chapter historical description will only be carried up to a point preceding the establishment of physical theories firm enough to remain valid into modern times. The particle view can be traced back to at least two kinds of early speculation, that on the elementary units of nature, and that on how terrestrial and heavenly bodies moved.

Many early cultures assumed there were a small number of building blocks in the natural world that they observed and used. Babylonians and Egyptians may have been the first to suppose elementary units called earth, water, air, and fire, in approximately 600 B.C. Greek schools of thought, known as Pythagoreans and Atomists in 500 to 400 B.C., elaborated on this theory and treated these elements as "atoms" (in the old sense). Major contributors were Leucippus and Democritus, who established for atoms the requirements of indestructibility, universalness, indivisibility, and presence in an infinite number. The primitive elements were treated in ways similar to our modern chemical elements. Elaborate theories were constructed whereby every observed object, living or not, was somehow composed of these elements. A metaphorical approach was taken, so that, fire, for example, could be identified with "spirit." These elements were thought to be indestructible and therefore always present if not obvious. They could be combined to make forms which seemed to be unlike earth, air, water, or fire, just as the chemical elements discovered later could be combined to form new substances. Aristotle added a fifth element, an ethereal substance, from which the heavens were made. In India the same five elements were considered as atoms. In China the list was earth, water, wood, metal, and fire, but there was no interest in an "atomic" theory. The Islam world worked with the original four elements of the early Greeks.

Early experiments with matter that exposed its particulate nature were done by Strato of Lampsacus, about 270 B.C. He studied charcoal and partial vacuums and was led to suppose that matter was made of particles ("atoms") and voids (empty space). Alchemy, a precursor of chemistry, began about 100 B.C. in the Roman empire. The principle goal became to create gold from other materials rather than to study matter for the sake of knowledge. Some alchemist philosophers, called Stoics, believed that all matter lived and grew and had gender. This was not unlike the theories in China starting about 300 B.C., where people were working quite in ignorance of Europe. Their goal was to create mercury rather than gold. There was more emphasis on a dual set of "influences," Yin and Yang, somewhat analogous to our modern concept of "force." They did not consider their elements to be building blocks as the Atomists did. They did have a strong interest in magnetism, a precursor of the interest in fields in modern physics, and developed the compass. Some cultures, such as China and Islam, believed certain materials, usually mercury and sulfur, were "influences" or "principles," affecting the behavior of matter rather than being elements themselves.

The study of matter did not seem to progress further until the end of the middle ages, when magnetism became of interest in the western world and alchemy was banned for concentrating too much on "spirits." Gradually the earlier Greek theories were "rediscovered" in medieval Europe. An interest in medicine and biology helped revive a study of materials, and empirical methods—emphasizing observation and experiment—were more respected. Helmont in the 17th century set the stage for the beginnings of chemistry and of the physics of gases by proposing that air consisted of several distinct kinds of gases. But unless religious doctrines were incorporated into the newer theories, there was strong opposition from the Christian religion.

Meanwhile, in many early cultures, another form of the particle view was developing. The planets and stars that seemed to move in the sky were seen to be of small size and to move as "bodies" or particles. While some cultures may have imagined them as large in some sense, almost no one realized that astronomical objects were at great distances and must be very large. Only the sun and moon showed discs of some size in the sky. It was only known that they were well out of reach from the highest available mountains. Foundations for modern astronomy began in cultures as diverse as Sumer, Egypt, Greece, China, and native America. Either an explanation of how the planets and stars were related to the creation of the world, or a systematic study of their motions was made, or both. Sometimes the results were used for religious or philosophical purposes only. Impressive monuments were constructed that supposedly marked sightings of the motion of the sun, moon, planets, and some stars. In some cases precise observations stimulated further developments in mathematics and the sciences. What was lacking for thousands of years was physical theory to explain planetary motions and other aspects of the meticulous observations.

Astronomical observations accumulated for many centuries in the Mediterranean area, western Asia, China, and India. Basic astronomical information had been established by Babylonians. Eudoxus of Cnidus (Greece), around 400 B.C., proposed that the sun, moon, stars, and all the other planets revolved around the Earth using circles and spheres. Archimedes of Syracuse made a model of the solar system (a particle model). Aristarchus in Alexandria proposed rotations and revolutions of the planets around the sun almost as we now know them, but his view was not accepted. He also estimated sizes and distances for these bodies. Hipparchus at Rhodes tried to improve the earth-centered system with eccentric orbits. Ptolemy at Alexandria (in approximately A.D. 150 and perhaps the best remembered Greek philosopher of the heavens) expanded upon this view with epicycles. Ptolemy was one of many who also tried to estimate the size of the Earth, indicating that it was thought to be a huge sphere. The Chinese worked on two different views—that the heavenly bodies were in infinite space and that they moved on shells surrounding the Earth. In India early Vedic texts placed the sun at the center of the solar system and identified it as the

source of energy. Later, the Hindus adopted the Greek view that the planets orbited around the Earth, including epicycles. The Muslims preferred Eudoxus' earlier view.

In the middle ages, under the influence of Christianity, the view of the Earth returned to flatness and a central position in the universe. That the Earth could move as a particle seemed unacceptable until much later. Copernicus in northern Europe in the early 16th century put the sun back in the center of the universe and restored rotation to the Earth as it moved around the sun. This *heliocentric* view, named for the Greek "helios" for sun, is an early step in breaking away from a human-centered view of the universe, so important in modern theories of how the particles of the universe must behave. Later that century, building on new more accurate observations of the solar system by Tycho Brahe and further simplifying Copernicus' model, Kepler provided a mathematically-based description that set the foundation for modern physical theories. His laws were published in 1609 and 1618, some of the earliest in physics or any science. They stated that the planets follow elliptical paths and gave a simple relation between a planet's distance and its period of revolution. All of these natural philosophers had explicitly or implicitly assumed that the planets and stars were particles which remained discrete and followed definite predictable paths.

Throughout this history there were occasional developments in how ordinary terrestrial bodies (unconsciously idealized as particles) were thought to move. Aristotle was among the first to explore this area. Philoponos in 6th century Alexandria reaffirmed the idea that terrestrial or heavenly bodies could move on their own, not needing the constant guidance of "angels." The "impetus" theory was born: a push (or force) would start a body in motion, giving it a supply of "impetus" that would keep it moving for a while until the supply was exhausted. On the other side of the world the Chinese had invented gunpowder and the rocket (about 1000) and had the first self-propelled freely-moving bodies. In the middle ages actual measurement of motion was begun, and ways to show it on paper, with graphs, were developed. William of Occam (14th century), while believing in impetus, also proposed the possibility of "action at a distance", a modern concept that ultimately helped undermine the older theory.

The impetus theory was well-entrenched until definitive experiments led to the first mechanical theories in physics. It has been recently realized that the "impetus" theory is still widely believed by college students today, including some majoring in physics. The reader is cautioned that some effort and an upgrading of "common sense" may be needed to master the modern theories given in later chapters. It is understandable that earlier theories about motion and other aspects of nature were first invented because they seemed intuitively correct. The processes of discovery and education include the altering of one's intuition in accord with the results of observations and intellectual understanding.

1.3 MEASURING LENGTH, TIME, AND MASS

Size and certain other properties are very important if matter is composed of particles. Particles have definite effective sizes, however small. The size is the same everywhere for the same kind of particle. Continuous matter, if it were to exist, would appear the same no matter what size piece is examined.

Traditional units of length such as inches, feet, and miles are cumbersome to deal with, both in converting sizes and in dealing with the extremely large and small. What is the real meaning of 0.0000000000001 inch, still a rather large size compared to some known phenomena?

In addition to better units, a reference size is needed, one that ideally anyone can obtain and use to any degree of accuracy. Our current system of scientific measurements is based in part on the kilogram. To compare with a standard kilogram, one must handle it in some way. This is bound to rub off a few of its tiny particles, diminishing the size. Many other things can go wrong with such

a standard, and a better one is needed. For physicists' purposes it is quite a disadvantage when the standard does not involve particles directly.

We will want to describe particles in terms of the most basic units of measurement, involving length, time, and mass. We will want to know their size, weight or mass, how fast they move, and other measurements to be introduced later. Each property will be measured in units based on a universal standard, just as a conventional yardstick is useless unless it is the same length wherever purchased. Ideally, the standards should be available everywhere, should be inexpensive and indestructible, and should permit great accuracy. The trend is to use standards based on atoms, since they are cheap, tiny, very durable, and give exactly the same results no matter which particular atoms of a given element are chosen.

The units of measurement should form a consistent set. English units such as feet, inches, gallons, pounds, and so forth all had different origins but modern necessity has forced them to be defined consistently. For example, a gallon should contain so many cubic inches of volume or hold so many pounds of a given substance such as water. Scientists this century have developed a highly consistent system of units called SI (for the French "Systeme Internationale"). It was adopted fully at a conference in France in 1971 but is continually updated by further international conferences. Some of the units in the system have been in use for centuries. For example, the meter was first defined in France in 1795 as the ten-millionth part of the length of the meridian line from the north pole through Paris to the equator.

The universally accepted units for length, time, and mass (together with their abbreviations) are now the meter (m), second (s), and kilogram (kg), respectively. These may be referred to casually as "metric" units, but the term "SI" distinguishes them more carefully from another set using the centimeter and gram. Today only the U.S. and a few small countries are not using some form of metric units on a daily basis, aside from science and technology.

From 1875 to 1960 the standard meter existed as a platinum-iridium bar with two marks on it, stored in Paris. The modern standard for length was redefined in terms of a certain tiny length traceable to fundamental properties of a certain kind of atom (of the element krypton). Ideally these atoms should behave as simple particles, but they can be made to emit light. The light is so pure that a length property of it (the wavelength, to be explained later) can be measured very accurately. Just to illustrate the accuracy possible, the meter is defined as 1650763.73 of those wavelengths. However, science and technology have a habit of needing ever greater accuracy, so work on standards such as these continues.

Just as particles come in a variety of sizes (measured as length), so particle behavior involves a wide range of time scales. Until recently the unit of time, the second, has been defined as 1/86400 of one solar day, the time for the Earth to rotate once to the same position with respect to the sun. In 1967 the second was redefined in terms of an atomic standard. Microwave radiation from atoms of cesium vibrates very fast, and the second is defined to equal 9192631770 of those vibrations. This number of about 9 billion vibrations in a second indicates how rapidly events occur within atoms.

To specify how much material is packed into a particle, and to discuss its weight, we need to know its mass. The official unit is the kilogram, an unfortunate name since "kilo-" is a special prefix denoting 1000, as we shall see soon. But the gram is not the SI unit. The kilogram has been in regular use since 1795, and the most recent standard was constructed about 1875. It is a platinum-iridium cylinder kept in Paris. It can be compared to other masses to one part in fifty million, not quite as good as the standards based on atoms. Measurements are traced back to this standard by means of secondary standards made and then located in each country.

There is another kind of mass standard which is pertinent to particle physics, the one based on the mass of a certain kind of carbon atom called ^{12}C. Using techniques to be described later, the masses of other atoms can be compared with ^{12}C

very accurately. However, the comparison of ^{12}C to the standard kilogram is not so good. We do not know the mass of a ^{12}C atom better than about 1 part in ten thousand, when measured in kilograms.

It will be useful to combine two of the basic units in order to define another property of particles. This procedure can be carried out with many of the units in many combinations, but only a few combinations are helpful in an obvious way. The first concept of interest is *density*, defined as the mass contained in a unit volume. We are more familiar with density in an ordinary way than we are with ways of measuring it. A box of air weighs less than a box of sand (as illustrated in Fig. 1-2 where the weight of the boxes should be ignored), yet the use of the same size boxes shows that we are comparing the same volume. A kilogram of air weighs the same as a kilogram of sand. We know that sand seems compact or dense and that air seems thin or tenuous. A lot of air is needed to make a kilogram. How is "a lot" to be measured? We need a certain and much larger vol-

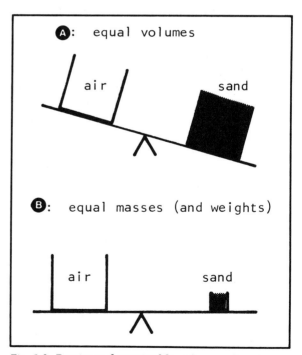

Fig. 1-2. Density as determined by volume and mass.

ume of air to possess the same amount of matter as a certain volume of sand.

What we wish to compare are the densities of substances such as air and sand. To do that we need to define volume in terms of basic units. A cubic meter can be a cubical volume measuring one meter along each of its three dimensions. To get a feel for this basic volume, imagine a box you can barely get your arms part way around. One of the most common metric units in general use (not SI) is the liter, a volume one-thousandth of a cubic meter. The technical way to express this would be: 1 liter $= 10^{-3}$ m^3.

Density can be defined as kilograms per cubic meter, or kg/m^3. Water, many of whose properties are intentionally round numbers in our unit systems, has a density of 1 pound per pint, or 62 pounds per cubic foot, or 1000 kilograms per cubic meter. Air at sea level weighs .075 pounds per cubic foot and has 1.29 kilograms of matter per cubic meter. We shall learn that elementary particles are extremely dense, a result of combining their tiny mass with their even tinier volume. Later we shall be more particular about the distinction between weight and mass; strictly speaking, density should be in terms of mass, the amount of matter.

1.4   SCALES OF SIZES IN NATURE

The unit of a meter is less than ideal for working with the smallest particles because it is so much larger than them. It is also not expressed directly in terms of particles. Scientists wish it were possible to say that, for example, 156784736506793 protons (a very common elementary particle) placed in a row would be the length of a meter. But the size of the proton is not known very well, and available methods of counting protons, not to mention lining them up or an equivalent procedure, are quite inadequate.

Whether we use meters or feet, we will have problems with very small and large objects. Instead of using lots of zeros to express, for example, the fact that the universe is about 100000000000000000000000000 meters big at present, scientists have developed a simple and elegant way to write large and small numbers. We shall find it essential for dealing with particles (and much else). The mathematical basis for expressing size in what is called *exponential* or *powers-of-ten* form is the fact that 10 to a whole number used as a power expresses a 1 followed by that number of zeros. A simple case is 10^1, which equals 10. The above stated size of the universe can be written with just four digits in the appropriate arrangement: 10^{26} An less obvious case is 10^0; this is equivalent to a 1 followed by no zeros. For very small numbers we have the rule that putting a minus with the numerical power or exponent is equivalent to a decimal point with one less than that number of zeroes before the 1. For example, 10^{-2}, which is read as "ten to the minus two" has the decimal value 0.01. The smallest particles for which size is reasonably well-established are about 10^{-15} meter or .000000000000001 meter across.

There have also been defined ways to express these numbers with words. Table 1-1 lists the prefixes and their abbreviations that express powers of ten. The traditional names are also given. The prefix is attached to a unit such as meters. For example 10,000 meters can also be written 10^4 meters or 10 kilometers or 10 km. Prefixes are provided only for each group of three powers of ten, giving fewer ones to remember. One of the most useful for discussing particles is the femtometer (fm) because it is approximately the size of a proton.

We shall usually find it sufficient to talk in terms of round powers of ten, also known as *orders of magnitude*. But when specific basic facts about nature must be given, a series of digits must be expressed. To state, for example, 45 km in powers of ten, one would write $4.5(10)^4$ m, or 4.5×10^4 m (or 4.5E4 in computer notation). To express a small size, for example:

$$7.5 \ \mu m$$

Strictly speaking, one is permitted only one digit before the decimal point to avoid difficulties in expressing accuracy. Using the powers-of-ten notation provides an easy way to indicate the ac-

Table 1-1. Powers of Ten Notation.

Power	Prefix	Symbol	Old name	Explicit form
10^{18}	exa-	E	quintillion	1000000000000000000
10^{15}	peta-	P	quadrillion	1000000000000000
10^{12}	tera-	T	trillion	1000000000000
10^{9}	giga-	G	billion	1000000000
10^{6}	mega-	M	million	1000000
10^{3}	kilo-	k	thousand	1000
10^{0}	- - -	- - -	unit	1
10^{-3}	milli-	m	thousandth	.001
10^{-6}	micro-	μ	millionth	.000001
10^{-9}	nano-	n	billionth	.000000001
10^{-12}	pico-	p	trillionth	.000000000001
10^{-15}	femto-	f	(etc.)	.000000000000001
10^{-18}	atto-	a		.000000000000000001

curacy. The number of digits given is called the *significant figures* for the number, and the requirement of only one digit before the decimal point eliminates any ambiguity about the significance of those that come after. For example, 4.5000 $(10)^4$ shows that we know this number to an accuracy of one part in nearly a hundred thousand, since the final zero occupies the fifth place in the number 45000. Using just 4.5 indicates an accuracy of one part in a hundred.

Calculations with powers of ten are easy when an order of magnitude estimate is wanted. When numbers are multiplied, the powers are added; when numbers are divided, the power used in the denominator is subtracted (algebraically) from that in the numerator. For example, if about 10^4 kilograms of sand are in a volume of about 10^{-7} cubic meters, the density can be calculated to be $10^{-4}/10^{-7} = 10^{-4-(-7)} = 10^3$ kg/m^3. In calculations involving squaring or cubing, the powers of ten are multiplied by the appropriate exponent. For example, a cube 10^2 meters on a side has a volume of $(10^2)^3 = 10^{22 \times 3} = 10^6$ m^3. Numbers can be added and subtracted only if they have the same powers.

$$10^6 m^3 \times 10^2 \times 3(10^2)^3$$

Looking at a particular range of sizes is known as examining the *scale*. We humans, who are about 1.5 meters in size, cannot take in at one glance the full range of sizes that nature seems to have. The stars we see in the sky are far from being the largest objects that exist. Instruments and photographs are needed to show us galaxies of stars and groups of galaxies. If we did have a picture of a galaxy, we would barely be able to see individual stars and could not see any details of the stars or their planets. If we limit ourselves to viewing the Earth, either personally from space or in a photograph, we cannot make out the numerous cities and roads that humans have laid across the surface. If we are viewing a city of buildings and houses from afar, we cannot make out the faces of its inhabitants, much less the print in the books some of them are reading. This restriction continues to smaller and smaller scales. In a picture of a bacteria, the smallest object that can be photographed well with ordinary light, we cannot see the atoms and molecules that comprise it, nor can we see the incredibly fast blur of activity that goes on within those atoms. Only instruments can inform us about the realm of the tiny, where the elementary particles are.

When given a picture, we make assumptions about its scale unless the scale is provided numerically in a form we can understand. As the structure of the universe has been studied, it has been noted that matter is more apparent and more densely packed at some scales than others. In Fig. 1-3 the *heirarchy* of the organization of matter is illustrated, from the very large to the very small. The sizes of particular objects are shown in powers of ten, with the units in meters. It should be noted that such pictures are printed at some size other than the size indicated numerically in the picture; they are scaled.

Some of the objects depicted are also listed in Table 1-2, where additional facts are given to better orient the reader to the enormous variation in the numbers. Most of the quantities are outside ordinary human experience. The mass, density, and number of protons in it are given for typical objects. The latter fact is to remind us of our interest in the fundamental building blocks of matter (principally protons). The density—telling how tightly matter is packed—will be seen to vary from very tenuous at large scales to very dense at small scales. Ordinary matter, and even stars (including their dense cores), have about the same density, compared to the extremes. (Objects such as neutron stars and black holes have extreme densities and defy the general pattern.)

In a few mighty leaps, as if the viewer had zoomed toward each picture in turn, the scale of the universe is illustrated in Fig. 1-3. We see that everything cannot be shown in one picture. An object filling one diagram is a minute speck in the next. On the largest scale the universe is populated with clusters of galaxies of stars. Typical clusters are about 10^{24} m across. There even seem to be superclusters of clusters forming the whole known universe (about 10^{26} m). The widely separated galaxies are about 10^{21} m in size. Stars are separated by enormous amounts of relatively empty space and are typically about 10^{17} m apart (0.1 exameter, using the largest prefix defined so far). Our planet, about 10^{7} m in diameter, seems rather solidly filled with matter, as does a human (about 1 m) and even a grain of powder or a bacteria (about 10^{-6} m). When we reach the scale of atoms (about 10^{-10} m), there again seems to be mostly empty space. At the smallest scale elementary particles seem to be hard and dense; their effective size is about 10^{-15} m.

For very large scales distances are also given in the popular unit *light year* (L.Y., not an SI unit), the distance light will travel in one Earth year. One L.Y. is nearly 10^{16} m. The stars are separated by about 10 L.Y., and organized into galaxies, whose size is typically about a hundred thousand L.Y. About ten million L.Y. separate the galaxies in a cluster. The observable universe is found to be about ten billion L.Y. in extent, a size difficult to conceive regardless of the units used.

The reader may wonder why the discussion often involves galaxies when we should be considering particles. Particles and galaxies are related in three ways: (1) Matter, however it

Table 1-2. Hierarchy of Matter.

Object	Size (m)	Mass (kg)	Density (kg/m³)	Number of protons	Time scale (s)
Universe	10^{26}	10^{51}	10^{-27}	10^{77}	10^{18}
galactic cluster	10^{24}	10^{43}	10^{-25}	10^{69}	10^{16}
galaxy	10^{21}	10^{41}	10^{-21}	10^{67}	10^{15}
star (normal)	10^{9}	10^{30}	10^{4}	10^{57}	10^{11}
planet (Earth)	10^{7}	10^{25}	10^{4}	10^{52}	10^{7}
mountain (big)	10^{4}	10^{15}	10^{4}	10^{42}	10^{3}
human	1	10^{2}	10^{3}	10^{29}	1
bacteria	10^{-6}	10^{-15}	10^{3}	10^{12}	10^{-6}
atom	10^{-10}	10^{-26}	10^{4}	1 to 100	10^{-14}
proton	10^{-15}	10^{-27}	10^{18}	1	10^{-23}

Notes: All quantities are shown approximately, within an order of magnitude. The time scales are for typical motions or processes; in some cases a wide range of possibilities exist. They are not lifetimes.

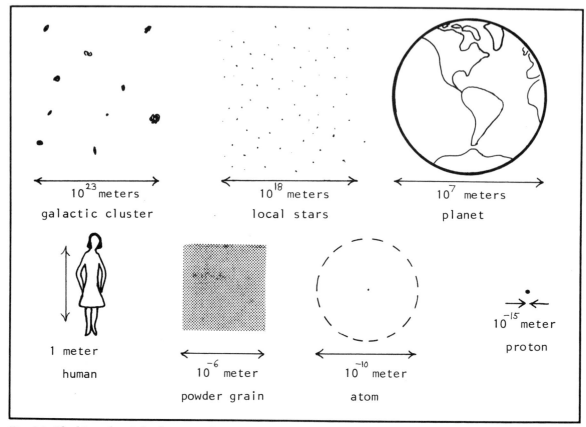

Fig. 1-3. The hierarchy of the clustering of matter at many scales of length.

clusters at different scales, is made of particles. (2) Each cluster or object can behave as a particle if considered at the appropriate scale. (3) The properties of the smallest particles determine the properties and behavior of the largest clusters of matter, even the universe itself. A major theme of this book will be to show how this can be so.

We can perceive the behavior of matter over a limited range of time scales, from seconds to decades. All else depends on instruments. Indeed, physics regularly manages to measure times down to 10^{-24} s. Near the end of the book we shall see that particle behavior in 10^{-43} s is important to the universe. At the other end, the universe has been evolving for about 10^{18} s (about 10^{10} years) and physicists are discussing up to 10^{120} years! Figure 1-3 also indicates typical times for the movements characteristic of the objects shown,

over the whole range of the universe. At the small end, particles move very fast and take very little time to cross the distances pertaining to them. On our scale of ordinary life, speeds and times are moderate. At larger scales, planets and stars move much faster, galaxies faster still, and the universe is expanding in a very rapid manner. We do not see much action because it takes so long for an object to cross the immense distances. The visible stars change their arrangement only slightly in thousands of years (about 10^{11} s), although they are moving more than 10^4 meters each second in various directions.

1.5 LIMITS TO HUMAN PERCEPTION

It has been indicated that we are limited in our ability to perceive directly the enormous range of sizes and time scales that characterize the universe. We have other perceptual limitations, and most progress in science has hinged upon the development of suitable instruments and procedures. We can never experience directly the enormous weight of the Earth or Moon, although we feel their influence directly as gravity all the time. We do not dare put our hands in very cold liquids or in fires, but simple instruments tell quickly and precisely their temperatures. We cannot sense electric charge or current very safely, especially not over the enormous range found in nature, but scientists and engineers are very good at measuring these. We can only see colors of light from deep red to deep violet. Other kinds of light and radiation exist and are very useful but are either completely imperceptible or would cause us injury. Scientific instruments, when used properly, are said to be *extensions* of our senses.

We are not so limited as to have only the five senses that are commonly cited. Besides sensing visible light, hearing a certain range of sounds, tasting sweet, sour, salt, and bitter, and distinguishing tens of thousands of substances by smell, we can sense pressure, cold, hot, vibration, our orientation and motion, electric current, gravity, the positions of our limbs, the tension of most of our muscles, and many subliminal aspects of our functioning bodies, such as whether there is too much carbon dioxide in our lungs. There are at least thirty distinct senses, and possibly more to be discovered. Admittedly we cannot make good numerical measurements with any of them, and a branch of psychology called psychophysics has been devoted to finding ways to quantify what seems subjective to us. Science requires *objective* measurements—ones which can be agreed upon or found independently by anyone. Indeed, a robot could make objective scientific measurements. But human senses produce only *subjective* data; it is very real in some ways but nevertheless cannot be accurately shared between any two people. A flower will look, feel, and smell differently to different people.

Information in our minds that seems very precise, such as numbers or simple symbols, can be shared only by a four stage process. First it must be represented or coded in terms of some physical variable. Then it must be transmitted as some disturbance in the physical world, for example, sound vibrations in the air during talking. The physical disturbance must be carried to the other person. There, sense organs (ears) receive the signal and change it into a form that then—somehow—becomes part of consciousness and is interpreted. Note the intended use of the word "interpret." What the other person receives is never exactly what was thought to be sent. In the purest case, speaking a number in English to another English-comprehending person tells the other person very little until the context is known. The number must refer to objects, have units, or otherwise be in context. Here is where the possibility of misunderstanding comes in. Scientists, of course, do not rely simply on oral communication to acquire and expand the body of knowledge. Scientific knowledge is recorded in the most unambiguous ways possible, usually in books and journals written in relatively unambiguous styles and accessible to all who have the assumed background.

Besides the fact that some particles are too big and too far away to study directly (galaxies), or that some are too small (protons), there are other more perplexing limitations on direct human observations. At one level of complication, the use of measuring instruments tends to change that which is being measured. A thermometer that was initially at room temperature cools off the hot drink into which it is immersed and reads something less than the original temperature of that drink (Fig. 1-4). There are ways to make instruments that cause very little change in what is being measured. But there are fundamental limits to how accurately anything can be measured, and these limits are often caused by the particulate nature of matter. If, as we shall learn, warm materials possess warmth by virtue of their atoms dashing about, then a very tiny thermometer that is designed to change things imperceptibly, encounters individual atoms striking it and responds erratically rather than smoothly.

Nature is even more peculiar than that! Attempts to measure one aspect of an atom or

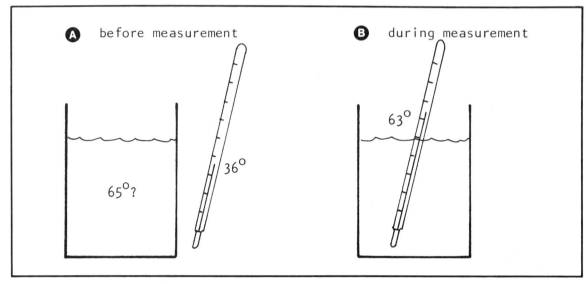

Fig. 1-4. *An instrument changing the property being measured.*

particle very accurately—a goal that can be achieved—cause another aspect to be totally unmeasurable. We need not be thwarted in our attempts to measure the speed of a thrown baseball as it crosses home plate. But in the microworld such a procedure of getting the speed and position simultaneously is absolutely impossible for a tiny particle. In another realm, we have no trouble keeping our watches in rather good synchronization when one person is on the ground and the other in a jetplane. But when much higher speeds are involved, strange difficulties arise whether we are talking about atoms, watches, or galaxies. We shall see later, when the modern view of atoms and particles is covered, that there are unresolved problems as to whether the presence of human consciousness—in the form of a subjective human observer—affects the way supposedly objective nature behaves.

1.6 SCIENCE AND ITS METHODS

The relative position of physics among the sciences is shown in Fig. 1-5. While the exact arrangement cannot be fixed, or shown in a simple manner, the major sciences are built on one another. Physics is at the base, establishing the fundamental laws and theories for matter and energy.

Chemistry studies the interactions of the different materials and elements. It assumes the laws of physics but develops knowledge at a more complex level of interaction. Physics has been gradually taking over chemistry as theories of matter have become more elaborate and inclusive. Biology needs the laws of physics and chemistry but works with still more complex and abstract forms of matter. There is a major step in the way matter is organized in the leap from nonliving to living material.

The global aspects of biology blend into earth science, the study of all aspects of our complex home world. But the Earth is hardly more than a particle in the field of astronomy, which explores as far as space seems to go and finds many uses for physics. Beyond biology in a different direction comes psychology, the study individual human behavior. Beyond that, humans organized as a society give rise to new phenomena not apparent at lower levels and studied in sociology. Names and organization may change, but this hierarchy or tree of sciences is expected to keep growing and branching, with ever more interconnections. Even when no connection is obvious, methods and mathematics first used in physics (or another field) often are found to apply elsewhere.

No attempt is made here to show how the technologies branch from the sciences. Mathematics is not shown because it is not a science in the usual sense.

A preview of scientific methods should help the reader in thinking about the process by which knowledge of particle physics is established. Sometimes called *the* scientific method, it was never explictly established by some committee or individual. Many philosophers and scientists have, in their own independent ways over many centuries, arrived at a consensus of how science should be done. It is comforting—indeed should be inspiring to the human species—that this is one endeavor where people can pursue their own ends, and the result is progress rather than chaos and anarchy. The essence of nature seems to be that, when every effort is made to understand nature, a process of unification and ordering comes about. This order not only has an intrinsic beauty but also has an effect on other human affairs. Unfortunately, the effect is not pervasive, given the current levels of greed, violence, and ignorance in all parts of the world.

Some scientists achieve prominence, winning prizes, advising governments, and carrying on socially beneficial projects beyond the call of duty to science. Many others are never heard about but seem to make altruistic contributions to science and society. Science itself can be a lonely or a selfish pursuit. Rarely do teams of scientists actually do the innovative work. Some scientists have the philosophy and integrity of science in their hearts as they work. Some simply do what they have learned to do. One usually cannot tell from a published research paper what kind of person did the work, and often only the scientist's peers can tell if the work was good or bad, useful or irrelevant or wrong. A few scientists go astray and intentionally publish falsified work. Regardless, the way science works as a community enterprise invariably filters the good from the bad, avoiding blame but honoring new and useful work. Every scientist has made mistakes along the way to truth; the quality of the work of recent scientists is yet to be fully judged. Nature is the final arbitrator. Usually, but not always, those who have made any lasting contribution will be remembered.

1.7 MODELING NATURE: LAWS AND THEORIES IN SCIENTIFIC METHOD

The scientific process starts with *observation*. In the past it was sufficient merely to name and describe what was found in nature. Scientists classified the information, looking for patterns, more so in the biological than the physical sciences. New phenomena must be measured very accurately in every way that seems important. The amount of experimental error must be ascertained. Results, even before they are understood, must be *replicated*—that is, repeated by other independent scientists. Comparisons can be done meaningfully only if the amount of error is known for each measurement. The first new but careful and reasonable observation is likely to attract fanfare when published. But if no one duplicates the result, that is the last one may hear of it, despite the

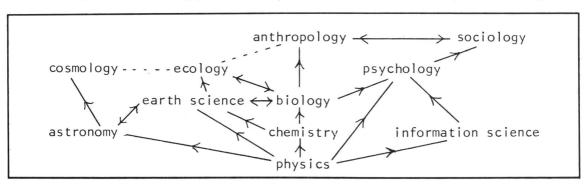

Fig. 1-5. The hierarchy and the interconnectedness of the sciences.

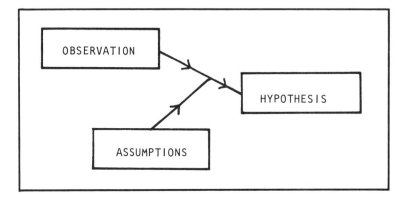

Fig. 1-6. A simple scientific method.

fact that journals show little interest in publishing replicated experiments or those with null results.

As has been discussed, observations must be objective. A person who thought (subjectively) that a certain elusive phenomenon was seen will get no scientific notice. If a photograph is provided, or a graph of data taken under controlled conditions, then it is publishable. Occasionally good observations are overlooked. Several scientists of repute who each witness some new elusive phenomena are likely to start a systematic investigation, especially if they know of each other's experience. Sometimes useful observations accidentally get notice even though their actual value at the time is unrecognized. One individual's evidence cannot help being subjective and therefore may be wrong. But finding objective tests can be a challenge. Kuhns wrote a study of science introducing the notion that "revolutions" are part of science. There is some inertia in science. When the basic prevailing theories seem well-established, groups and journals are biased against giving attention to some new contrary fact. The existence of "conventional wisdom" or "normal science" is a debated area in the philosophy of science.

As observations accumulate in a certain area, a working *hypothesis* will be proposed (Fig. 1-6). This hypothesis states that under certain conditions such-and-such always happens. Some of the assumptions that are needed to state the hypothesis are recognized. The process of jumping from observations to hypothesis is called inferring. The result can be called an *inference*, but if further testing is planned it is better known as an hypothesis. If the exceptions to the rule are none or minimal, work proceeds. *Experiments* may be done, in which the conditions are artificially arranged and observations are made. The inferring or hypothesizing process is mainly one of *deduction*, a rational process where the conclusions follow from the initial data and assumptions. It is cut-and-dried and has even been done by computer.

For example, suppose that people working in different places all find that an ice cube weighs less after it melts to water, and that the accumulated results average to some number larger than the weighing error. The inference is that melting ice leads to or is associated with—note that no one has said "causes"—a weight decrease. An average numerical change can be stated, and can be expressed independently of how much ice each scientist started with. The implicit assumption is that no one did anything else to cause the weight change and that no one observed some material to leave the containers used. The scientists should make their assumptions explicit and state either that they observed no material to go elsewhere, or that they were unable to check this aspect (perhaps it cost too much), or perhaps that they forgot (!) to check it. For those who feel further checking is justified, the inference is an hypothesis, and experiments are arranged accordingly. Accuracy is improved and all stated assumptions are examined.

If the expected outcome occurs, and if enough data is accumulated from experiments and observations, a *law* may be proposed, as part of a more thorough scientific process (Fig. 1-7). Physicists and most other scientists prefer to state laws in mathematical symbols where possible. The exactness is justified in physics but may be misleading in the "softer" sciences. The goal of experimental work is to state a natural law based on evidence. Such natural laws are not irrefutable. Anytime a number of scientists can independently agree that the exceptions are the basis of some other better law that covers both the original inference and the exceptions, or that most of the original experiments were wrong, the law is replaced or eliminated. In the above example, it might be observed during the experiments that there is condensation on the containers, or that there is more weight change the longer one waits. In the former case, the weight loss expressed in the law seems dead wrong, since additional material condenses from the air onto the experimental container. Rethinking and new work is recommended. In the latter case, someone might examine the previously overlooked assumption that there was no evaporation and redo the experiment in a sealed container. If that shows the weight change to be less than the magnitude of experimental errors, the law is again dead.

The reader may have realized that nothing has been said so far about *explaining* what was observed. No mention has been made of theories. A law is usually not much of an explanation. We have laws for gravity in elegant mathematical form. Except for what one might read into the symbols or quantities used in the laws, they do not explain how gravity works. A full *theory* for gravity will include not only the mathematical law but also the description of how to use it and the background connecting it to other physical concepts. This is a fuzzy area to consider, because what constitutes an explanation to one person, physicist or not, may not be an explanation to someone else. A law only summarizes what is objectively known. A theory goes further and contains possibilities not yet discovered or verified. A good theory not only explains all that we know within certain stated restrictions but also predicts some new phenomena. If experiments are done and the new predictions are confirmed, the theory is verified. It stands until someone comes along with a better one, usually goaded by new observations that were not predicted by the old theory.

The characteristics of a good theory are more elusive than the methods of experimental science. The theory often is *simple*. When given a choice between a complicated and a simple theory, scientists and others are likely to choose the simple one, a process called using "Ockham's razor." As discussed, the theory should be *predictive*. We shall find that the current revolution in particle physics is due partly to the finding of new particles based on predictions from perhaps the most abstract and complicated theory ever created. (No simpler one has arisen yet to do the same job.) Also, a good theory should explain things in a way that seems intuitive to an experienced scientist. It would be nice if the explanation is accessible to everyone else as well, but this is less and less easily accomplished in physics.

Where do theories come from? Nature only

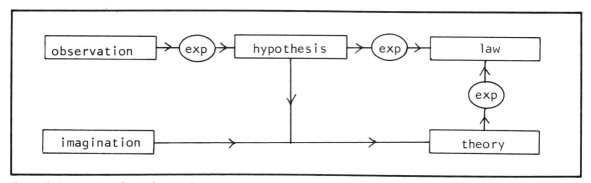

Fig. 1-7. A more complex and more thorough scientific method.

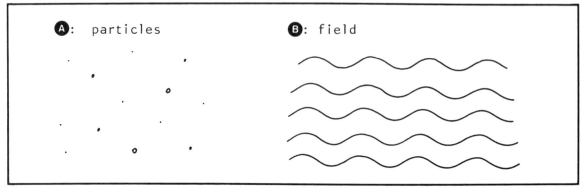

Fig. 1-8. Particle versus field models.

hints at possibilities and does not provide theories. They are literally dreamed up by human imagination (see Fig. 1-7), in a process called *induction*, which is a somewhat organized form of intuition much informed by experience. The human mind is a great hypothesis generator, coming up with all kinds of nonsense based on incomplete evidence. This is the origin of superstitions and myths and the basis for the apparent irrationality of mental processes. Subjecting the new ideas to experiments and logical examination helps select plausible theories.

Some very important work has come from "thought experiments," where one carries out a mental experiment, usually testing an extreme case. Einstein made enormous breakthroughs as a theoretician. His mental "experiments" used hypothesizing, metaphors, extreme cases, reasoning, and other less tangible ingredients. For example, in his youth he wondered what would be the consequences if a scientist moving at high speed had to observe exactly the same phenomena as one supposedly stationary. His working this out mathematically resulted in a major revolution in science and human affairs.

1.8 PHYSICAL MODELS

The general theoretical results from applying several centuries of scientific method in physics are known as *models*, a term different from "theory." This introduction to the physicist's tools is concluded with a brief description of the prevail-

ing models. We are most concerned with the particle model. The other major model is the *field* model, which we will not be able to avoid discussing as it pertains to particles (see Fig. 1-8). It is sometimes called the *wave* model. One commonly experienced field is the gravitational field for the Earth. It cannot be seen or touched, yet its influence is felt everywhere in a rather consistent way. While particles and fields are a duality and have well-established validities, there is no reason why other major models might not arise to accompany or replace these two.

Models should not be confused with the languages used in science. The reader probably handles most information in general English, perhaps aided with some diagrams (the level of communication of this book). Physical scientists use more abstract diagrams (which will be shown), a language filled with jargon (names for new concepts such as the ones in this book), many shorthand symbols (more names), and most importantly, mathematical description. Where the government likes to use three- or four-letter acronyms to name agencies or programs (for example, NASA), physicists like to name each quantity with a single letter (m for mass) and then calculate with these symbols. The mathematical part of the discussion in this treatment will be limited to showing what major physical laws look like in mathematical form, and to explaining what they are trying to say.

The particle and field models have subtle connections. Beginning students of physics may have learned that waves are the "opposite" of particles. But the complement to particles is actually fields, and the notion of "opposite" has no clear meaning in this case. We shall see that fields are the more fundamental concept and are the medium that "waves" as certain waves travel. Some waves occur in a collection of particles, or in seemingly solid matter. Particles and fields are dual views, corresponding respectively to discontinuous and continuous views of matter. In reality, matter can only be discontinuous, although it may be modeled as continuous. Fields are defined as continuous, although certain discontinuous aspects will arise. Particles may be abstractly represented as dots or points, fields may be represented as lines, straight or curved (Fig. 1-8).

Taking the particle view does not give us much of an explanation of how things work, so particles in themselves are not theories. A particle by itself is simply an abstract concept. It is true, for example, that we can predict some new properties of air if we view it as made of idealized particles, but the theory lies in the process of working out the results. It would best be said that the particle view of air is a good model for air. The reader may have heard the term "particle theory" applied to elementary particle physics. In this sense the term refers to the models used, together with the processes assumed and their mathematical description. Rather than continue here with trying to define laws, theories, models, and so forth, we would best proceed with the body of laws and theories that are physics and learn further by example.

Chapter 2

Forces, Motion, and Energy

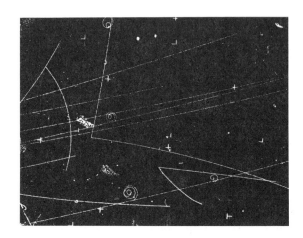

T HE STAGE WILL NOW BE SET FOR DISCUSSING the frontiers of modern particle physics by providing the basic physical background needed to understand the more advanced theories. This will require several chapters, each building upon the preceding. This chapter covers, literally, the "mechanics" of physics—what forces are, and how they are related to motion and energy. The subject is called "classical mechanics" to distinguish it from more recent forms of physics. Some useful modern principles for particle physics will emerge from this basic "bread and butter" physics.

2.1 THE FUNDAMENTAL FORCES

By the mid-20th century four fundamental forces in nature had been well established. These are, in order of historical discovery, gravity, electromagnetism, the strong nuclear force, and the weak force (see Fig. 2-1). Each force is now well understood, the first two better than the last two. For gravity and electromagnetism we have es-

pecially good theories—that is, there is a mathematical law describing each force and a broad explanation of how it works. The theory for each force is well-verified by a long history of experiments. A simple law for the strong and weak forces is not yet explicitly available, nor is it likely to be. However, we shall encounter new developments later that change the whole situation.

The reader may wonder about the other forces experienced every day. Why did only gravity make the list of the fundamental ones? Our personal experience includes pushes and pulls we can make with our muscles and with our engine-powered vehicles. These are legitimate forces in physics but are not fundamental. The four fundamental ones will be seen to apply to four very distinct realms of nature. As we learn more about the structure of matter we shall see that electromagnetic force is the basic one at work when one object pushes or pulls another. The same story applies to friction, the force which seems to keep us from starting a heavy object sliding along a surface.

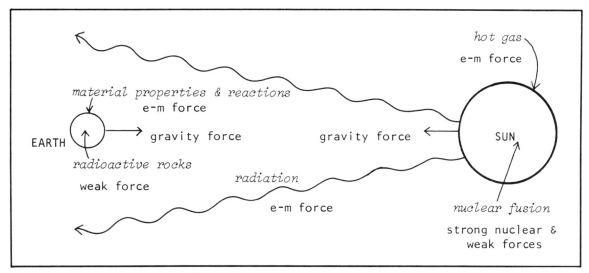

Fig. 2-1. All four fundamental forces illustrated in the Earth-Sun system.

We have all experienced certain aspects of the electromagnetic force more directly. In regard to the electric part of it, we have noticed that a comb drawn through hair then attracts bits of paper, pulling them with some force. In the magnetic realm, we have played with magnets, compasses, nails, and certain other materials and noted again that some "influence" or force occurs between magnet and object. Despite the presence of nuclear radioactivity in our everyday world, none of us have ever experienced directly the strong or weak forces, for reasons to be given much later. We may have observed the consequences of those forces—for example, film fogging or a Geiger counter chattering. The strong and weak forces do have very great indirect effects on us.

While there has been speculation, there is no firm evidence for any more fundamental forces. There are other derived forces used in physics, besides friction and muscle action, which will be described as needed. Physics does not have anything to do with the unsupported claims of mental "force" or the "force" of good and evil, etc. Technical terms such as "force" used in physics have come from, or have been adapted to, many nonscientific uses.

Let us look at the general characteristics of forces, particularly the fundamental ones. Even if we cannot perceive the "push" that is involved, the force should always be measurable by finding out what it does to matter. We cannot detect gravity unless we hold an object in its influence and measure the pull on the object. A force can be either a push or pull. Gravity only attracts, but electromagnetic forces can push or pull. Forces always have direction, although not necessarily a direction exactly toward or away from the object causing the force.

The fundamental forces seem to occur without the intervention of matter. This is called "action at a distance." There are about 400,000 km of vacuum between the Earth and Moon, yet these bodies continuously attract each other by means of gravity. Somehow the Earth knows we are present and attracts us so that we feel it. Different forces have different *range*, the distance over which they act. We are influenced here by the gravity of distant galaxies, and we can feel the electric field on the ground from clouds charged during a thunderstorm. The theoretical range of gravity and electromagnetism is infinite. But the strong and weak forces have very short range and cannot be felt by a particle any farther than about 10^{-14} m (10 fm) from the particle that is causing the force. That is one reason we do not experience them directly.

Table 2-1. The Fundamental Forces or Interactions.

Interaction	Relative strength	Source	Range (meters)
strong nuclear	1	baryons	10^{-15}
electromagnetic	10^{-2}	charges	infinite
weak	10^{-5}	leptons	10^{-17}
gravity	10^{-38}	mass/energy	infinite

Note: Strength and range are approximate. Strengths are given relative to the strong force, but all vary with distance.

Forces cannot exist without the presence of matter somewhere, however far away. Every force has as its *source* some type of matter. Table 2-1 summarizes the basic properties of the fundamental forces, and more explanation is given below. Sometimes the fundamental forces are called *interactions* to signify that the force is detectable when there is interaction of two pieces of matter.

2.2 GRAVITY

Gravity has been perceived ever since humans became aware of their orientation on the Earth and noticed that objects and themselves can fall. After the Renaissance but before the beginnings of modern science, the nature of gravity was poorly understood. Things were imagined to push an object that fell down or was thrown. Galileo Galilei in 17th century Italy studied the fall of projectiles (inspired by military problems, as many early physicists unfortunately were). He learned how to analyze motion and pondered the role played by weight during falling, using thought experiments and real experiments. He made many discoveries in the physics of motion, and the one of interest for the moment is that gravity acts the same independent of the amount of matter (mass) in an object.

It remained for Isaac Newton in 1666 to find the source of the gravitational force we experience. The legend of the apple falling on Newton serves only to illustrate that he drew a connection between the way in which the moon "fell" toward the Earth during its orbit and the way local projectiles fell to the Earth. The motions of the planets

were better understood at this time than everyday motions, and it was suspected that some force kept the planets moving about the sun and the moon moving about the Earth. There is some dispute, but Newton or one of his contemporaries, deduced that the force was an "inverse square" one and applied to the Earth and all other bodies. Later work began on explicitly measuring gravitation forces. Nevil Maskelyne in 1774 estimated the mass of the Earth from how much a mountain deflected a plumb line. In 1798 Henry Cavendish constructed a laboratory experiment whereby gravity was measured between two pairs of balls of known mass. One pair was suspended on a thread as shown in Fig. 2-2.

The modern form of the gravitational force law incorporates all the separately established earlier laws into one mathematical package thus:

$$F_g = \frac{Gm_1 m_2}{r_{12}^2}$$

Here F_g is the symbol for the force that results from the indicated calculation. Masses m_1 and m_2 are the masses of the two objects between which the force occurs. Distance r_{12} is between the centers of the two objects. The inverse-square part of the law is shown by the power of 2 on r_{12}. The amount of force is proportional to either mass. The constant G is needed to make the units agree. When m_1 and m_2 are in kilograms and r_{12} is in meters, G must be $6.6732(10)^{-11}$ for F_g to come out in units of force called "newtons." (Newton did not invent this unit and had to use very cumbersome units, but, like many physicists, his contributions were honored with this unit name in 1938.) G is known accurately to only about six places (significant figures).

An order of magnitude estimate of a person's weight can be made with this law, to illustrate both the method of estimating and the use of the law. To do this, we must know the mass of the Earth to be about $6(10)^{24}$ kg, and that we are located about $6.4(10)^6$ m from its center. These numbers and G will be rounded to the nearest power of ten. Then the weight of a person whose mass is 100 kg is:

$$F_g = (10)^{-10} (10)^{25} (10)^2 / (10)^{7 \times 2} = (10)^3 \text{ newtons.}$$

A simpler force law for this situation will be given shortly. The rounding off in this case has still allowed a more accurate result than one might expect.

Although this law describes the motions of the planets and stars very well, it has not been verified more accurately than indicated by G. For extended bodies such as books or planets, the gravitation interaction is understood to occur between the centers of the masses. This is what makes a body equivalent to a particle; it acts as if all its mass were concentrated at a special central point.

The inverse-square behavior of a force means that as the distance between the two masses increases, the force becomes weaker at a certain rate. The effect of distance is "squared," so that, for example, doubling the distance reduces the force by a factor of four. Such a force theoretically never diminishes to zero, but for practical purposes it may be assumed zero depending on the circumstances. If we travel a good part of a light year away from the sun, its gravity is negligible. However, we would not have escaped the aggregate gravitational force due to all the stars in our galaxy. The gravity law applies to all known particles, as well as to bodies of any size and mass. It does not work perfectly, and for special applications we shall learn about a law by Einstein that supersedes it.

2.3 ELECTROMAGNETIC AND NUCLEAR FORCES AND THEIR RELATIVE STRENGTHS

A century later Charles Coulomb established (in 1785-1789) the force laws for electricity and magnetism. They were also found to be inverse

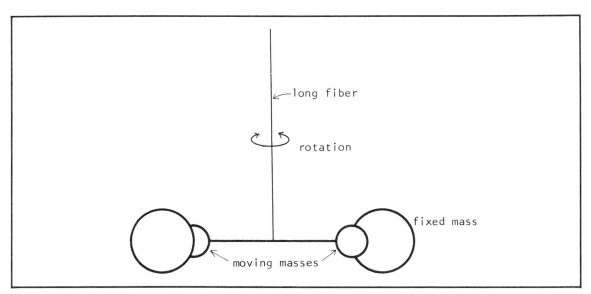

Fig. 2-2. The Cavendish gravity balance.

square laws. The electric force law has very much the same form as Newton's gravity:

$$F_e = \frac{kq_1q_2}{r_{12}{}^2}$$

This force, while electric, is understood to be measured in newtons again. The sources of the force are charges now, not masses, and symbolized with qs. There is a different constant k characteristic of electric forces to match the charge units to force units. (We will discuss charge further in later chapters.) Constant k is $8.9875543(10)^9$ in SI units, and is known to eight significant figures accurately. The direction of the force (attractive or repulsive) depends on the kinds of charges. The range is again infinite because of the inverse square, but in practice charges manage to shield each other and the force may not be felt very far away. That the force is "action at a distance" was first proposed by Franz Aepinus about 1759. The inverse square part has been tested to extreme accuracy. By 1971 it was known that the power could not differ from "2" by more than a few parts in 10^{16}.

The electric force law given above is not the whole electromagnetic force law. We also need the magnetic part. One approach is to find a similar force law for magnets, and this has been done historically. The total electromagnetic force is the sum of forces due to charges and magnets, but such a law is unnecessarily cumbersome to state in mathematical form. A better approach is to learn about electric and magnetic fields in Chapter 4. The electromagnetic theory was, and still is, our best and most complete theory for a fundamental interaction. As with gravity, the above simple law applies to particles as well as to larger bodies of any shape. The calculations become difficult or impossible if the shape is not spherical, however.

The state of understanding is not so good in regard to the strong and weak nuclear forces. The properties of the strong nuclear force have been gradually pieced together since the work of Ernest Rutherford. In 1911 he concluded that an atom had to have a very small and massive core called the nucleus, containing protons. The mutual repulsion of the protons is very strong, and knowing it gave an estimate of the strength of a new nuclear force that had to hold the protons together. The source of the strong nuclear force is most matter; the elementary particles are called "baryons" and include the proton. The range over which the force could act was found to be always less than $(10)^{-14}$ m (10 fm). Some significant progress was made in 1935 when the theory of Hideki Yukawa gave a simplified calculation for nuclear force and an explanation for it. Details are reserved for the chapter on the nucleus. Numerous physicists have struggled with the theory of the force, making little headway until a major theoretical revolution occurred in the 1960s.

The first major evidence for what came to be known as the weak force (or weak interaction) was the discovery of radioactivity by Antoine Bequerel in 1896. Certain chemical elements emitted tiny particles with substantial energy, and in so doing changed into other chemical elements. Many decades of patient experiments passed before it was suspected that a new force was at work. New particles (which we shall meet later) were discovered as a result of studying interactions involving the weak force. Like the strong force, the weak one has a range so small (now known to be about 10^{-17} m) that it has an effect only when elementary particles are very close together. Only a few kinds of elementary particles (called "leptons") seem to be sources of the force, and it is best not thought of as a "nuclear" force. Substantial theory for the weak force had to wait until the 1970s for the revolution in particle physics.

At this point a little can be said about the relative strength of the four forces. In any unit system gravity appears to be much weaker than the electric force. In SI units G is much smaller than k. There are other considerations, such as the question of how to compare masses and charges, but comparisons can be made for elementary particles. For example, two protons repel each other about 10^{36} times more strongly because of their charges than they attract because of their masses, regardless of their separation. (Their magnetic interaction also exists but is weaker.)

The use of two protons allows us also to compare the strength of the strong nuclear force. It is aptly called "strong" because two protons attract each other with a nuclear interaction over 100 times stronger than the electromagnetic. The "weak" force is so-named because it is no better than a thousandth of the electromagnetic force, depending on the conditions used. Table 2-1 summarizes the four fundamental forces, which are listed in decreasing order of strength. The arbitrary strength of 1 is assigned to the strong nuclear, and the others are listed relative to this. Having forged this far into nuclear physics, let us return to more classical physics.

2.4 SOME COMMON FORCES

Friction and muscle force have been mentioned as everyday consequences of the electromagnetic force law. A mathematical law for friction is available but belongs more to technology than to particle physics. We shall be needing some acquaintance with another force law, one that holds for springs. Whether Robert Hooke discovered the law which bears his name is not known, but he did work with weights vibrating on springs to measure time and had to understand the mechanics of springs. The law states simply that:

$$F_s = kx$$

where F_s is the force needed to stretch a spring a distance x from its relaxed position, and k is a new constant to fix the units, measured in newtons per meter. We will be interested later in how waves get started, and a force such as this one will be needed. The strength is proportional to the amount of stretch.

Because it is an important part of elementary physics, the simplified form of the gravity law that holds near the surface of the Earth will be given. Near the ground, so that the Earth seems flat rather than round, the downward force on a mass m is given by:

$$F = mg$$

where g is another constant, approximately 9.8 newtons per kilogram. This F is also known as the *weight* of the object of mass m. Whether it is falling or standing still on the ground (or on a scale) the Earth pulls with this force on objects.

2.5 MECHANICS

The earliest sub-field of physics to be securely established, mechanics is the study of the motion of objects, including particles. Physicists have tended to call objects under consideration "bodies," which may or may not act as ideal particles. Galileo and Kepler were major influences in mechanics. Galileo's work was mainly in what we now call "one dimension." Moving objects were constrained to a single path, not necessarily straight. Kepler, perhaps unknowingly, introduced two dimensions by his assertion that the planets move in elliptical paths. Such a path is necessarily in a flat plane, but there are two independent directions in which the planet or other body can move. This was not clear at the time, so work by Newton in regard to combining forces was the first hint of the need to take two or three independent directions into account.

A force always requires two or three numbers to describe it fully. In a plane, such as assumed for Fig. 2-3, we need the strength (called the *magnitude*) and direction of the force, or we need the *components* of the force in two independent directions. The directions of the components must be at a right angle, or "orthogonal." When a rectangle is formed with a scaled diagram of the two components, the equivalent force lies along the diagonal. Force is an example of a mathematical quantity called a *vector*. We are familiar with simpler quantities called *scalar*, which do not have direction, just magnitude—for example, mass or temperature.

There are at least two procedures for working with force vectors: drawing them to scale in diagrams, or using trigonometry to relate magnitudes and angles. Every force pointing in some direction can be replaced in its effect with two

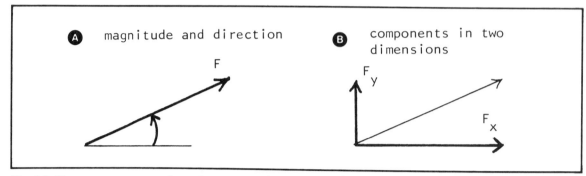

Ⓐ magnitude and direction

Ⓑ components in two dimensions

F

F_y

F_x

Fig. 2-3. Two ways to represent forces (or any vectors), using two independent quantities in two dimensions.

forces at right angles. These components will have some definite and smaller magnitudes. For example, the effect of the downward gravitational force on the block on an inclined plane in Fig. 2-4 is the same as if gravity were replaced by a certain force along the plane and a certain force against the plane.

Regardless of their positions and directions, any two vectors $\vec{A}$ and $\vec{B}$ (Fig. 2-5A) can be added by the parallelogram method. First slide the two vectors, without changing their directions, so that the tails of each coincide at the same point P (Fig. 2-5B). The heads indicate a parallelogram, which can be constructed as shown with lines parallel to each vector. Then draw the diagonal from P across the parallelogram. The length and direction of this diagonal represents the magnitude and direction of the resulting vector $\overrightarrow{A + B}$. To subtract vectors, draw the vector to be subtracted in reverse, shown as $-\vec{B}$ in Fig. 2-5C. Then proceed with addition. Vector addition is useful for finding the net result of several vectors such as forces working together on one body. It can be carried out with any number of vectors.

In three-dimensional space either a magnitude and two directions, or three components, are needed to specify a vector. Again, a force pointing in some direction can be replaced in its effect with three forces all at right angles as shown in Fig. 2-6. This diagram attempts to show on flat paper the spatial relationships of the three components. Any body or particle moving through space will require consideration of three dimensions to study its behavior. Particles are said to

have three degrees of freedom. This is a physical law about the kind of space we live in. Any object can act or move in three independent ways.

To discuss particles further, we need to know more language for describing where they are, how fast they move, and so forth. Position is indicated in terms of *coordinate axes*, commonly labeled X, Y, and Z as in Fig. 2-7A. A zero position must be defined for each axis, and it should be placed where the axes cross at O, the *origin*. Distance from O in the direction of the arrow is recorded as a positive number, and the other direction is negative, as shown. Positions should be measured in length units such as meters. Each axis is a straight line, and the three together form a "right-handed" system. If the thumb points in the direction of the X-axis, and the first finger points in the direction of the Y-axis, then the second or middle finger will be forced to point in the direction of the Z-axis as shown (Fig. 2-7B). It is important to

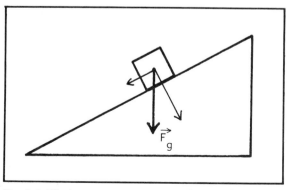

$\vec{F}_g$

Fig. 2-4. The force components most relevant when gravity acts on a mass on an incline.

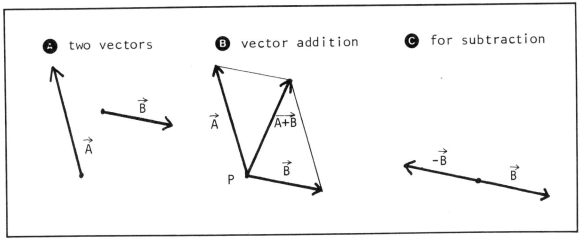

Fig. 2-5. Adding vectors by the parallelogram method.

consistently use a right-hand system because all quantities in physics are defined and measured in reference to it.

Sometimes it is useful to consider a position as a vector. In Fig. 2-7C the point P with coordinates (x,y,z) is shown. A position vector for this point must be drawn from O to P, and is labeled $\vec{r}$ by convention. The arrow on a symbol denotes that a quantity is a vector and has length and direction. In this case the length $\vec{r}$ can be found from the coordinates by the Pythagorean theorem:

$$r = \sqrt{x^2 + y^2 + z^2}.$$

If we were interested in just one dimension, such as the motion of a train on a track, we would choose one axis along which to measure position, say the X-axis. The object would be said to have position $x = 3$ if it were seen to be adjacent to the value $x = 3$ along the numbered axis. If we really want to know where it is, we must specify the units for the axis as well (for example, meters). If the track were to curve, we may or may not be able to make use of an unorthodox axis that curves along with the track. It is better to measure the position in two dimensions by adding a second axis (Fig. 2-8).

To describe motion we must know positions of the object at certain times; then we can discuss its speed and acceleration. Figure 2-8 shows an object's positions at several successive times, thus indicating the direction it moves as well. *Speed* is the distance a moving body covers in a certain time. For example, in Fig. 2-8 the object moves 5 m in the X-direction in 1 second. A change in position can be specified with the symbolism Δx, where Δ simply denotes a small change in x. Similarly a time interval is called Δt. Speed v is calculated in one dimension with:

$$v = \Delta x/\Delta t$$

The units of speed must be meters per second (m/s) if meters were used for x and seconds for t. In the example, the object is moving 5 m/s between

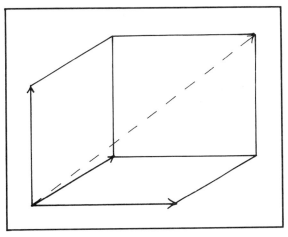

Fig. 2-6. The three independent components of a vector in three dimensions .

25

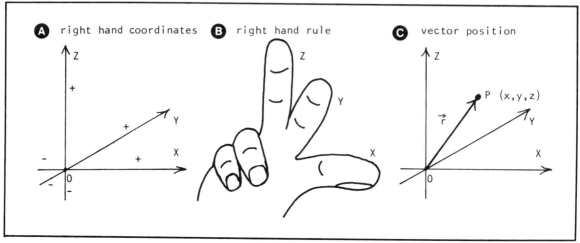

Fig. 2-7. The right-handed three-dimensional coordinate system.

$t = 2$ s and $t = 3$ s. Speed is positive if the direction of travel is the direction of the X-axis and negative in the other direction.

If the object happens to be speeding up or slowing down, then its speed is changing. A change in speed is denoted Δv and, if known in a given time, can be used to calculate its *acceleration* (symbol *a*). Even if the body is changing speed, we can find its speed at given times or at given positions. Acceleration is calculated in one dimension by:

$$a = \frac{\Delta v Gv}{\Delta t}$$

The units must be meters per second divided by seconds, or meters per second-squared (m/s²). In Fig. 2-8, the object has slowed after $t = 3$ s, reducing its speed to 3 m/s by $t = 4$ s. Its speed has been reduced by 2 m/s in about 1 s, corresponding to an acceleration of -2 m/s². Unlike speed, acceleration could be negative even if the direction of travel is positive. In such a case the object

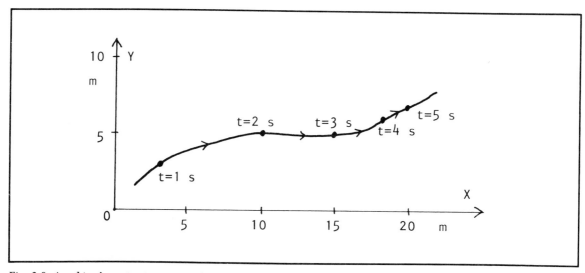

Fig. 2-8. An object's motion in space and time.

Table 2-2. Table of Motion for a Moving Particle.

Time t (s)	Position x (m)	Speed v (m/s)	Acceleration a (m/s²)
1	6		—
		− 2	
2	4		1
		− 1	
3	3		2
		1	
4	4		3
		4	
5	8		

is slowing down (popularly called "decelerating"). When an object moves in two or more dimensions, these calculations of speed and acceleration can be done independently in each dimension, along each axis.

We can practice calculating the motion of an imaginary particle with the aid of an arrangement such as that shown in Table 2-2. Suppose that the positions x of the particle have been measured at one second intervals (at $t = 1$, $t = 2$, and so forth) and are given as shown. The distance the particle went from $t = 1$ to $t = 2$ is $6 - 4 = 2$ m. The speed v is 2 m divided by the 1 s interval, or 2 m/s. Since position is decreasing, this v must be negative: $- 2$ m/s. This speed is most accurately known between times $t = 1$ and $t = 2$. Therefore the v column is stepped down one-half unit and the speed entered to correspond with $t = 1.5$. Similar work gives the other speeds shown. Note that after $t = 3$ the position is increasing and v turns positive.

A similar process is used to calculate accelerations. The first known change in speed is from $- 2$ m/s to $- 1$ m/s. The change in speed is $\Delta v = 1$ m/s, positive since the speed is increasing (even though the speed itself is still negative). The acceleration is 1 m/s² and is entered halfway between the speeds, in the $t = 2$ row.

This numerical calculation method is called "numerical differentiation," and is the same process used in computer programs to find speed and acceleration from position data. Accuracy is achieved simply by knowing positions at shorter time intervals. There exists also a reverse procedure called "numerical integration," which allows the calculation of positions, given speeds, and the calculation of speeds, given accelerations. [Calculus students can do this algebraically, but

anyone can do the equivalent of calculus with numbers in the way shown.] The procedure can be extended to two or three dimensions simply by adding columns for y and the corresponding v_y and a_y and for z and its v_z and a_z. The original v and a must of course be thought of as v_x and a_x. Motion calculations are done for each dimension independently of the other dimensions. This branch of physics that analyzes motion is sometimes called "kinematics."

Just as forces in two or three dimensions can be considered as vectors with components, so can position and motion. If an object is moving southeast with a speed of 10 m/s (Fig. 2-9A), it can be said to be advancing about 7.07 m/s in the south direction and about 7.07 m/s in the east direction, simultaneously. If the object was once at the origin and is now 10 m to the southeast, the change in location can be represented by a translation of about 7.07 m to the east, followed by 7.07 m south, or by 7.07 m south followed by 7.07 m east, as shown in Fig. 2-9B. When considering speed in two or three dimensions, it should be called *velocity*, a term that signifies that the speed has direction. Velocity and acceleration can have components written (v_x, v_y, v_z) and (a_x, a_y, a_z), respectively.

2.6 NEWTON'S LAW OF MOTION

We have discussed forces, which are in some sense *causes* of motion. We know that the harder we push on something, the more motion we get. But we have also observed that big heavy objects do not respond so well. Early thinkers noticed this too, and assigned the concept of "inertia" to bodies. Objects which weighed more were found

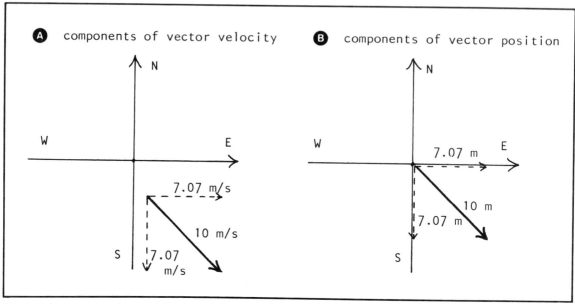

Fig. 2-9. *Components of vector velocity and position in two dimensions.*

to have more inertia, and so weight and inertia came to be equated. But after we discuss the law of motion, we shall see that the distinction must be retained. For the moment let us identify inertia with *m*, the mass of a body.

Isaac Newton studied carefully the motions of objects under various conditions in the late 17th century. In many ways he can be characterized as one of the last great natural philosophers. For example, his major published work *Philosophiae Naturalis Principia Mathematica* (1686) had to be written in Latin and was more of a philosophical treatise than a work in mathematics or physics. In it he laid out a broad philosophical approach to studying nature: to describe motion, then to find the forces that cause it, then to study what gives rise to those forces. Motion was simply the clue to grander processes in nature. Observation and experiment led him to formulate three laws about forces and motion. The work was difficult because not only could practical systems not be simplified or isolated easily, but also many of the results seemed to defy intuition.

In a paraphrase of his original words, the main conclusion was that a body in motion remains in that motion in a straight line unless acted upon by an outside force. If there is a force, it is related to an acceleration of the body. The modern statement of the law of motion in mathematical form is:

$$\Sigma \vec{F} = m\vec{a}$$

The symbol Σ shows that there are a number of forces acting upon a body of mass *m* and that their effect is to be summed to a net vector force before proceeding further. Another way to read ΣF is to say "the net force." The result of adding all the forces on the body is a single force in a certain direction (Fig. 2-10A). The object then experiences a vector acceleration in that same direction. The arrows show that vector directions are an important part of the law. If there is only one force *F* acting on a body, the law may be written: *F* = *ma*, and the vector symbols dropped in one dimension.

There are two ways to consider what is being defined in the law of motion. One is that mass *m* is the appropriate constant needed to make the units of *F* (newtons or N) and *a* agree, as measured. The units of *m* must then be newtons seconds squared per meter. We also know *m* as being measured in kilograms, and indeed kg and Ns²/m are the same. The equation thus can define m. If by experiment we find a and the F that causes it, then m is given by F/a. The other way to look at it is that F is being defined by the equation. For a known m undergoing a known a, a net F is found from calculating ma. Then the units of F, newtons, are derived units, defined from a combination of more basic ones by $N = kg \ m/s^2$. One cannot have it both ways at once; Newton's relation defines either F or m.

Table 2-2 can be extended. We can find the behavior of the mysterious force causing the motion as described numerically. Let us assume that $m = 2$ kg. Then a column of forces can be calculated from the acceleration column. At $t = 2$ we find that $F = 2$ N, in the positive direction. At $t = 3$, F is 4 N, and so forth. The physical interpretation is that a steadily increasing force caused this particle to cease its negative motion, reverse, and move positively along the X-axis.

Let us consider some special cases of Newton's law of motion. Suppose there is no acceleration. This does not necessarily mean the body is not moving; it may be going at constant speed. Then the net F must be zero. Notice the importance of "net." If a body is being pushed equally from both sides, as in Fig. 2-10B, the forces cancel to nothing. We also should be careful about the notion of constant speed. If there is no net F, so that the velocity is unchanged, then the direction cannot change. A very special case occurs when there is no velocity and no F. Then the body stays at rest. If

net F is not zero and there is no motion, then the body is acting strangely, as if it had infinite mass. Indeed this is the case if one pushes on an object fastened firmly to the Earth (Fig. 2-10C); nothing happens except fatigue. The effective mass is enormous, as long as the contact with the ground remains unbroken. (It does not matter here if the pushing agent happens to stand on the same ground.)

Suppose that $m = 0$, or, more realistically, m is very small, perhaps the mass of an elementary particle. If we apply an ordinary human-sized force, a huge acceleration is obtained. This is best seen by trying to calculate $a = F/m$. A large F divided by a tiny m gives a very large result. Elementary particles will acquire huge speeds under the huge accelerations caused by ordinary forces.

Newton gave us another law, one basically about forces. In a paraphrase this law of *action-reaction* states that for every action (really, a force) there is an equal and opposite reaction (a force). Thus two bodies act upon each other in equal and opposite ways. This law helps us interpret what happens in the example of an object resting on a table. Gravity exerts a force (weight) on it, downward. We note that the object does not accelerate downward despite the law of motion. We deduce that the net force on the book must be zero. The other force that counteracts gravity is a force applied by the table pushing upward. It is equal and opposite to gravity. We must be careful in drawing the vector forces on a body. The body

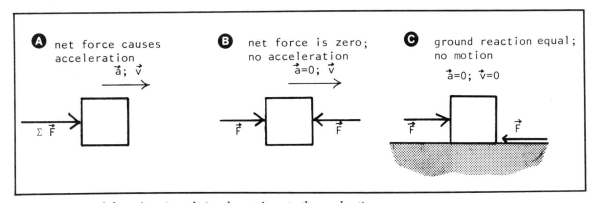

Fig. 2-10. Newton's law of motion relating the net force to the acceleration.

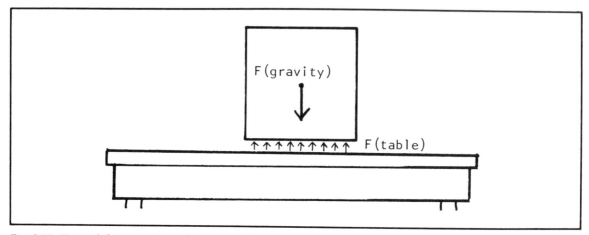

Fig. 2-11. Newton's law of action and reaction at the interface between a block and the table supporting it.

must be isolated from all other bodies, and only "pure" forces may be drawn, as in Fig. 2-11. The body is shown to remain in place with F(gravity) acting downward on all parts of the body, and F(table) acting upward on the bottom of the body where it touched the table. Newton's law is best applied to "free-body" diagrams such as this, where a body is shown, surrounded by the relevant forces. The action-reaction law is like a force law, allowing us to deduce certain needed forces to analyze the situation.

2.7 ENERGY

Energy has become a popular subject, due in part to its economic scarcity, but physicists have always considered energy as part of their domain and have defined its many forms very carefully. We should suspect that a body in motion possesses energy, because heat (one form of energy) is released when the body is stopped, for example when a baseball is caught or an object is dropped without bouncing (Fig. 2-12). We also know intuitively that we have put energy into a body when we push it with a force and it flies away at a certain speed. The energy possessed by a moving mass can be calculated to be:

$$KE \equiv \frac{1}{2} mv^2$$

where KE is the abbreviation for kinetic energy,

and the triple equal sign shows that this statement is a definition of KE. The units of KE, as for all energies, are defined to be joules (abbreviated J, equivalent to kg m²/s²). Kinetic energy was first discussed by Christian Huyghens in the 17th century, when it was called "vis viva" (which reminds us of v-squared!).

Since a dropped object arrives at the ground with a certain speed and therefore KE, the fact that it may do damage when it strikes shows that it possessed energy. Before it was released, did it

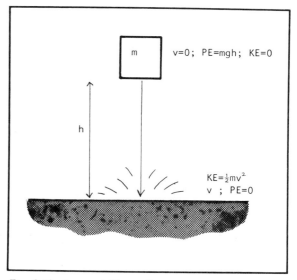

Fig. 2-12. A mass falling a distance converts its potential energy into kinetic energy.

possess energy? The convenient approach in physics is to say that it did and to find some rational way of defining what energy it had. The clue is to look at what caused the motion. Gravitational force seems related to the KE the body acquired. The farther it falls, the longer gravity acts on it and the more KE it has. It is postulated that the body had a certain gravitational energy before it fell and a different and lesser gravitational energy when it reached the ground. Such an energy is called a potential energy (PE). Before it fell it had a certain value of PE. During falling some PE is converted into KE and it arrives with less PE.

Potential energy must be defined with respect to some reference. If the ground is taken as the position of zero energy, we have the basic definition of gravitational PE near the Earth's surface:

$$PE = mgy$$

where y is the height of the body above the ground. PE is measured in joules. PE can be negative, as would occur if gravitational PE is measured in a hole below ground level (where y is negative, too).

Let us do a numerical calculation of motion for a falling object. In Table 2-3 a body of mass 3 kg is shown to be at height $y = 80$ m. From the simple gravitational force law we know that $F = mg \cong (3)(10) = 30$ N, where the value of g has been rounded to a convenient 10 m/s^2. This force of 30 N is downward and thus shown negative. Height y is measured positively upward from the ground. Using Newton's law of motion we can calculate the downward acceleration to be 10 m/s^2. It is no accident that a_y comes out to be the same as g, as will be discussed shortly. We can find the downward speed in a simple way after 1 second has elapsed by considering that -10 m/s^2 acting during 1 s results in a speed of -10 m/s. After 2 seconds the body picks up an additional 10 m/s of speed and this is added to the preceding speed. Similarly we can also find its position as it falls. During the first second the speed has an average value of -5 m/s, as shown in the v_{avg} column, having increased from 0 to -10 m/s. This average speed results in a fall of 5 m downward, putting it at 75 m. At $t = 2$ s, the body has moved an additional 15 m down, leaving it at 60 m. This process, the reverse of the table of motion previously calculated, is continued until the ground is reached, conveniently arranged to be at $t = 4$ s, when $y = 0$.

Besides illustrating the use of Newton's law of motion to calculate motion in the way a computer would do it, our detailed breakdown of the motion allows the illustration of a major principle in physics pertaining to energy. Included in Table 2-3 are tabulations of the falling body's PE and KE, second by second. At $t = 0$ it started with PE = (3)(10)(80) = 2400 joules. At $t = 1$ the height is less and PE is calculated to be 2250 J. At the bottom PE must be zero. Meanwhile the KE has been increasing. At $t = 1$ the data in the table allow the KE to be calculated to be KE = $(.5)(3)(-10)^2$ = 150 J. At the ground the speed of 40 m/s gives a KE of 2400 J. The point of this exercise is to examine the sum of KE and PE, moment by moment. The TE (Total Energy) column shows that the sum remains constant at 2400 J. This quantity of energy for this system never changes during the motion.

Table 2-3. Table of Motion and Energy for Falling Body (Mass m = 3 kg).

Time t (s)	Force F (N)	Acceleration ay (m/s²)	Speed vy (m/s)	Average speed v_avg (m/s)	Position y (m)	Energies: PE (J)	KE (J)	TE (J)
0	−30	−10	0	0	80	2400	0	2400
1	−30	−10	−10	−5	75	2250	150	2400
2	−30	−10	−20	−15	60	1800	600	2400
3	−30	−10	−30	−25	35	1050	1350	2400
4	−30	−10	−40	−35	0	0	2400	2400

The PE for the more general gravitational force law can most easily be found with the aid of calculus, and the result is:

$$PE(gravity) = \frac{-Gm_1 m_2}{r}$$

The reference for this PE is taken at the only sensible place possible, infinity. When r becomes very large, PE approaches zero. Therefore PE must be defined as negative, so that as one body approaches another, the PE can decrease in the absolute (algebraic) sense. A large negative value is less than a small negative value. This PE behaves unpleasantly when r approaches zero and in fact cannot be used there. There is no serious problem because the PE at the center of a mass becomes zero for another reason. How the simple form PE = mgh can be obtained from the more general PE(gravity) is another matter not needed in this book.

The PE for the spring force will be needed later and can be found to be:

$$PE(spring) = \frac{1}{2} kx^2$$

The spring's PE is zero when it is not stretched, and increases as the square of the stretch x. The PEs for other fundamental forces will be given later.

One of the greatest principles of physics is the law of *conservation of energy*. This law implies that when one considers all relevant forms of energy, the total energy is always the same. Energy is neither created nor destroyed. It is always present in some form and may be converted to another form. The law of conservation allows some difficult questions to be answered easily. During a long complicated physical process, we do not have to know the details in order to predict the outcome from the starting conditions, when energy is involved. This allows a black box approach to physics. A particle of known energy can strike a complex nucleus and, no matter what happens, we know what the PE change in the nucleus was after we account for the energies of any particles that emerged.

"Conservation" in practical terms in our technological society has come to mean not wasting energy—to get as much useful work or heat out of fuel as possible. But there is never a loss of energy; it is just turned into nonuseful heat or other forms when it is "wasted." Other forms of energy will be defined as other branches of physics are explored. We shall use electric, magnetic, nuclear, and spring energy; each of these forms of potential energy corresponds to a force law. We shall also meet heat energy and light energy. Chemical energy, the most commonly used form, is really an application of the electromagnetic energy released when atoms react.

The range of energies found in nature is enormous. It will not be attempted at this time to express the total energy of the universe, all of which presumably came out of the "Big Bang." We shall confine ourselves to some ordinary energies and the energy of particles. The lowest energy something can have is probably zero, although this will never be achieved. There is always some force pulling on any given body. There is also, as will be seen, trouble with claiming that empty space has zero energy. To give a few examples in ascending order, the KE possessed by a typical molecule in the air is about 10^{-20} joule. 10^{-10} J can be extracted from one uranium or plutonium nucleus. A person running has about 1000 J of KE, about the same energy as a small flashlight battery can produce over its lifetime. It takes about 2 million joules to vaporize 1 kg of water. The U.S. uses over 10^{20} J per year, converting it mainly to heat. The sun puts out 10^{27} J each second. The amount emitted by what seem to be very distant exploding galaxies is absurdly bigger (10^{50} J?).

2.8 CONSERVATIVE FORCES

To call a force *conservative* is not to comment on its politics but to express a fundamental point of view about forces. We shall find this concept elusive at first but important for introducing modern topics about particle physics. A conservative force is said to have acted if the same energy is given to a particle in moving it along one path as by moving it along another path. The energy imparted as a particle is moved by a force called *work*. It is defined by:

$$W \equiv \Sigma F(x)\Delta x$$

where the symbol Σ (Greek upper-case sigma) signifies the addition of all the small parts that are to follow. $F(x)$ is notation that says use the value of the force F at location x. F may vary from place to place, but F must be only the component that points along the direction of travel. Δx is the distance over which the value $F(x)$ holds. [Calculus students will note that a dot product is being set up for integration.] W of course should come out in energy units. It corresponds to ordinary notions of work under certain conditions.

This definition of work is useful for finding a PE, given a (conservative) force law, or it can be "inverted" to find a force law, given a PE. The method is to realize that the work done in moving a body against the force pulling it back—for example, a rocket leaving the Earth—is equal to the PE so gained. Calculus may be needed to carry out the calculations.

A conservative force does the same work no matter which route is taken to the same endpoint. Figure 2-13 illustrates the process. A body is moved from point A to point B by two routes. Route 1 shows that the body is lifted up 7 m, then lowered 2 m, all the while moving aside on an erratic path. Work was done against gravity in lifting it, but some work was recovered in lowering it. (It is against the rules to drop it or otherwise let it arrive at B with any excess speed.) Along Route 2 the body was slid (without friction or perhaps carried) along the floor, then raised straight up to B. No work was done as it moved sideways because the gravitational force was at a right angle to the direction of travel. The same work was done raising it straight up 5 m as to raise it 7 m, then lower it by 2 m. This procedure works for any complex path. Work only accrues from the component of motion in the direction of the force, and this adds up the same for any path!

Since this analysis has proven to be correct for gravity, gravity must be a conservative force, as all the fundamental ones seem to be. The spring force is conservative, but friction is not. If one slides a box on a rough surface from one point to

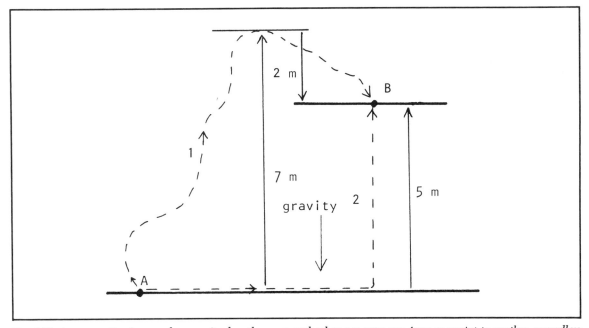

Fig. 2-13. A conservative force such as gravity does the same work when a mass moves from one point to another, regardless of the path taken.

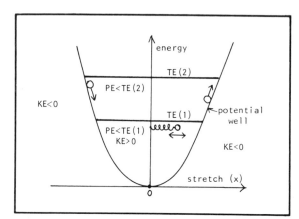

Fig. 2-14. *The potential well for a particle bound by a spring* (*Note: TE = PE + KE*).

another, more work is done if a long roundabout path is taken.

A graph of PE (spring) versus stretch x illustrates a specific and important model in physics, the concept of the *potential well*. The shape of the graph (Fig. 2-14) is one form of potential well because it appears to be able to contain something that cannot escape. It represents the constraints on a particle connected to the spring. The idealized spring has no length except when stretched, and the particle moves back and forth along the X-axis, hindered by the spring. If it is given a certain speed to start with, it has a certain total energy TE(1). That can never change, nor can its KE be negative. Thus the graph of the potential shows how far from $x = 0$ the particle can move. With larger TE(2) it can move farther from 0 before its KE would become negative. Another way to visualize the potential well is to imagine the particle rolling (without friction) on the sides of the well. If it starts with PE = TE(2), then

it can roll down to the center, converting all its energy to kinetic, then roll up the other side, stopping and turning around when it again reaches energy level TE(2). This energy graph is a good way to picture system behavior as long as conservative forces are acting and TE is conserved.

2.9 THE INVARIANCE PRINCIPLE

To set the stage for some ideas of modern particle theory, a viewpoint originating in mechanics and called the *principle of invariance* will be introduced. For most basic variables in physics, such as position, time, charge, and so forth, there are related conserved quantities and conservation laws. The conserved quantity corresponding to time is energy. The total energy of any system and of the universe never changes over time. There is also a symmetry involved with this and other applications of the principle. Invariance in time is a symmetry; physics, or at least energy, should be the same no matter what time an experiment is done. Therefore, as time passes, energy remains invariant.

With the invariance principle comes a mathematical method for calculating what is conserved for a particular variable. The procedure involves what aspects are minimized as a system changes over time. When a physical law is required to be invariant with respect to time, the calculation method produces as a result the quantity $\frac{1}{2} mv^2$, which we know as KE. It is included in a sum which contains whatever PEs are appropriate to the system. Later we will see what conserved quantity is paired with position, and so on, and how such complementary pairs of quantities are at the heart of modern physics.

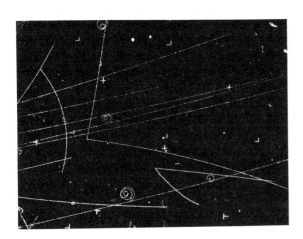

Chapter 3

Rotations and Interactions of Particles

SOME FURTHER ASPECTS OF THE MOTION OF PAR-ticles are discussed here. We need these concepts not only to understand better the mechanics of particles but also for later discussion of the design of particle accelerators.

3.1 MOMENTUM

We have introduced the law of conservation of energy based partly on evidence and partly on a consideration of what must be conserved over time. If another important physical variable, length, is examined, one might wonder what is conserved in regard to length. The principle of invariance shows a way to calculate a quantity that never changes, no matter what position a moving particle has. That quantity is the particle mass times its velocity and called *momentum*. It is calculated thus:

$$\vec{p} \equiv m\vec{v}$$

The momentum $\vec{p}$ is a vector quantity, defined to point in the same direction as the velocity. The concept of momentum has a long history, traced back to Newton who called it "quantity of motion." He suspected its relevance to dynamics from experiment and intuition.

We know from our ordinary experiences that something besides velocity seems to matter in motions and collisions of objects. One everyday experience involving momentum, whether or not we suspect it, is the dynamics of stepping from a cart (Fig. 3-1A,B). If the cart has low-friction wheels, we know it rolls in a direction opposite to the motion of the person leaving it and the resulting confusion may lead to an unexpected fall. We know further that if one steps from a massive vehicle, it moves slowly and the person moves almost as fast as expected. In contrast, if one steps from a light object such as a skateboard, it moves away rapidly and the person hardly moves (but stumbles nevertheless). Both the mass of the objects involved and their speed seem to count in what happens. The light object moves away with high speed, the massive object with low speed. It is the product

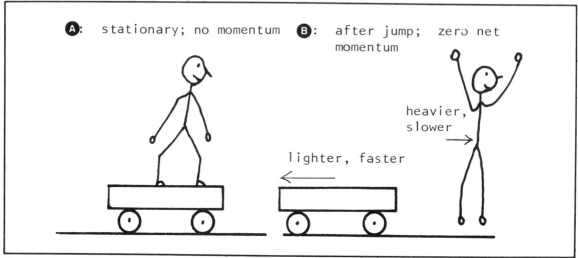

Fig. 3-1. *Conservation of momentum in one dimension.*

of mass times speed that does not change in these cases. We might also suspect a conservation law. Before the action neither body had momentum. Afterward, the momentum of the cart is equal and opposite to the momentum of the person, resulting in a net momentum that remains zero.

Another relevant observation is that a struck pool ball or a thrown baseball seems to bounce from surfaces at predictable angles, a kind of "reflection." In Fig. 3-2 a ball is shown approaching a flat wall with a velocity (and momentum) which has two components, one toward the wall and one parallel to it. The interaction at the wall is such that the only force the wall can apply to the ball is perpendicular to its surface. By Newton's law of motion, the acceleration of the ball during collision is in that direction; its velocity toward the right is reversed to become a velocity to the left (assuming a very elastic superball). The result of an unchanged velocity parallel to the wall and a reversed perpendicular component is a ball that appears to have reflected from the wall in a symmetrical manner.

Let us consider the effect on momentum in this collision. Since the mass did not change during this collision, the momentum must have undergone the same effect as the velocity: the component perpendicular to the wall was reversed.

The momentum of the ball in Fig. 3-2 has changed direction so that the momentum has changed from $\vec{p}$ to $\vec{p}'$, although the magnitude is the same. If there is to be a law of conservation of momentum, we must account for this change in momentum. The missing momentum was transferred to the wall. As a result of the collision the wall is now moving slowly to the right. If it is firmly attached to the massive Earth, then the Earth is moving very slowly to the right. Moreover, if the wall is now moving, we ought to explain how the motion

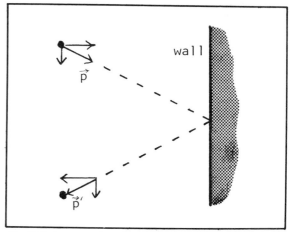

Fig. 3-2. *A component of momentum reverses when an object rebounds elastically.*

was started, using Newton's law which pertains to all motion. The force which set the wall moving rightward was the equal and opposite reaction corresponding to the force the wall applied to the colliding ball.

The law of conservation of momentum simply states that the total momentum of a system is conserved—that is, the total is unchanged no matter what happens. The crucial provision is that the system one is talking about be fully considered both before and after an event. One cannot look at the ball alone if it has an interaction with anything else. As soon as one considers ball and wall together, both before and after collision, then the momentum of both added together (vectorially) is unchanged. The alternative is to consider the wall as applying an external force to the system (the ball alone). Then the ball's momentum is not invariant. If momentum is not conserved, there must be regions of space where the physics is not invariant—namely, behind the wall.

That conservation of momentum results from an invariance of position can be seen in the case of a person stepping from a moving cart. Regardless of where the cart is, or how fast it moves, the changes in momentum of cart and person balance out. It is unnecessary, perhaps impossible, to know the absolute total momentum of the cart before and after. We live on a rapidly rotating Earth (surface speed at mid-latitude almost a meter per second), yet we do not need to account for this and other motions of the Earth to find that Newton's

law of motion always seems to work on the Earth. to find that Newton's law of motion always seems to work on the Earth.

Newton's own refinement of his law of motion can now be stated (in modern symbols):

$$\Sigma \vec{F} = \frac{\Delta \bar{p}}{\Delta t}$$

The net applied force to a body of mass m results in a change of momentum $\Delta \vec{p}$ in time Δt. The equation is a vector one, as before. This law does not contradict the conservation of momentum; $\Delta \vec{p}$ can change if an outside force is applied.

3.2 CIRCULAR MOTION

To proceed further we need to add the description of rotational and circular motion to our understanding of dynamics. Newton himself stated that a body would proceed in a straight line until acted upon by an outside force. Suppose, as in Fig. 3-3A, a particle is moving with velocity $\Delta \vec{v}$ and a force $\Delta \vec{F}$ is briefly applied at an angle to its motion. By Newton's law, the body should accelerate in the direction of the force, gaining a velocity component $\Delta \vec{v}$ in that direction. Its total vector velocity will have changed direction, to become $\Delta \vec{v}'$ in Fig. 3-3B. It will also be moving with a different speed (magnitude of $\Delta \vec{v}'$).

Now suppose that a steady and carefully selected force is applied to a moving particle such that the force is always at right angles to the path of motion. This would be easily done, for example, by twirling an object at the end of a string (Fig. 3-4, position a), or by considering a planet in a circular orbit around the sun. We have assumed the path is circular, but can we find a constant force that will assure it is circular for a given speed? The velocity of the body continually changes direction as it goes around. However, a force always at right angles to the path should not cause any change in speed. The magnitude of the velocity should be constant, and just the direction should change. The answer is yes, and a simple calculation [see any physics textbook] gives the required relation between speed, radius, and central force:

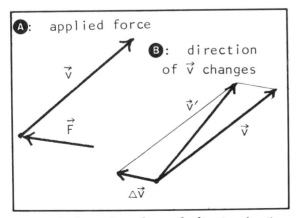

Fig. 3-3. A sideways force changes the direction of motion.

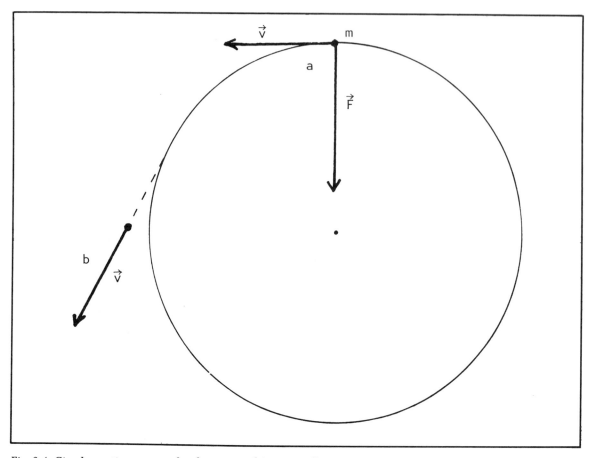

Fig. 3-4. Circular motion occurs only when a central (centripetal) force acts on a moving body. (The radius of the circle depends on the force and the tangential speed of the body.)

$$F_c = \frac{mv^2}{r}$$

Here v is the speed, r is the radius of the presumed circular path, and F_c is the symbol for the *centripetal* force needed. The equation is not in vector form, but it is understood that the vectors are continually rotating around, as shown in Fig. 3-4. Another trick to keeping F_c from increasing the speed is to do the calculation for very small changes in direction [using calculus].

The explanation for what keeps an object on a string (or planet) in a circular orbit is the centripetal force applied by pulling on the string (or by the sun's gravity). If the string is let go or breaks, then at that instant the object obeys

Newton's law that says an object must move in a straight line when there are no applied forces. At that moment it had a certain velocity, and it keeps it, moving away in a straight line tangent to the original circle, as shown in position b of Fig. 3-4. Under no conditions does there ever exist such a thing as "centrifugal force." No force pulls the object outward, nor does it move outward when let go.

3.3 ANGULAR MOMENTUM

We can sense intuitively that an object constrained to a circular path somehow conserves a quantity resembling momentum, yet we know now that its velocity and therefore its momentum

38

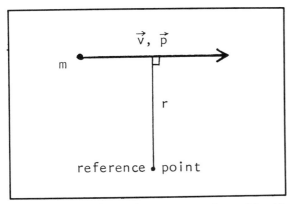

Fig. 3-5. Defining the angular momentum of a particle with respect to a point.

$\vec{p}$ are changing direction. We can deduce that $\vec{p}$ itself is not conserved because an outside (centripetal) force is acting on the object. When we consider a special characteristic of circular or rotational motion, that the motion consists of regular changes in angular position, then a new conserved quantity is revealed—the *angular momentum*. The magnitude of this vector quantity $\vec{L}$ is calculated by:

$$|\vec{L}| = L = mvr$$

where the absolute magnitude symbols show that the magnitude of the vector is being calculated. This equation is for a mass m traveling past a point at radius r with speed $\vec{v}$ (Fig. 3-5). This definition is not limited to circular motion but can be used (with care) for any moving particle with respect to any chosen reference point. The particle need not change direction. For example, a car passing by a mile-post along a road has a certain angular momentum with respect to that point. It is necessary to define the direction of $\vec{L}$ as the direction perpendicular to the plane of the motion. This is the only direction that does not change. Indeed, common experience is that a rotating top or wheel attempts to keep its plane of rotation fixed despite disturbances (sometimes called the "gyroscopic" effect).

Angular momentum can be defined and calculated for a solid object. For a rotating sphere (Fig. 3-6A) the axis of rotation tells us the direction of $\vec{L}$. Since an axis points two ways, the direction can be found exactly by using the right-hand rule. If the direction of rotation is shown by the curled fingers of the right hand, the thumb points in the direction of positive $\vec{L}$. The calculation of angular momentum in this case is not so simple.

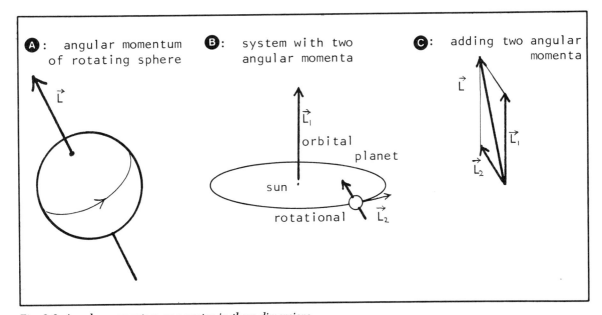

Fig. 3-6. Angular momentum as a vector in three dimensions.

The speed of the surface (and interior) of a rotating ball varies from a maximum at the outside on its "equator" to zero for parts near the axis and the poles. These variations in speed are sometimes avoided by defining an angular speed for a rotating object. Calculus is needed to find a $\vec{L}$ for a rotating sphere. We will not need the result, but we will be considering particles as tiny spheres and referring to their angular momentum as "spin."

A solid rotating object illustrates how the particle view is applied to extended bodies. Indeed, the rotating body would fly to pieces if there were not electromagnetic forces holding its atoms together. The body can be considered as composed of discrete chunks or particles as small as one wishes—at least down to the size of atoms. Each chunk follows Newton's law of motion by itself, when all forces from neighboring chunks are considered. Calculation of the behavior of the whole rotating body is actually carried out by this method. The use of calculus for an extended body automatically supposes it is composed of very small mutually interacting particles.

The conservation of a vector quantity such as angular momentum means that both direction and magnitude remain unchanged, unless there are outside forces. A spinning top is acted upon by gravity, an external force (really a torque), and we should not expect its $\vec{L}$ to remain constant. Indeed, the reader may have observed the axis of a spinning top to rotate slowly, a motion called *precession*. We will encounter precession again when we consider atoms and nuclei as tiny tops. In another example, the angular momentum possessed by a rotating skater is approximately conserved on low-friction ice. If the skater moves some mass outward by an extension of the arms, thereby increasing the radius for that rotating mass, then some other quantity must change in compensation. The speed decreases, preserving constant angular momentum. In a sense, the moving arms slow down, or speed up, the rest of the body—an application of Newton's law internal to the system.

One of Kepler's original laws for planetary motion was a statement of the conservation of angular momentum. Kepler's way to express it

was that the radius from sun to planet sweeps out equal areas in equal times. [Physics students should be able to connect these two statements.] Orbiting bodies, whether planets or particles, possess two kinds of angular momentum: that for their orbital motion around a central point, and that for their rotation or spin about their own axes, as shown in Fig. 3-6B. Each individually is conserved, unless there is interaction ("coupling") between the two (as happens for atoms). Two or more angular momenta are added just as any other vectors, as shown in Fig. 3-6C. The total angular momentum $\vec{L}$, which is the vector sum of $\vec{L_1}$ for the orbital and $\vec{L_2}$ for the spin, is conserved.

A central force such as the sun's gravity cannot change orbital angular momentum. Angular momentum is changed by a *torque*, a combination of a force F acting in some direction not passing through the center of motion and the perpendicular distance r from F to the center (Fig. 3-7A). The magnitude of the torque τ(tau) is calculated from the product of F and r thus:

$$\tau \equiv Fr$$

Torque is always a vector, with a direction perpendicular to the plane containing F and r. Thus torque can have the same direction as angular momentum. Usually we will be interested in cases where a couple of equal and opposite forces act, as in Fig. 3-7B. No linear motion results, and all the effect is rotational. There is a version of Newton's law for finding the change in angular momentum for a given torque, but we will not be needing its explicit form. Torque has units of newton meters, but they should not be confused with energy units.

3.4 ROTATIONAL AND ORBITAL ENERGY

The material flying around an axis in a rotating body such as that of Fig. 3-6A has speed and therefore kinetic energy. If the body is not moving as a whole, it possesses no kinetic energy corresponding to straight line motion. But it possesses some sort of energy of motion. The

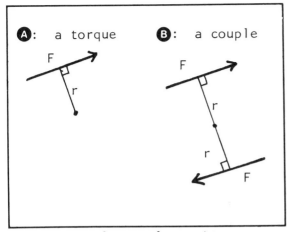

Fig. 3-7. Defining the torque about a point.

mathematical description of that energy will not be needed here. The rotational kinetic energy is to be considered as part of the total energy package in considering conservation.

For orbiting particles, the KE is simply $\frac{1}{2}mv^2$ as before, where v is understood to be the tangential speed along the curved path. Speed v is not constant unless the orbit is a circle. Thus an orbiting body has a varying kinetic energy, more near the center, less farther away. For a non-circular orbit the central force can have a component in the direction of the particle's motion. Sometimes the force speeds up the particle and sometimes slows it down.

If one particle passes near another, we can consider one particle fixed and examine the motion of the other. What happens depends on the total energy TE of the moving particle. For a system such as the sun and a planet, TE consists of the (negative) gravitational potential energy PE and the KE. The sum TE = PE + KE must be constant. There are four possible kinds of orbits, depending on the value of TE (Fig. 3-8). If TE is zero, then the particle (planet) will have zero speed when infinitely far away, out of the sun's "potential well." The path it follows is a parabola as shown. If TE is less than zero, the planet is bound; it can never have enough KE to escape the sun as PE is too large and negative. It follows an elliptical orbit with the sun at one focus, as shown.

This is the path first predicted by Kepler for motion near an inverse-square force. We can see that KE should vary, being large when the planet is in the part of the orbit near the sun, and smallest when the planet is farthest away (and having the largest PE). A component of the gravitational force speeds the planet up as it approaches and slows it down as it recedes. A special case of an ellipse is a circle, when the speed and distance are just right, as discussed earlier. At the other extreme, the planet (more likely a passing comet) can have too much TE to be captured. It has net positive KE even when infinitely far away and follows a hyperbola as shown. In all these cases angular momentum is conserved along with TE. Each is unvarying at any point of the orbit.

3.5 COLLISIONS

A one-dimensional collision of two balls (or

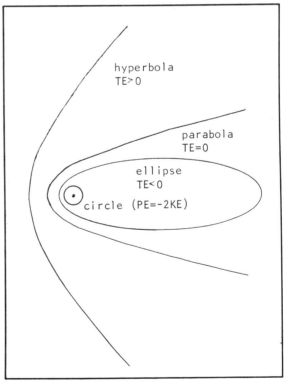

Fig. 3-8. The four kinds of paths of motion possible in a gravitational or other inverse-square field. (The path depends on the value of TE = PE + KE.)

two particles which act as small hard round balls) gives us few surprises. It also is rarely encountered in nature. This is a head-on collision in effect because the colliding bodies enter and leave along exactly the same straight line. Figure 3-9A shows the before and after of a head-on collision. We usually will not be interested in what happens during the collision. Thus we may not need to know about the forces involved in the collision, even if they act over long distances between the bodies.

When the forces are not of interest, we can step far back from a collision and explain what happens in terms of the conservation laws for energy and momentum. In effect an input-output analysis is done, with the details of the collision hidden in a "black box." This is the way nature forces us to proceed when dealing with atoms and elementary particles. Given the input kinetic energy and momentum, and some knowledge about the motion afterward, we will be able to predict other aspects of the motion of the departing particles. If the masses m_1 and m_2 are given and the incoming speeds v_1 and v_2 are known (Fig. 3-9A), then it remains to find two outgoing speeds

v_1' and v_2'. These are called two "unknowns." To find them mathematically two equations involving them would be required. Fortunately we have one from the law of conservation of energy that says:

$$KE(1) + KE(2) = KE(1') + KE(2')$$

This equation is a way of writing that total initial KE is equal to total final KE. We should recall that KE(1) involves the square of speed v_1' and so forth. It is assumed here that there is no potential energy; that is, no forces are acting at the places we measure speed and KE.

The other equation comes from the conservation of momentum before and after the collision, giving in one dimension:

$$m_1 v_1 + m_2 v_2 = m_1 v_1' + m_2 v_2'$$

Again, this law holds because there are no forces where the measurements are made. We need not worry about vectors in one dimension, but we should realize that a speed can be positive

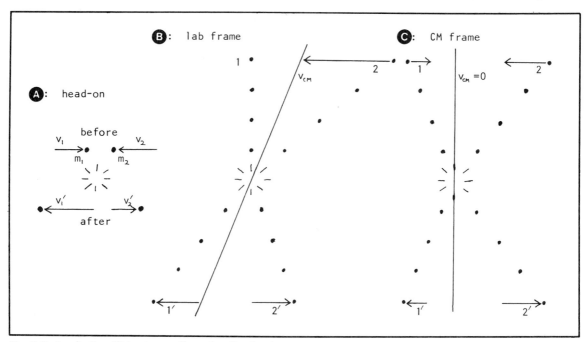

Fig. 3-9. An elastic collision in one dimension (head-on).

or negative, depending on its direction. If directions toward the right are positive in Fig. 3-9A by the usual convention, then speeds v_2 and v_1' are negative.

We will not be concerned with the algebra of solving such a collision, but we can guess some of the results in extreme cases (a common practice by physicists exercising their trained intuition). Properties of the particles such as the mass are usually considered to be *parameters*, values which can be fixed for a given problem but changed in different problems. The best procedure is to consider different values of the parameter first, then consider the effect of varying a variable such as speed.

Suppose first that the masses are equal $(m_1 = m_2)$ and that m_2 is at rest $(v_2 = 0)$. Then afterward m_1 will be at rest and m_2 will have acquired all the energy and momentum, having speed $v_2' = v_1$. The particles have exchanged momenta. This holds true even if v_2 is not zero. The process of "exchanging" will arise repeatedly in particle physics.

Suppose next that m_2 is huge, so that it acts like a brick wall. Whether or not m_2 is moving, m_1 will bounce from it, acquiring a negatively-directed speed that is increased by the amount of v_2. The speed of m_2 is essentially unchanged when a tiny particle bounces from it. If m_2 is motionless, then m_1 reflects from it. We should not overlook an important symmetry in these collisions. The same results occur when the roles of m_1 and m_2 are reversed.

We have presented this collision as if some sort of action-reaction force occurred only when the balls touched; there could just as well have been a strong electric repulsion between two like charges that never touch. The particles will approach to the point that both stop, having converted all their KE temporarily into electric PE. Then the repulsive force reverses their motion and they fly apart. The collision could also be a close gravitational encounter between two masses, except that the masses never stop but continue on their hyperbolic paths.

For collisions with contact, we have assumed that they are *elastic*, where all the energy that goes into "squashing" the balls is regained when they rebound. No energy is lost to heating the balls. When an *inelastic* collision occurs, an amount of energy—usually unknown—is lost to heating the bodies or particles, sticking them together, or breaking them up to form new particles. The incoming particles can no longer be said to be *scattered*, with the same ones coming out as went in. A better term is that they *"reacted."* Energy conservation cannot be used unless one knows more about the reaction. It certainly cannot be said that the total KE input is equal to the total KE output. (Sometimes more KE can come out, for example when one of the particles "explodes," releasing previously stored energy.)

There is an important and different way to look at the one-dimensional collision discussed above. Suppose that m_1 is moving slower than m_2 and may be stationary. Mathematically one would say $v_1 < v_2$, disregarding direction. For any two masses, moving or not, there is a center of mass. In this case it is moving, to the left at a slower speed than either v_1 or v_2. Figure 3-9B shows a series of snapshots in time of the particles as seen in our ordinary "lab" frame of reference. The center of mass moves along the line shown. We can calculate the speed v_{CM} of the center of mass by:

$$v_{CM} = \frac{m_1 v_1 + m_2 v_2}{m_1 + m_2}$$

Speed v_{CM} is an algebraic average of v_1 and v_2, weighted by the masses. As far as the momentum is concerned, the total momentum can be calculated just as well from the sum of the masses and v_{CM}. In two or more dimensions, the equation for v_{CM} can be considered a vector equation.

The reason for considering the motion of the center of mass is so that we can travel along with it and look at the collision from a new perspective. The law of conservation of momentum guarantees that the center of mass never changes its motion, so that the observer is in steady motion at speed v_{CM} during and after the collision.

Regardless of the initial speeds of the individual particles, they come together in a symmetric manner, at the same speeds ($v_1 - v_{CM}$, and $v_2 - v_{CM}$. After the collision they depart with the same speeds as if they had reflected from each other. A series of snapshots of the two particles colliding, as seen by a camera moving with speed v_{CM} is shown in Fig. 3-9C.

For several reasons physicists find it easier to consider collisions in what is called a "center of mass frame of reference" (shortened to "CM frame") and then to transform the results to the "lab frame," which is where speeds and masses can be measured. The transformation for any (low) speed has the form:

$$v_L = v_C + v_{CM}$$

where L and C refer to lab and CM frames, respectively. This equation works in two or three dimensions when used with vectors.

The lab frame is considered stationary and the particles do all the moving. In the CM frame the physics of the collision occurs regardless of the outside world and nature is seen at her simplest. What happens in the collision depends on how the two particles move toward each other and nothing else. More accurately, the CM frame is known as the "center of momentum system," since the common parts of the particles' momenta are what are eliminated from consideration. A collision can be analyzed without regard to energy in the CM frame. Thus inelastic collisions are easier to study.

In two and three dimensions collisions are much more complicated. The incoming bodies define a plane, and the outgoing bodies could be in a different plane. Sometimes one or both have unusual spins that interact (as sometimes happens in billiards, causing one ball to leave the plane of the table). Fortunately, there is a trick for simplifying three-dimensional collision of two bodies to two-dimensions. It might seem that the billiards table is an ideal model for collisions, since the balls are hard, rather elastic, and normally move in only two dimensions. Billiard balls also rotate as they move. Elementary particles have spin, too,

but the spins do not interact like billiard ball spins. Frictionless hockey is no better model. Only models using extremely small round bodies minimize the effects of mechanical spin on collisions.

Difficulty arises as soon as we count equations in trying to solve a two-dimensional collision. Now there are four unknowns, consisting of the two components of each velocity $\vec{v_1}'$ and $\vec{v_2}'$ shown in Fig. 3-10A. We can get one equation from the conservation of energy and two from vector momentum conservation in two dimensions. We are short one equation, and some other means must be used to measure at least one fact about the outgoing particles or the collision cannot be fully solved.

In some cases we could know the total angular momentum of the particles before the collision and use its conservation law to predict it afterwards. No new unknown variables are introduced, so this provides one more equation. But if the particles can spin, and most do, that introduces new variables, one for each spin, and we are worse off! Physicists working with collisions of elementary particles are left with incomplete information no matter where they turn. They must measure as many variables as possible, both before and after the collision.

The CM system is quite helpful for studying two-dimensional collisions. The $\vec{v}_{CM}$ is now a vector which can be calculated from the incoming velocities. It represents the total momentum available, as shown in Fig. 3-10B. In the CM frame the two colliding particles appear to reflect from each other, as shown in Fig. 3-10C. Each particle is deflected through the same angle θ. The collision appears symmetric.

The force of interaction is often electric, nuclear, or gravitational. From far away a comet appears as if it were moving in a straight line toward the sun. If it were to continue in the straight line as in Fig. 3-11A, it would miss the sun by a distance b called the *impact parameter*. As far as we and the comet are concerned, it has an initial angular momentum $L = mvb$, using the sun as the reference point. This L cannot change even though the path curves (as an hyperbola). It is assumed that there is no touching, so that comet and sun

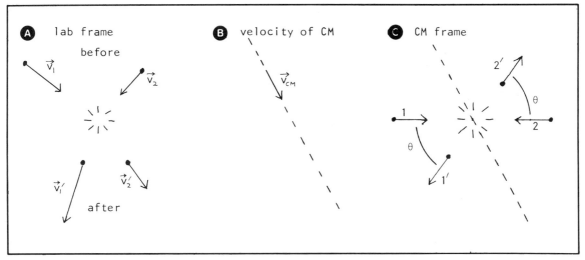

Fig. 3-10. An elastic collision in two dimensions.

cannot exchange angular momenta, and further-more that our comet comes back intact. After this "collision" the comet approaches another straight line in a different direction but the same distance b from the sun.

Another useful parameter to examine in collisions is *cross section*. If the two identical balls in Fig. 3-11B each have radius $b/2$, then there are four different kinds of collisions possible. Each ball has a cylindrical region of diameter b around it, as indicated. If the two regions do not overlap, there will be no collision (case 1). In case 2 the

regions just touch and the collision will be grazing—a slight touch but no noticeable change in the motion. In case 3 there will be a glancing collision and both balls will depart at certain angles from their original paths. In case 4, a head-on collision is obtained. The impact parameter for case 1, measured from center to center, is greater than b. For case 2, it is equal to b. For case 3, it is less, and for case 4 it is zero. Since the size of b determines whether the two particles will ever touch, the cross section is the cross-sectional area of a cylinder of radius b, or πb^2. The effective

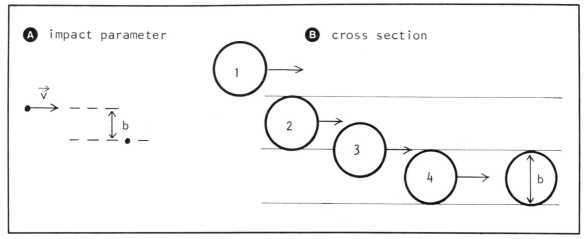

Fig. 3-11. Describing a collision with impact parameter and cross section.

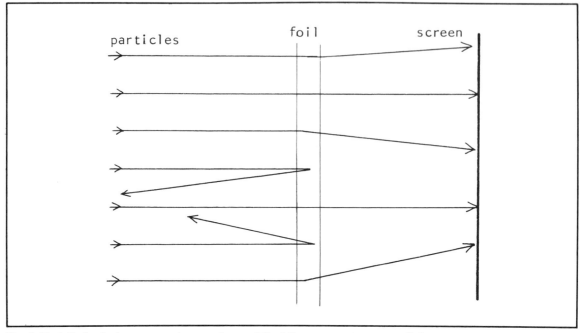

Fig. 3-12. A Rutherford-type scattering experiment. (The foil must contain very small dense scattering centers.)

cross-section as far as collisions are concerned is four times the cross-sectional area of one of the particles:

$$\frac{\pi b^2}{4}$$

When the particles do not touch, as in the case of the comet passing the sun, there is still an effective cross section. Complex calculations are needed to obtain it in terms of the impact parameter, the energy, and the strength of the interaction. The cross-section depends on the energy, since, for example, a comet passing far from the sun is deflected little, and one passing close by seems to rebound strongly. In the case of elementary particles, we often have little idea of their actual diameter and cross-sectional area. Indeed, for reasons to appear later, such a model of particles is necessarily inaccurate and misleading. Nor can we measure the tiny impact parameters. We do not know where the particles are, except in a statistical sense. Any experiment must have a statistical result—that is, an average result from many repetitions of the collisions, each one unpredictably different in itself.

One thing we can observe is the pattern that results when a spray of particles passes near another particle, and from the pattern deduce the effective cross section. The equivalent experiment was first done by Ernest Rutherford in the early 20th century. He observed the points of light that appeared on a screen when high speed positively-charged particles from a radioactive substance impinged on thin gold foil. A few of the outgoing particles after collision were scattered very far from their initial line of travel, but most traveled straight through as if there were no foil in the way (Fig. 3-12). He deduced in 1911 that gold atoms must consist of mostly empty space and very small heavy cores, called "nuclei."

Statistically, the same results are obtained whether a sleet of particles passes near one heavy gold nucleus or through a foil randomly filled with them. The occasional repulsion of a particle can be sufficiently explained by supposing the nucleus has a large positive charge. With several more facts about the experiment, Rutherford even calculated the effective cross-section of the nucleus, indicating its size. Many refinements of this sort of experiment have occurred since, but physicists are always interested in finding the effective cross-section as various other parameters, particularly the energy of the particles, are varied. The scattering of particles by other particles, with electric forces providing the interaction, has been given the generic name of "Rutherford scattering." Later we will be interested in scattering involving the strong nuclear interaction.

Chapter 4
Fields

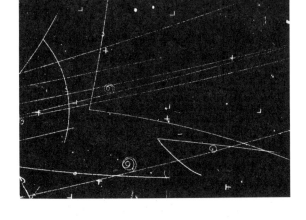

WE MUST INTRODUCE THAT OTHER MAJOR model used in physics, the concept of a field. Very different from particles, fields are nevertheless quite essential to our later discussion of modern theories about particles. Moreover, the modern approach is to use fields instead of forces when talking about how one body or particle interacts with another.

4.1 FIELDS AND FORCES

A *field* might be defined as "something" that exists in space around a source of a force. The field has direction and therefore must be described with vectors. At every point in space around a source, there is a certain strength and direction of field. The field is in some way a fiction because we have to put some form of detector there to know there is a field. But it also has some reality; for example, we can send radar signals across the solar system and receive their echoes. The signals are fields which carry energy toward a distant body, and bring back the tiny amount reflected. Fields

can be either stationary (*static*) or moving (*dynamic* or *radiation*). We will discuss radiation fields later.

The gravitational field is referred to so widely that the reader is likely to be already familiar with the concept. Also in one's common experience might be the electric field near something charged and the magnetic field near a compass or magnetized iron. Figure 4-1 represents the gravity field around the outside of the Earth in schematic fashion. The Earth is really spherical and has radial field in all directions in three-dimensional space, but just a cross section can be shown on paper. The presence of the otherwise invisible field is signified with thin lines. These extend radially outward to infinity, but in the case of gravity the direction of the field is inward. The reader has perhaps traced the field lines of a magnet using iron filings, as represented in Fig. 4-2. This method gives the lines more reality than is merited.

For a source of gravitational field such as the Earth (presumably spherical) and for a source of electric field such as a charge, the "lines" are not only radial but also uniformly arranged. However, they do not remain at constant spacing but diverge. They are most concentrated near the source. Before we discuss finding the force from a field, we can use the field concept to understand why certain forces have an inverse-square dependence. Consider an imaginary sphere drawn around the Earth, as in Fig. 4-3. Although the lines are just a mathematical model, every line starting at the Earth must pass through sphere around it. The density of lines—that is, the number that pass through a unit area—must decrease as the imaginary sphere is made larger and farther away. There are available only so many lines, although the exact count is irrelevant in physics. The area of the sphere increases with r as r^2, so the density must decrease as r^2. There must be something about the gravitational field that decreases as r^2,

and indeed we have learned that its strength so decreases. This is the same as saying that gravitation force is an inverse-square force. A field that is becoming weaker in the distance is represented with less closely spaced lines.

The force is found from the field simply by putting a test particle in the field and measuring the force on that test particle. The force can be calculated by multiplying the field strength times the appropriate characteristic of the test particle (its mass or charge, for example). Of course, a different value of the force will be found for a different size of test particle.

4.2 THE GRAVITATIONAL FIELD

In the case of gravity, the force is equal to the test mass times the field strength. The gravitational field strength at a distance r is given by:

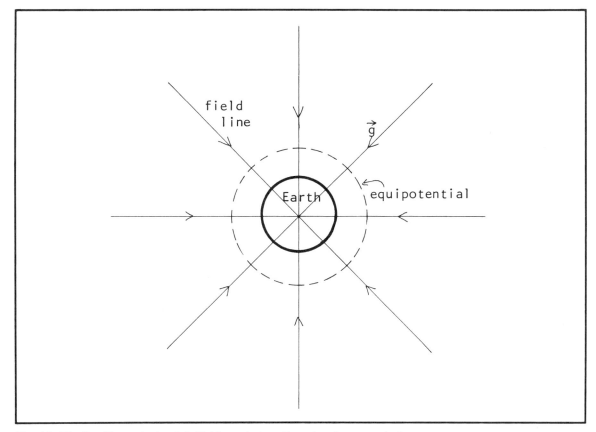

Fig. 4-1. The Earth's gravitational field (shown in two-dimensional cross section).

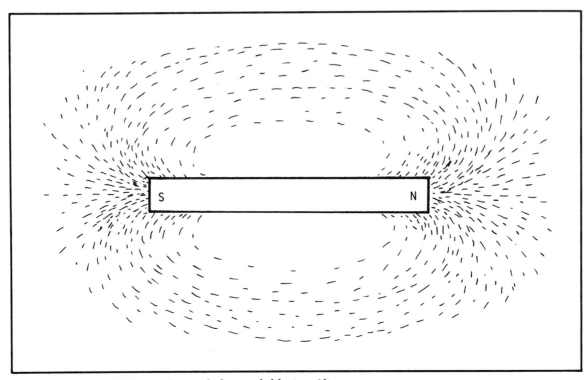

Fig. 4-2. *Magnetic field lines as they might be revealed by iron filings.*

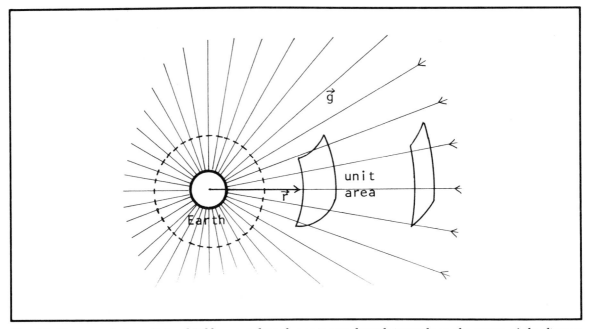

Fig. 4-3. *The amount of gravitational field passing through a unit area depends inversely on the square of the distance to its source.*

50

$$\vec{g} = \frac{-GMr}{r^2}$$

where $\vec{g}$ symbolizes the vector field, G is the usual constant, and M is the source. The r is called a *unit vector*, telling that direction of positive field is radially outward from the source. Equations such as these must have vectors on each side to be meaningful. Since gravity really points inward, a minus is used. The force calculated with $\vec{F} = m\vec{g}$ appears the same as the simple equation for gravity stated in Chapter 2, but vector arrows have been added and the field is understood to vary with radial distance. To make the units come out, the field strength must be measured in newtons per kilogram.

We have discussed gravitational potential energy and found that anywhere around a mass another test mass has the potential to do work. At every point there must be an energy in the field corresponding to the PE of a mass. A useful result from more advanced physics is that a field strength must be squared to calculate an energy. Since there is no limit to the space occupied by the field of one mass, the energy would be infinite unless we defined the gravitational field energy to be a density, an energy per unit volume. The field energy density then is:

$$GE = \frac{\vec{g}^2}{8\pi G}$$

The square of the vector field $\vec{g}$ is a new operation that makes a scalar field for GE. GE depends on the square of the magnitude of the vector field strength. To avoid there being an infinite total energy, the GE must decrease with distance from the source mass; it does so because $\vec{g}$ decreases. GE appears in a different form in Einstein's theory of gravity, covered later.

It will be useful to describe how the motion of a particle is calculated for a gravitational field. The results should agree with what Kepler found for the planets. We will also learn more about how physicists use laws to obtain useful results. When Newton's law of motion is coupled with the gravitational force law, the following equation connecting the properties of gravity and the motion (acceleration) results:

$$\frac{GmMr}{r^2} = m\vec{a}$$

This equation describes motion in three dimensions. [Physics students who have had calculus will recognize that the $\vec{a}$ should be written as the second time derivative of $\vec{r}$.] A major restriction on this procedure is that we must use only low speeds. Speeds which can approach a substantial fraction of the speed of light require use of Einstein's relativity (a later chapter).

An equation of motion has one or more solutions which give the results in terms of how the coordinates of the moving mass m vary in time. The exact solutions depend on exactly what initial conditions are chosen at a given starting time— for example, what velocity the planet has at $t = 0$ and where it was located. We will not get into details of which coordinates to use except to say that the most convenient coordinates turn out to be ones involving the radial distance of m from the central mass M, and two angles.

The solution of this equation involving the two masses breaks into three parts, the motion of the center of mass and the motion of each mass around the center of mass. For planets moving around a sun, the sun is so much heavier that the center of mass hardly moves. To a good approximation the solution of interest is that of one planet moving around a fixed sun. After the mathematical work is done the solution gives a path lying in one plane that is a conic section—a circle, an ellipse, a parabola, or an hyperbola as were shown in Fig. 3-8. The different possible paths depend on the total energy and angular momentum that the planet has; these are both conserved, as discussed earlier. The position at any time of the planet or other particle along the path can be found from the solution if the initial position is given.

4.3 ELECTRIC FIELD, CHARGE, AND CURRENT

Before we can assemble electromagnetism out of electric and magnetic fields, we need to learn more about the electric one. We must necessarily consider only a static field involving stationary sources (charges). Any motion would connect electric and magnetic effects together.

Charge is a fundamental property of matter and has its own units. These are called *coulombs*, after Charles Coulomb who first found the laws governing charges about 1785. A charge of one coulomb has no simple relation to our experience. It is a very large amount of charge, containing about 10^{19} bits of elementary charge, and would be rather dangerous to get near. Charge comes as two kinds only, positive and negative. This was discovered by Charles Dufay in the 18th century. Most elementary particles have one unit of charge, either positive or negative. That unit of charge is very small, being $1.6021917(10)^{-19}$ coulomb, as best can be measured thus far.

To discuss charge further, we must introduce another elementary particle, the *electron*. Its name comes from the Greek word for "amber" because the ancient Greeks were familiar with the static electric properties of amber. By 1881 the mysterious phenomenon of electricity was suspected (by Hermann Helmholtz) to be made of discrete bits of charge. They were named electrons by Johnstone Stoney in 1891 and found by J. J. Thompson in 1897. It was established by Philipp Lenard, among others, that electricity, presumably composed of electrons, could be made to pass between two plates in a partial vacuum. One of the plates emitted something that acted like a particle and was charged, since other charged plates affected its motion. Thompson was able to measure the ratio of the charge of individual particles to their mass—a ratio called $\frac{e}{m}$, where e denotes the electron charge. Lenard received the Nobel Prize in 1905, and Thompson in 1906.

It remained for experiments by Robert Millikan in 1917 to measure the forces on elementary charges and show that the charges come in discrete units. He obtained and measured different amounts of charge on oil drops suspended between charged plates and was able to deduce the smallest unit and thus obtain the charge of the electron explicitly. Millikan received the Nobel Prize in 1923. All this work was done with the charge that could be made to flow through wires (now known to be electrons) before its properties were fully known. How the experiments were done can be better explained when electric fields are further discussed. The electron charge, defined negative, was also found to be equal in size and opposite in sign to the positive proton charge. This was in contrast to the fact already known that the proton must have over 1800 times the mass of the electron. The electric field, called $\vec{E}$, can be calculated from its source charge q by:

$$\vec{E} = \frac{Kq\vec{r}}{r^2}.$$

Again, we see an inverse square vector field, showing that the density of field lines must decrease at exactly the right rate with distance so as to continue radially forever. The direction of $\vec{E}$ depends on the sign of q. The law for the field is arranged so that a positive charge gives rise to a vector field that points radially outward (Fig. 4-4A). If q is negative, $\vec{E}$ points inward toward the charge (Fig. 4-4B) because it must be in a direction opposite to r.

As we examine the electric field law, we can discover some new peculiarities about the universe itself. For example, where do the field lines eventually go? If the universe is electrically neutral, then there must be negative charge somewhere else to balance the positive charge of Fig. 4-4A. In practical terms, isolated charges are obtained by rubbing or otherwise extricating them from matter, leaving negative charge behind. We have already learned that the electric force is extremely strong so that charges cannot be isolated very readily.

A law of nature is that every electric field line must connect two opposite charges. That is, every fictitious line leaving one charge must arrive at another opposite charge. Therefore let us consider

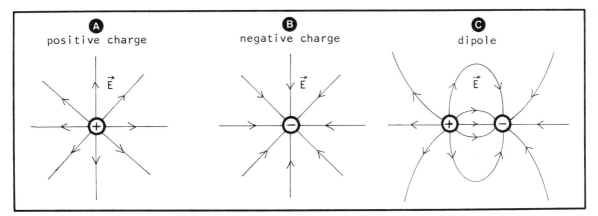

Fig. 4-4. Electric fields near point charges (shown in two-dimensional cross section).

two charges near each other, one positive, the other negative, in an arrangement called a *dipole* (Fig. 4-4C). Might they be satisfied with each other? Yes, in some respects. If the diagram could be drawn large enough, we could show that every line leaving the positive charge curves around to land on the negative charge, and does this in such a manner as to leave one and reach the other in a uniform manner. The lines bunch together between the two charges and are very sparsely distributed elsewhere. We have already learned that the electric field is strong where the lines are dense. If you imagine yourself to be very far from the dipole, then the two charges will appear to be very close together and in fact partly cancel each other's effect. Calculation would show that the field at large distances grows weaker by r^3, a faster rate than the inverse square law.

There is something tense about the space between this pair of charges. Wherever there is field, there is energy. A third positive test charge placed between the two in Fig. 4-4C will experience a force to the right, repelled by the positive and attracted toward the negative charges. It would have a potential energy decreasing steadily to the right. The electric field energy density EE can be calculated by:

$$EE = \frac{1}{2} E_0 \vec{E}^2.$$

This law contains several aspects of describ-

ing physics mathematically. First, there is the energy density. In each cubic meter near the charge(s) there is a certain amount of energy, more near the charge, less farther away. Since $\vec{E}$ decreases by r^2, the energy density decreases by $(r^2)^2 = r^4$. Secondly, we need another special constant ϵ_0, called the *permittivity* of empty space. It tells how much electric field is "permitted" from a charge in space, and has the value $8.8541853(10)^{-12}$ in SI units. We are also now in position to remove some of the arbitrariness from the electric force law. The constant k that was used is defined to be

$$\frac{1}{4\pi\epsilon_0}$$

(Why there is so much complication to the constants is another story.) Thirdly, the equation for EE shows a squared vector field, which is a scalar field and intensifies the effect of large values of $\vec{E}$.

The force on a charge in an electric field is readily calculated with the vector relation:

$$\vec{F} = q\vec{E}.$$

If q represents positive charge, the force is in the same direction as the field. If q is negative, the force is in the opposite direction.

We will be needing to know the patterns in space, or the "shapes," of the fields due to charges in a few simple arrangements. If mass could be arranged similarly, the resulting gravitational fields would have the same shapes. We have discussed the fields for single and pairs of charges. If the pair consists of two or more charges of the same sign, the field is rather complicated near the charges but at large distances appears to be coming from a compact cluster of charge which acts like a single point charge of the appropriate size (Fig. 4-5A).

The next complication is to place charges in a spherical arrangement, such as on the surface of a metal sphere (Fig. 4-5B). Although they want to repel each other away from the sphere, they will not leap away unless there are too many. This arrangement of charges also results in a radial field and appears as a point charge if one is far away.

In Fig. 4-5C the charges are arranged in a sheet, something charges do not like, but it can be done if they are held by atoms in the sheet. The field from this arrangement points uniformly away in both directions from the sheet and is said to be *homogeneous*. The density of its lines does not change anywhere and so the field does not grow weaker at great distance. An infinite sheet of charge viewed from very far away still looks infinite and the field strength is unchanged. A more realistic sheet of finite size has a complicated field near its edges, and from very far away would appear as a point charge.

To prepare us better to understand how elementary particles can be moved, detected, and measured, we can now explore how an electric field affects charges. We have already hinted that a charge not in motion will tend to start moving if it is near other charges (which are the source of a field). The dipole shown in Fig. 4-4C would collapse, perhaps explode, if the charges were not held in place. We are restricted here to low speeds or Einstein's theory for motion becomes applicable. We must also neglect for the present the field that a moving charge "carries" with it.

A familiar field is the radial field. If our test charge is negative, and attracted to a fixed positive charge, the motion is quite similar to motion in a gravitational field. The charge can have a circular, elliptical, or hyperbolic orbit in one plane (recall Fig. 3-8). The circular and elliptic orbits, with their negative total energy, will remind us later of an oversimplified model of an atom. The hyperbolic orbit describes what happens if, say, an electron is shot near a proton (Fig. 4-6A). It starts in a straight line, curves around a certain amount, and comes out in a curve that straightens. We say that the electron is **deflected** by the proton; the latter will not move much since it is much heavier. If the incident charge is a light weight positive one (the positron), then the deflection will work as in Fig. 4-6B. The two charges can never orbit, but the repulsion results in an hyperbolic path similar to that for the electron but turned the other way.

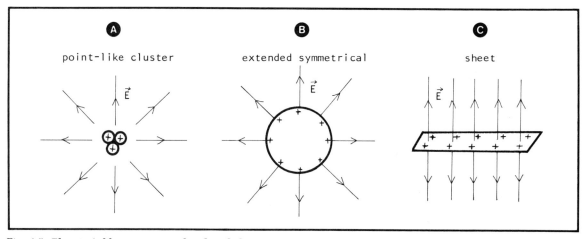

Fig. 4-5. Electric fields near various distributed charges.

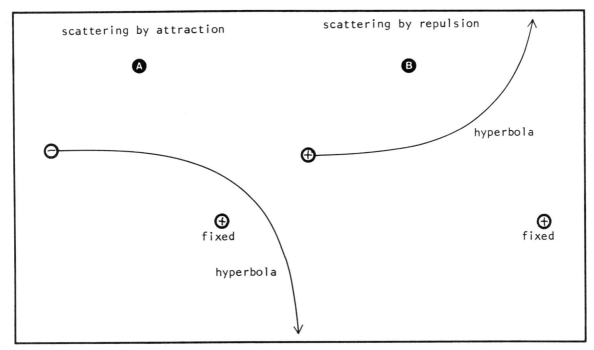

Fig. 4-6. Motion by charges in the electric fields of other charges.

A useful arrangement of charges giving rise to a uniform (homogeneous) electric field is made with two parallel conducting plates, often called a "capacitor" (shown in cross section in Fig. 4-7). Field lines start on the positive charges and end on their negative counterparts. A negative charge such as an electron will experience an upward force in the downward field, as shown. When the plates are oriented horizontally, an experiment similar to Millikan's oil drop experiment can be done. For a given negative charge (one or more electrons residing on an oil drop), there is some strength of electric field which will provide an upward force $\vec{F_e}$ to exactly balance gravity $\vec{F_g}$. Other procedures enable finding the mass of the drop, and then its charge can be found. Millikan was thus able to measure the smallest unit of charge.

Another tool of historic and modern importance for studying matter is the use of electric currents. We will discuss electricity only enough to see the particle aspect of it and to mention some generally useful concepts. Any moving charge, whether in space or in a wire (a conductor), can be viewed as a *current*. Current I is defined as the amount of charge Δq passing a given point in a given time Δt. It is calculated by:

$$I = \frac{\Delta q}{\Delta t}$$

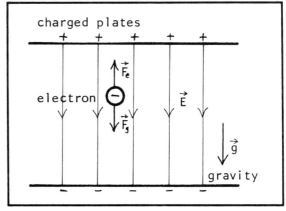

Fig. 4-7. Balancing electric and gravitational force, such as for a Millikan-type experiment (shown in cross section).

I has the units of coulombs per second, which has been defined as the *ampere*. The ampere rather than the coulomb has been taken as a basic SI unit for electric concepts. (We will meet Msr. Ampere later.) The current in a wire consists of moving electrons, but positive charges also constitute a current in space or in an appropriate material. Positive current, which is how I is defined, moves in the opposite direction from a current of electrons.

What makes current flow? For charges to move, they must be attracted or repelled, finding themselves in a "downhill" situation in regard to their PE. The amount of PE per charge is defined to be the *potential* or *voltage*. The potential V can be calculated from:

$$V = PE/e \ ,$$

where e is the charge of an electron. V has the units of joules per coulomb, which has been defined as another common electric unit, the *volt*. The volt is named after Alessandro Volta who, in 1799, made the first "voltaic cell," a chemical arrangement that produced electric current and potential energy from chemical potential energy. These cells, usually called "batteries" now, were once the major source of voltage and current for experimenters. The flow of currents in solutions, whether in batteries or other electrolytic cells, was another source of early information about the size of the electric charge.

The preference of charge to reside on the outside of metal containers, an effect noticed by Benjamin Franklin about 1755, was first seen as a proof of the inverse-square form of the electric force in 1767 by Joseph Priestly. Extremely sensitive tests of the law are obtained by putting charge on the outside of a metal shell and observing if any flows to the interior.

The principle of invariance can be used to obtain the law of conservation of charge. The invariance or symmetry involved is called one of "gauge." This idea has proven to be very important in modern particle theory and will be discussed further in a later chapter. The conservation

law has been experimentally verified in many cases. All particle interactions, involving any of the four forces, always preserve the total charge count. As far as we know the universe as a whole is electrically neutral. Any small imbalance anywhere would lead to a flow of charge to correct the imbalance. Therefore the number of positive unit charges in any situation, minus the number of negative units, must be zero.

4.4 THE MAGNETIC FIELD

Looking for the source of the magnetic field, in analogy to the charges that give rise to the electric field, leads us into some very modern physics indeed. At first it might be thought that "magnets" are the source of the field. But these are large material objects, and we must find the field based in elementary particles or abandon the concept.

Historically credit can be given to Hans Oersted for showing in 1820 that electric current is a source of magnetic field. A straight wire carrying a current I has a magnetic field around it (readily detected with a tiny compass as in Fig. 4-8A). The field has a circular form and a direction; it is designated with the symbol $\vec{B}$ and is a vector field. Andre Ampere showed at the same time that a circular coil of wire carrying a current had a rather uniform straight magnetic field inside it (Fig. 4-8B) and exhibited magnetic *poles* at the ends. Compasses and other magnetized pieces of iron have poles of two kinds, called North and South from use as direction finders in the Earth's magnetic field. The elementary laws for poles were that "like" poles—that is, a North and a North, or a South and a South, repelled each other—and that "unlike" poles—a North and a South—had a strong attraction (indeed, followed approximately an inverse square force law as shown by John Michell in 1750). A magnetized bar of iron (recall Fig. 4-2) exhibits the same vector field as the coil of wire carrying current, with the direction of $\vec{B}$ coming from the N pole.

A familiar "righthand rule" relates the directions of current and field. If the thumb points in the direction of positive current, then the curled fingers give the direction of the resulting magnetic

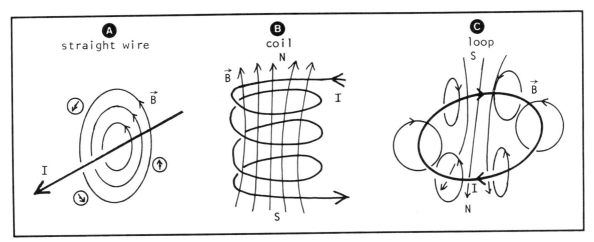

Fig. 4-8. Magnetic fields from currents.

field (Fig. 4-8A). If the thumb is pointed in the direction of a known straight field, then the curled fingers show which way positive current must be traveling in a coil (Fig. 4-8B). To use this convention correctly, we must recall that the direction of positive I is opposite to that of electrons in a wire.

Both the analogy of poles to electric charge and the relation of current to magnetism have intrigued physicists. Even a single loop of current is known to act as a magnet (Fig. 4-8C). Ampere did not take long to suppose an atomic theory for magnetism: that tiny electric currents in "magnetic particles" gave rise to the magnetic field of a chunk of matter composed of those particles. He was partly correct. The established facts that charge could "induce" charge in other materials and that magnetic poles could magnetize certain other materials caused a search for how currents could "induce" currents in other conductors. The successful results of this search are a basis for the later chapter on electromagnetic effects.

With the magnetic field we encounter different shapes of field. It is more difficult to obtain a uniform magnetic field, and there is a requirement for field lines to be closed curves. A magnetic field line does not seem to need a source. A magnet with its two poles is simply a material device for concentrating and continuing a closed field line. Physicists have long searched for a source of magnetic field lines called the "monopole," and this search has acquired an importance in elementary particle physics far beyond its relation to magnetism.

The magnetic field strength $\vec{B}$ has SI units of tesla (T), named after the 19th century experimenter Nikola Tesla. The Earth's field is less than 10^{-4} T. The largest practical fields available in iron are about 2 T. By analogy to electric field energy density, there is a magnetic field energy density which can be written mathematically in a similar form:

$$ME = \frac{1}{2} \vec{B}^2/M_0$$

The force which a magnetic field can apply to a particle depends both on the two magnetic poles, now called a *magnetic dipole moment*, of the particle, and on its charge. We shall learn that most elementary particles have both charge and magnetic dipole moment. The dipole moment is like a small pointer in space, and the forces on it in a magnetic field tend to align it with the field, as in Fig. 4-9A. The forces act as a torque, which only can turn the particle or body, not move it. The dipole moment has minimum energy when aligned with the field. The magnetic dipole moment is assigned the vector symbol $\vec{\mu}$ (Greek mu). The potential energy of it in a field is given by:

$$PE = -\mu B \cos \theta ,$$

where θ is the angle between $\vec{\mu}$ and $\vec{B}$. The connection between μ and a tiny loop of current I is given by:

$$\mu = IA ,$$

where A is the area of the loop.

The magnitude of the magnetic field at the center of a loop of current I of radius r is given by:

$$B = \frac{M_0 I}{2\pi r} .$$

The permeability constant M_0 should not be confused with the magnetic moment μ. This relation is useful for estimating the strength of the field near a particle of known magnetic moment. The effective I can be estimated if the approximate cross-sectional area of the particle is known.

In regard to the charge of a particle, the magnetic field causes it to move in an unusual manner. The force exerted by the field is at right angles to both the field and the direction of motion (Fig. 4-9B). Furthermore, the force depends on the charge and speed of the particle. If there is no motion, there is no force, and nothing happens. The right hand rule gives the relationship of the directions of force $\vec{F}_m$, velocity $\vec{v}$, and field

$\vec{B}$ as shown. If the right hand thumb points in the direction of $\vec{v}$, and the first finger points along $\vec{B}$, then the force is in the direction of the second finger. This rule is for positive charge. If q is negative, the force is reversed.

For this simple case involving only right angles and a uniform field, the magnetic force can be calculated with:

$$F_m = qvB.$$

No attempt is made here to give the vector relation mathematically because it involves further symbolism we will not be needing. When right angles are not involved, the component of $\vec{v}$ perpendicular to $\vec{B}$ is used. The force will always be perpendicular to the plane containing the field lines and the velocity vector. If $\vec{v}$ happens to be in the direction of $\vec{B}$, then there is no force. A magnetic field does not "see" a particle traveling along with it.

The motion of charged particles in magnetic fields can be quite complicated. We can consider only one special case, involving a uniform field. The particle experiencing magnetic force at right angles to its motion will not continue in a straight line. Such a force causes it to pursue a circular path, as in Fig. 4-9C. Such a motion is very useful in designing machines to produce rapidly moving particles. Since the force is always at a right angle

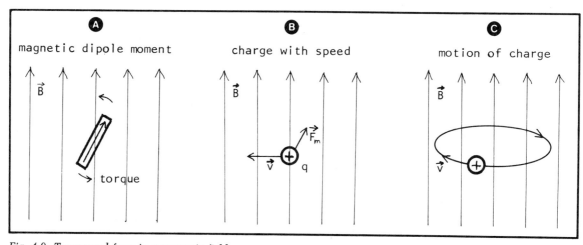

Fig. 4-9. Torque and force in a magnetic field.

to the motion, it can never accelerate the particle, however.

4.5 INTRODUCTION TO PARTICLE ACCELERATORS

A "particle accelerator" is a machine for accelerating particles, whether elementary ones or atoms, to high speeds, giving them enough kinetic energy to disrupt other particles and atoms or otherwise find new phenomena. Electric and magnetic fields are very useful for accomplishing this objective, provided that the particles are charged. (We shall meet some neutral particles later, and there is difficulty controlling them.) Here we discuss some of the simpler forms of particle accelerators.

The common energy unit used for discussing the energy given to particles by accelerators is called the *electron volt* (eV). One eV is the energy a charge acquires as it passes through a potential difference of one volt. A potential difference of one volt occurs in an electric field of a certain strength. For example, a field strength of a million newtons per coulomb, which is equivalent to a million volts per meter, gives a potential difference of a 1000 volts in 1 mm, almost enough to break air down (ionize it and allow a spark to discharge).

In Fig. 4-10 an electric field is shown between two charged plates in cross section. To be charged, there is a potential difference between the plates. A positive test charge, as it is moved toward the positive plate, has higher and higher potential energy and therefore greater potential. The location of zero potential can be chosen anywhere. Some regions of higher and lower potential are shown (as dashed lines), numbered with typical potentials in volts. These lines are called "equipotentials" since a particle moved along one would not change its electric potential energy. A positive charge tossed into the central part of Fig. 4-10 and left on its own would accelerate downward, toward the negative plate. There would be a conversion of its PE to KE. It would collide with the plate if left alone. A charged particle or body would acquire the same kinetic energy regardless

of its mass. A massive particle would simply take longer to accelerate from one equipotential to the next.

To set the scale, one eV is equivalent to $1.602(10)^{-19}$ joules, using the basic unit of charge. At room temperature the air molecules are carrying about 1/40 eV as they dash about colliding with us. So 1 eV is quite a bit of energy for a tiny particle such as an electron, but it is a trifle on our scale. In modern particle physics, not much happens until particles are given over one million electron volts (1 MeV), the energy a charge would acquire in falling between two plates with a potential difference of a million volts. MeV (mega-electron volts) is a popular unit, as is GeV (giga-electron volts). A particle with 1 GeV is probably moving near the speed of light and carrying enough KE to break up most other particles. Physicists are now reaching a thousand times higher, to 1 TeV (tera-electron volt).

A *linear accelerator* is an extension of the method used to accelerate electrons in a television tube. Only electric fields are used. The method is the oldest used, first introduced in the 1920's. It gradually replaced the reliance on radioactive sources and cosmic rays for high speed particles. A van de Graaf generator, which works by carrying electrons on a moving belt and collecting them on a charged sphere, can produce up to ten million volts and remains a popular source of high voltage. The potential from it is put at one end of an evacuated tube in which charged particles are in-

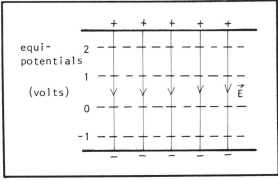

Fig. 4-10. Equipotentials (surfaces of constant voltage) in an electric field (shown in two-dimensional cross section).

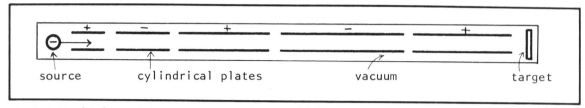

Fig. 4-11. Simplified form of a linear accelerator, using potential differences to accelerate charged particles (shown in cross section).

troduced, and they simply dash toward it. Another form of linear accelerator uses an electrical method to produce high potential. It is named after John Cockcroft and Ernest Walton of the 1930s, who used it to earn a Nobel Prize. It is difficult to reach higher than 10 MeV, since the ten million volts involved wants to discharge in air and insulators. Both accelerators are still in use as sources of fast particles for injection into bigger accelerators.

A more advanced linear accelerator is shown in Fig. 4-11. Free charged particles, often electrons, are prepared at one end of a long evacuated tube. As much of the air is removed as is practical, usually leaving less than one part in 10^{15} air. Electrons are attracted to positive potentials, so the first conducting cylindrical plate (shown in cross-section) is charged positive. After the electrons pass through, that plate is changed to a negative potential (voltage) to "push" the electrons, while the next plate is made positive. The electrons accelerate more and acquire a higher speed. By switching voltage rapidly to each successive plate, the electrons continually find themselves with negative behind and positive ahead. The effect of the available voltage is added as many times as there are plates. This can go on for miles, and indeed a two-mile linear accelerator at Stanford University Linear Accelerator Center (SLAC) has been operating since 1966. The only drawback of this design is that the plates must be made longer and longer at the proper rate so that electrons of known but increasing speed get through them before the voltage is switched. The kinetic energy that can be obtained is limited by the length, and at the current level of about 30 GeV is no longer high compared to that available

from other types of accelerator. Nevertheless, the linear accelerator is useful because charged particles moving in a straight line do not lose much of their energy as radiation.

The use of magnetic fields to make charged particles travel in circular paths was the main technical breakthrough that led to modern research and theory with particles. The forerunner was the *cyclotron* by Ernest Lawrence in 1929, which used a powerful uniform magnetic field between two huge poles to keep particles in approximately circular orbits in an evacuated chamber (Fig. 4-12). The particles were accelerated by the potential difference between two half cylindrical electrodes shaped like the letter "D" and called

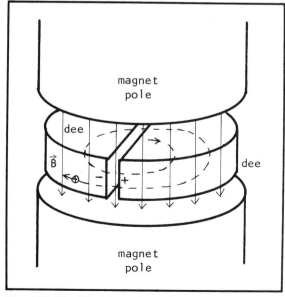

Fig. 4-12. Simplified form of a cyclotron, using a potential difference to accelerate charged particles and a magnetic field to guide them.

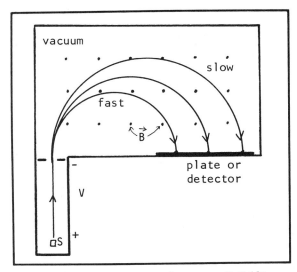

Fig. 4-13. *A mass spectrometer, using a magnetic field to separate particles of different mass.*

in 1939 (see Plate 1).

The history of accelerators will continue in a later chapter where the results of their use are described. The identification of charged particles is also greatly aided by the use of electric and magnetic fields together. It is possible to make a particle mass/momentum sorter, whereby charged particles with known kinetic energy take different circular paths in a magnetic field, depending on their speeds and masses. The device is called a *mass spectrometer* because it finds all the possible masses that the particles have. Figure 4-13 shows the internal arrangement. Ions (charged atoms or particles) from a source S are given a certain energy as they pass through the potential difference V. Heavy ions will have lower speeds than lighter ones. When they enter the region of uniform magnetic field, which is perpendicular to the apparatus, their different circular paths cause them to land at different places on the plate, depending on their speeds and therefore their original masses. The paths of fast charges are bent more sharply than the paths of slow ones. The particles can pile up on the photographic plate, exposing it, or be captured as actual piles of material (if there are enough), or be detected as a current. The method is very sensitive to mass, and we shall see need for it.

"dees." The potential was rapidly reversed as the charges passed from one dee to the other, so that they always were in an accelerating electric field. This design succeeded because the orbiting particles took the same time for an orbit, regardless of speed. They simply spiraled out, making larger and larger orbits in the magnetic field. The typical maximum energy is about 25 MeV. For this accomplishment Lawrence received the Nobel Prize

Chapter 5

Waves

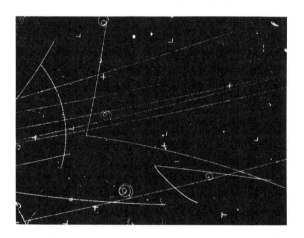

BEFORE REENTERING THE REALM OF PARTICLES, we must discuss one more area of basic physics, that involving waves. The wave model of nature is often thought of as the principal alternative to the particle model. We shall see that we cannot do without either.

5.1 BASIC PROPERTIES OF WAVES

The reader is probably familiar with waves as the back and forth motion of material. A disturbance produced in a substance such as air or water is able to travel or propagate itself over some distance. The material in which the wave travels is called the *medium* (plural: media). One important feature is that the atoms making up the medium do not move; only the disturbance moves. The wave carries energy away from the source of the wave and must act elastically, not dissipating the energy too rapidly. A medium is not always needed for waves, as we shall see in the case of radio and light waves in the next chapter.

As a first example, let us consider waves on

a stretched string (a rope, wire, or anything else that is flexible and elastic). To obtain a wave, we need merely pull a small portion of the string aside with a force, as shown in Fig. 5-1A. The resulting wave has a little more complication than might have been expected. It starts traveling as two "pulses" and travels in both directions along the string, as shown in Fig. 5-1B. It travels with a definite speed, which is determined by the tension in the string and the mass density of the string. The wave will not expire when it reaches the fixed ends but instead will reflect. Figure 5-1C shows the reflected waves returning in the opposite directions. The reason that an initially downward pulse is oriented upward after reflection is ascribed to the action-reaction at the fixed end. The downward pulse pulls on the fastener. The fastener may not move, but it pulls upward on the string in reaction. The result is a upward pulse. Momentarily the two pulses pass through each other (Fig. 5-1D), coalescing to one pulse before continuing (Fig. 5-1E). After another reflection at each end, the pulses are again downward (Fig. 5-1F) and

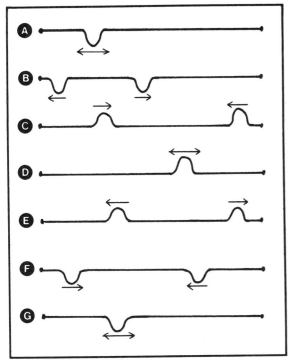

Fig. 5-1. *Pulses (single waves) traveling on a stretched string and reflecting from the fixed ends.*

momentarily merge to form a replica of the original pulse (Fig. 5-1G).

The time for a wave on a medium such as the string to return exactly to its original form is called the *period* (symbol T) of the wave. During one period the wave has traveled a distance twice the length of the string. The reader's experience may be that the form of a wave on a string is more like a smoothly curved bow, as in Fig. 5-2. The result of plucking the string indeed results in such a shape after a short time because the string is only partially elastic. The string has a strong preference to vibrate in the manner shown. Figure 5-2 gives successive carefully chosen pictures of a vibrating string at intervals of one-fourth period ($T/4$) after it has been plucked at $t = 0$. Instead of traveling back and forth in a regular manner, the wave stands in place and regularly repeats its motion. This form of wave is called a *standing wave*. A medium without boundaries, such as an infinite string, can only have traveling waves. A medium

with boundaries, such as a string between two fasteners or air in a room, will settle down to vibrating with standing waves, no matter what the original disturbance was.

The wave shown in Fig. 5-2 has only one of many possible forms. It is called the *fundamental mode*. Other modes, said to be "higher" because there are more bumps in the string, are shown in Fig. 5-3. There is no theoretical limit to the number of modes, but in practical terms a limit is reached when the string will not bend more sharply without using up the available energy. There are other points of the string beside the end point which do not move as the wave vibrates wildly up and down. These points are called *nodes* and shown with dots in Fig. 5-3. The third mode (also called the third *harmonic*) has two nodes, for example. For the case shown, each mode has two fixed ends, but different modes are possible if one end of the string is "free"—left to wave in the air.

The diagram for the second mode in Fig. 5-3 shows the shape of one complete standing wave. This shape is called a *sine wave* or "sinusoidal." This shape by itself is shown as a graph in Fig.

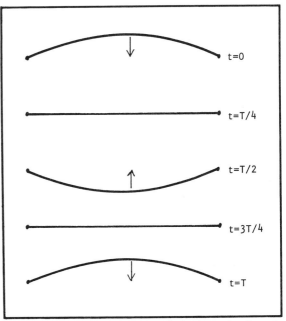

Fig. 5-2. *A standing wave on a stretched string, going through one cycle of vibration.*

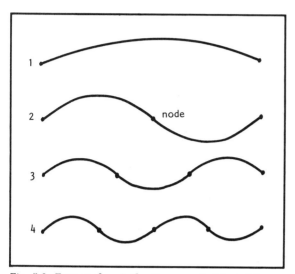

Fig. 5-3. Four modes of vibration on a stretched string.

The mathematical expression for a sine wave distributed along a string is:

$$y = A \sin \frac{2\pi x}{\lambda}$$

where y is the height at any point x along the string. The coordinate system has been placed with its origin at a convenient place to keep things simple. *Sin* is the symbolic abbreviation for a mathematical operation called "sine." The sine is calculated for the quantity to the right of *sin*. This expression does not describe a wave that moves. More complication would be needed. The sine wave shape is the smoothest possible shape that can vary regularly. It is the most gentle possible wave in nature.

Alternatively, the wave may be viewed as vibrating in place in time. The appropriate mathematical expression is then:

$$y = y_0 \sin \frac{2\pi t}{T}$$

where y_0 is the maximum height of the wave at a given place, t is the time, and T is the period. At $t = 0$ the whole wave is flat (all $y = 0$, as shown in Fig. 5-5). At $t = T/4$, the wave is at maximum

5-4 and can represent either a standing or traveling wave. The complete sine wave has a length called the *wavelength* (λ, Greek lambda). The central node occurs at the halfway point, as it should for symmetry. The maximum height of the wave is called the *amplitude* (A). If the sine wave is traveling, it moves along X-axis, keeping the shape shown. If it represents a standing wave, its height shrinks and grows periodically; every half cycle the wave is zero or flat all along the X-axis.

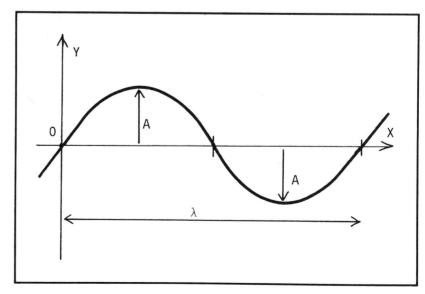

Fig. 5-4. Graph of a sine wave.

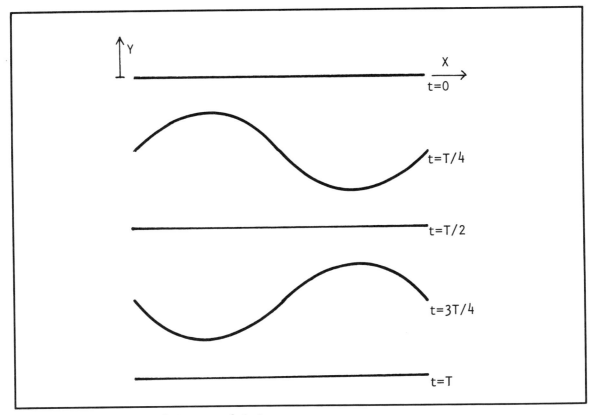

Fig. 5-5. A sine wave varying during one cycle in time.

height ($y_0 = A$). At $t = T/2$ the wave is again flat. At $t = 3T/4$ the sine operation reverses the wave and again provides maximum height. At $t = T$ the cycle starts over.

Related to the period is another quantity we shall need often, the frequency (*f*). It is measured in hertz (Hz), formerly "cycles per second." The frequency can be calculated from the period by:

$$f = \frac{1}{T}$$

When the period is long, the frequency is low. When the period is short, the frequency is high. These two variables are related reciprocally.

5.2 GENERATION OF WAVES

Besides plucking a string there are many other ways to generate waves. Some sort of force is required. To cause sine waves to propagate in a medium, a special restoring force is needed. The medium must pull back on the disturbance in proportion to the size of the disturbance. This exactly describes the spring force, which can also be called an elastic force. A weight suspended on a spring has a regular period and is called a *harmonic oscillator* because its motion is sinusoidal. Its motion can be fully described if the position and speed of the weight are specified at some initial time. The frequency of the motion is determined by its mass *m* and spring constant *k* thus:

$$f = \frac{1}{2\pi} \sqrt{\frac{k}{m}}$$

Almost any stretched string is capable of some elasticity and provides the required restoring force. So does any elastic solid material. Water and air provide the proper restoring force in different ways. All such media will carry sine waves. Of course, plucking or pulling a string too hard might cause a permanent stretch and so alter the intended situation. Other sorts of restoring forces might give different kinds of waves, for example, the "soliton" we shall encounter later.

There is a simple relation between circular motion and sine waves. Imagine twirling a weight on a string vertically as shown in Fig. 5-6. Also shine a uniform beam of light past this arrangement so that a shadow of it falls on a screen. The shadow of the weight will be seen to move up and down in a sinusoidal manner. That is, its height y will vary sinusoidally in time. The amplitude A is equal to the radius R of the string. The period and the frequency of the twirling and of the sinusoidal motion are, of course, the same.

One way to generate sine waves on a string is to tie one end to a rotating device. If the speed of rotation can be varied, then the frequency and period of the wave can be varied. For a given stretched string, the speed v of the wave is fixed. There is a simple relation between speed, frequency, and wavelength:

$$v = \lambda f$$

A units check should be made. Speed has units of meters per second. The right side of this relation has units of meters times reciprocal seconds, which amounts to the same units. For this thought experiment, as the rotation frequency is increased, the wavelength λ must necessarily decrease.

It may seem that rotating motors causing vibrations on strings are irrelevant to the forthcoming discussion of particles. But we are accumulating simple models which are good analogies to how atoms and particles behave. For example, the orbiting of an electron can cause a light wave to vibrate and be emitted. Conversely, a wave striking an atom will change its internal motion.

5.3 WAVE ENERGY AND RESONANCE

Waves are useful in physics because they carry energy as they travel, just as particles do. The amount of energy can be found by considering a wave when it has "stretched" its medium to the maximum. We should recall that the potential energy of an elastic material is proportional

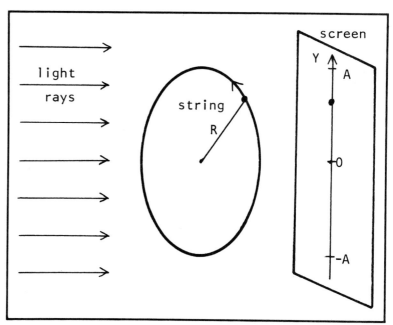

Fig. 5-6. The connection of circular motion to sinusoidal motion.

to the square of its extension. Therefore the energy carried by a wave is proportional to the square of its amplitude (A^2). At any moment a few parts of the wave are fully stretched to A and a few parts are zero, but most parts are somewhere between. The average energy stored in the wave (and therefore carried along) is best represented by $1/2kA^2$, where k represents the usual elastic property of the medium. A units check will show that this form has the units of energy.

A standing wave represents energy of another kind also. The string or other medium is in vigorous motion, possessing kinetic energy. The vibrating medium trades PE and KE, interconverting them at a regular rate. This can be seen for a mass bouncing on a spring, too. In time the energy is converted to heat because the material is not ideally elastic, and the vibration dies out or "decays." When an initial energy is given to a vibrating system, the time for it to decay to 37 % of its original amount is called the *time constant* τ. The time constant has the special property that 37 % of what is left is again lost in one more time τ. This decay would continue forever if it were not soon hidden by the random motions of atoms.

A given stretched string has a certain preferred mode of vibration, determined by its length, tension, and mass. If a rotating drive is connected to it, the vibration will be small and erratic unless the frequency of the drive is adjusted to match a natural frequency of the string. As the match gets closer, the string can absorb more energy from the motor. At an exact match the string could theoretically take up an infinite amount of energy over time. But various friction losses limit its vibration to some large amplitude. This process of matching the driving frequency to a natural mode is called *resonance* and is shown as a graph in Fig. 5-7. If the motor is speeded up or slowed, the match between its frequency and resonance is poor and the vibration nearly dies out. Often there are higher modes of resonance, and the second harmonic will start resonating when the driving frequency reaches the proper value. Forced vibration and resulting resonances can be found in almost any system capable of supporting waves.

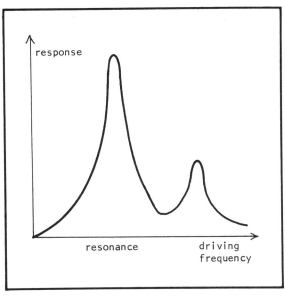

Fig. 5-7. *A resonant response graph: the amount of vibration depends on the driving frequency.*

There are systems which do not have harmonic modes. We shall rarely need to refer to them. When a mode does not have a natural frequency that is 2, 3, 4, and so on, times the fundamental frequency, the mode is said to be a "partial." Common examples are the modes of bells and struck bars. Piano strings are thick enough and stiff enough to act as bars of metal and produce partials when struck. Vibrating plates and spherical systems, including the Earth and atoms, have partials for modes.

5.4 PLANE AND SPHERICAL WAVES AND THEIR INTENSITIES

In air and other extended media (also in space) we can obtain waves spread throughout space. These three-dimensional waves can have two simple forms, plane and spherical. More complex forms are possible, but we shall see that complex waves can usually be made of simple ones. Since most sources of interest are small or point-like, it will be more realistic to consider spherical waves first.

A stone thrown into a lake causes circular waves to spread out. The three dimensional analogy is that any small disturbance in air, in space, or in a solid material causes spherical waves to travel outward. If there is no complication, the wave spreads symmetrically in all directions from source S, as shown in cross section in Fig. 5-8. A diagram such as this represents the traveling wave with a series of wavefronts. The wavefront is usually the region of maximum amplitude at a given instant. If the medium is air and the wave is sound, then each thin spherical shell shown is a region where air molecules are most compressed together. The distance between two wavefronts is the wavelength λ. In between two wavefronts is a region where the air is thinnest (*rarefacted*). The sine wave is involved in that a graph of air density going outward from the source would have a sinusoidal shape. It is important to note that each wavefront is at all points perpendicular to a radius vector from the source.

At this point more discussion of wave energy is needed. A new term is needed to describe a source that emits energy steadily. *Power* is a quantity that tells how much energy is passing by in a given time. It is measured in watts (w), after James Watt in the 18th century, who worked with mechanical power. One watt is equivalent to one joule per second. Since the power of a wave is spread all over the wavefront, we often are interested in *intensity*, the amount of power per unit area. The intensity is proportional to the amplitude squared, just as the energy is. The intensity decreases with radial distance from a source by the inverse square law, in order that the energy emitted be conserved over all space.

The usual way to obtain a plane wave is to go far from the source of spherical waves. Then a small section of the wavefront appears flat, as shown to the right in Fig. 5-8. Not much energy will be obtained from a given piece of a spherical wave because the energy from the source has spread in all directions, never to be seen again. With radio, light, and some other waves, there are other technical methods for obtaining nearly plane waves. A salient example is the laser, which can concentrate most of its energy into a narrow beam of good plane waves. A beam has a high intensity near one direction and weak intensity in other directions. We shall see later other shapes and beams of waves from particles.

5.5 SUPERPOSITION, INTERFERENCE, AND REFRACTION

In the case of two pulses that coalesced momentarily on a string, we saw an example of *superposition*—literally, putting one on top of another. When there are two or more sources of waves, the effect of each at a given point is sim-

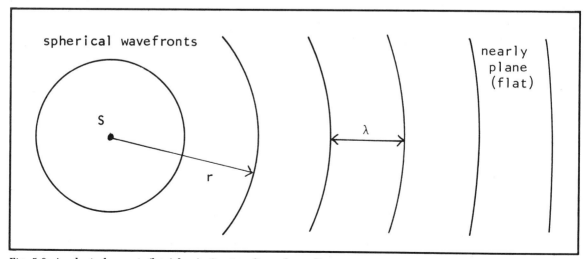

Fig. 5-8. *A spherical wave is flat (plane) after traveling a large distance.*

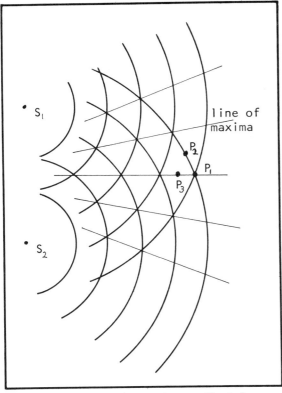

Fig. 5-9. *The interference of waves from two identical sources.*

times the average put out by one source, as it should. By the principle of superposition, the sine wave from each source adds to produce another sine wave, not some other shape.

The superposition of two waves produces an *interference*, giving a new pattern where maxima are preferred in certain directions, as indicated in Fig. 5-9. The pattern depends on the separation between the sources and their wavelength. This phenomenon is most readily demonstrated with water waves, either by throwing two stones in simultaneously or by using two identical vibrators.

Two other pertinent effects can be seen on water. Suppose there is one source S and a barrier with a hole, as shown in Fig. 5-10. The wave energy that arrives at the hole serves as a new source of "spherical" waves, which emerge through the hole. On the left of the barrier can also be seen waves that were reflected. Any barrier to waves acts as a mirror, changing the direction of the waves but preserving their form. Two holes in the barrier would give two sources of waves which could then interfere, as will be seen later with light.

Refraction of waves occurs when their speed changes. This can be done with strings by connecting a heavy and a light string as in Fig. 5-11A. For the same tension, each string propagates waves at a different speed. Since a heavy string will have a slower speed, the wavelength on it will be shorter. A vibrator connected to this hybrid

ply the sum of the effects of each separately. In Fig. 5-9 two identical sources S_1 and S_2 and the resulting spherical waves in cross section are shown. Observations are made at several points. At the instant shown, a maximum from each source is obtained simultaneously at point P_1. A wave detector there would show double amplitude (and four times the energy) as compared to one source. At P_2 the maximum from S_2 falls on the minimum from S_1. The correct interpretation is that the wave height or amplitude from S_2 is largest and positive, while that from S_1 is largest but negative. A detector at P_2 finds the resulting amplitude to be zero. At P_3 the minima from each source add to produce a doubly negative amplitude (and again four times the energy). The detector at any of these places would find a sine wave passing by over time with amplitude double that of either source alone. The *average* energy received by the detector would be two, not four,

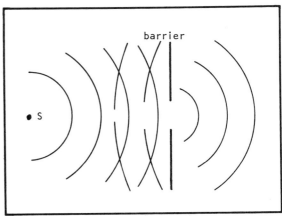

Fig. 5-10. *Diffraction through an aperture.*

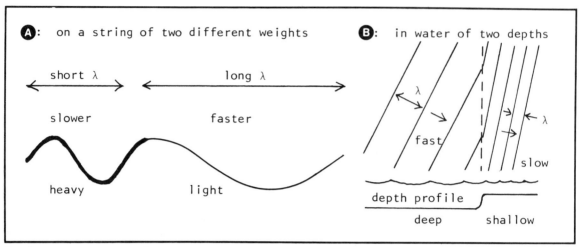

Fig. 5-11. Refraction in two different media.

string will produce shorter waves on the heavy part and longer ones on the lighter part (assuming parameters have been arranged to allow standing waves). A wave sent down this string changes speed at the junction. If the string were to change gradually in weight, the speed would change gradually.

With two dimensional waves in water, a new effect appears with refraction. In Fig. 5-11B waves are approaching the shore at an angle. The change in depth is shown in profile. In shallower water the waves move slower (an externally obtained fact needed for discussion). Again, their wavelength must be smaller. The only way to match the wavefronts as they cross the boundary from deep to shallow water is to change their direction, and this is what happens. We see a bending of the waves toward the boundary, the usual characteristic of refraction from a "faster" medium to a "slower" one. In getting this result we have

made use of an important procedure in physics—the matching of results at the boundary between one simple region and another. Such problems are called *boundary value* problems.

Another important effect occurs for the waves in Fig. 5-11A. There is a mismatch at the junction of the heavy and light strings. A wave traveling down one is partially reflected at the junction (or boundary) because there is a reaction force there. This partial reflection is what allows each section of string to acquire its own standing wave. On an infinite string with a junction in the center, wave energy provided on one side will not all travel to the other side. All of the energy will not pass through a mismatch. In regard to the water waves in Fig. 5-11B, wave energy is also reflected at the boundary. Some wave energy returns to deeper water, where it interferes with incoming waves. (For clarity the reflected wave is not shown in the figure.)

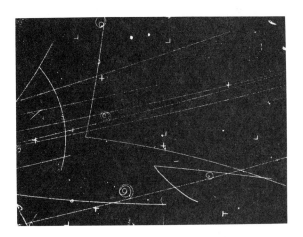

Chapter 6

Light and

Other Radiation

W E ARE READY TO EXAMINE AN ASPECT OF nature which combines the particle and field (or wave) models. Then we can proceed to the basic issues of particle physics. Visible light was the first part of the electro-magnetic spectrum to be shown to have such a dual aspect. After radio waves and x-rays were shown to be part of the same spectrum, their dualistic nature has also become important.

6.1 LIGHT

The history of theories of light shows a vacillation between particle and wave models. The Pythagoreans said that light was something particle- or fluid-like that was emitted in all directions by glowing bodies and bounced around the environment, finally entering the eye. Light was of special interest to the scientifically-inclined during the Renaissance, when analysis of the painting of pictures included studies of perspective, illumination, color, and other aspects of how the eye sees the world. Laws about the behavior of

light were found in the 17th century by the Dutch who invented the microscope and telescope, instruments which aided other beginnings of science. Kepler added to the knowledge of optics. In 1621 Willebrod Snell found the correct law of refraction. Called Snell's law, it gives a way to calculate by what angle light is bent as it passes from one transparent medium to another. As our earlier discussion of waves showed, the bending depends on the speed of the waves, although the wave nature of light was not suspected at the time.

Rene Descartes in 1637 attempted to explain the law of refraction not with waves but with particles. Having observed that light reflects as if it consisted of particles rebounding from a flat surface, he supposed that the light particles changed their direction by collisions with a denser substance. In supposing that light went faster in a denser medium, he was led astray. At the same time Descartes also started what would become a wave model for light. He supposed that light traveled by means of forces in a medium, a sort of action at a distance. The medium was called "ether" and was thought to exist everywhere. The ether remained a vital part of physics until the late 19th century.

Pierre de Fermat in the 17th century formulated a principle of nature which could be applied to light. Called "Fermat's principle," or the "principle of least action," it states that natural processes occur in the least possible time. When passing from one medium to another, light would take the path that required the least time. In Fig. 6-1 the straight line route from A to B would require that the light spend more time going slower in the denser medium. If it were to take the longer path ACB, it could travel faster and use less time along AC in the less dense medium (air), then save more time along path CB which is shorter. Path ACB describes how light actually travels to cross a boundary, and Snell's law can be calculated from considering it. Fermat also found that light had to travel slower in denser media in order to preserve what was known about reflection and refraction.

The wave theory of light is traceable to Francesco Grimaldi, who observed a diffuseness and coloration at the edges of shadows. He proposed in 1665 that light was a fast-moving fluid with waves included. Christian Huyghens decided by 1690 that the fluid or ether did not move but did have waves. Olaus Rohmer observed the motions of Jupiter's Moons sufficiently to obtain an approximate speed for light in 1676, showing that it was neither infinite nor slow. He also made the ether model into a sea of elastic particles which did not move.

Understanding of light moved forward in 1672 when Newton used a prism to demonstrate that ordinary "white" light could be refracted into a regular set of colors, now called a *spectrum*. Newton's theory for this favored the particle view (he called them "corpuscles"). He knew from experiment that light was both reflected and refracted at any boundary, but he had to assume that the ether vibrated (reminiscent of waves) in order to explain the observations in terms of corpuscles. In other experiments which produced bands of light and dark, which we will recognize as wave interference, he managed to stay with the corpuscular view.

Early scientists continued to alternate between wave and particle models. Even the principle of least action was applied to the particle view by defining action as speed times distance. It predicted (incorrectly) that light would travel faster in denser media. Clearly an experiment that measured the speed in media would decide between the two theories. Lacking this, the trend during the 18th century was to view light as particles. Thomas Young in 1801 revived the wave theory with the analogy between light and sound. Color was supposed to correspond to sounds of different frequencies. It was also known that sound traveled the same low speed, regardless of its intensity. Most particle theories had to suppose that more intense light (or sound) would have more and/or faster particles. Young was able to do the light experiment analogous to the interference of water waves and so measure the wavelength of light. This "Young's experiment" (to be shown later) found the wavelength to be very small, so that many phenomena explainable with particles could be explained similarly with waves. Light waves did not bend around obstacles nearly as readily as water and sound waves because of the small wavelength.

At this time the results of several investigators began to show that light was *polarized*—that is,

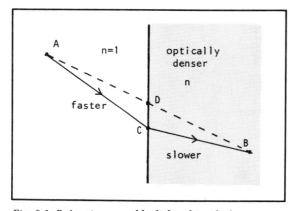

Fig. 6-1. Refraction caused by light taking the least time to travel between two points (Fermat's principle).

there were preferred orientations for light. In 1817 Young made a radical departure in regard to the way in which waves were thought to propagate. Sound waves were known to consist of air density variations that occurred along the direction of travel. Young showed that if light somehow vibrated perpendicular to its travel, polarizations could be explained with light as a wave. A fuller theory was developed by Augustin Fresnel.

Fresnel also studied diffraction, the bending of light around obstacles. He raised, in 1821, doubts about the ether, since it might have to be very stiff to carry such light waves. It was known that the planets move rather freely among the ether, if there exists such a medium. Research proceeded on what properties the ether should have in order to satisfy many conflicting requirements, aided by the growing research on waves in solid materials. Armande Fizeau and Jean Foucault measured the speed of light in various media in the mid-19th century. Fizeau used a rotating wheel with slots for his measurement in 1849. The speeds in space, air, and water confirmed the mathematical law of refraction quite well for the wave theory.

6.2 ELECTROMAGNETIC RADIATION

The foundation of the modern theory of electromagnetic radiation, including light, was laid by Hans Oersted in 1820 and Michael Faraday in 1831, among others. Oersted discovered that an electric current produced a magnetic field. Faraday worked with the converse effect and found that a changing magnetic field could produce a changing electric current in a conductor. The term *induction* in physics now refers principally to this effect. Figure 6-2 shows two of many ways to produce the changing magnetic field; the resulting voltage in the closed loop of wire can be measured with a voltmeter. The strength of the induced voltage and current depends on the strength of the magnetic field, how fast it changes, and the size of the coil. The magnetic field can be made to change either by moving it or by varying the current in the another coil. Faraday used the "lines of force" model in an ether to explain these effects.

A subtle contribution to the understanding of induction was made by Heinrich Lenz in 1834. The level of his reasoning in regard to the energy flow that must be involved in induction was ahead of his time and a foretaste of the principle of the conservation of energy, which came quite late in physics. He realized that an induced current represented energy and must come from somewhere, namely from whatever agent caused any changes during induction. In order to avoid getting something for nothing, Lenz found that the direction of the change and the direction of the induced current had a certain relation so that the change was opposed. Any other relation would re-

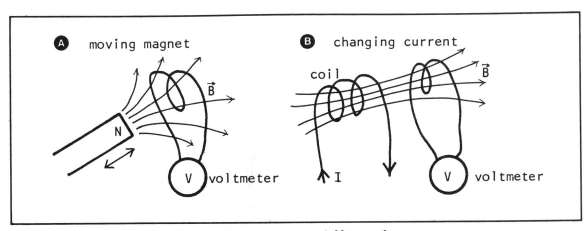

Fig. 6-2. *Two ways to induce a voltage by changing a magnetic field in a coil.*

sult in catastrophe! His law has been named after him.

Another contribution by Faraday was the observation (in 1837) that the type of insulator between two charged plates helped determine the strength of the electric field there. The "lines" supposedly crowded together better in some insulators (called "dielectrics"), increasing the intensity of the field. Some materials had a higher "dielectric constant" or "permittivity" and encouraged more concentration of charge and field. The same effect occurred with magnetism: magnetic materials concentrated the "lines." The better the material, the higher its "permeability."

Faraday's observation that magnetic fields affected light and Gustav Kirchhoff's work with the magnetic field from a known flow of charge led Clerk Maxwell to formulate a complete mathematical theory of electromagnetism in 1862. Maxwell was deeply committed to the ether model, yet managed to produce a theory so complete as to predict the velocity of light from measured physical constants. The theory is still so elegant after a century that it is worth presenting here in the modern mathematical form known as Maxwell's equations:

$$\vec{\nabla} \cdot \vec{E} = \varrho P/E_0 \qquad \text{(a)}$$
$$\vec{\nabla} \cdot \vec{B} = 0 \qquad \text{(b)}$$
$$\vec{\nabla} \times \vec{E} = -\dot{\vec{B}} \qquad \text{(c)}$$
$$\vec{\nabla} \times \vec{B} = \mu_0\vec{J} + \mu_0 G_0 \dot{\vec{E}} \qquad \text{(d)}$$

We might notice with these equations that more complex symbolism is needed, a common trend in modern physics. [The mathematics would appear so much more complex as we reach modern versions of other forces that showing their form is essentially impossible for this book.] The equations suppose that all possible phenomena are present and interacting in space: electric and magnetic fields, charges, and current. When some of these are not present, things are much simpler. In some ways these equations completely describe the fundamental electromagnetic force. They tell about the fields, and then the force can be calculated from:

$$\vec{F} = q\vec{E} + q\vec{v} \times \vec{B}$$

Maxwell's equations have some symmetry, but it is not complete. Equation (a) for the electric field looks like (b) for the magnetic except the source term ϱ (Greek rho) on the right for the electric is charge density and for the magnetic is nothing. Equation (a) tells mathematically that electric field lines start and end on charges, and (b) tells that magnetic field lines are closed. Equation (c) tells that something (a circular aspect) of the electric field is related to the change in time of the magnetic field (that is what the dot on top means). Equation (d) tells that the circular behavior of the magnetic field is related to the change in time of the electric field. The circular aspect of the magnetic field is also related to any current flowing, and $\vec{J}$ specifies the density of the current. We shall see later that physicists have strived for complete symmetry in the description of electromagnetism. As often happens, other physicists needed time to get used to such a major new theory, and sought other less disturbing models. We would now consider their concoctions to be crazier than Maxwell's.

An astonishing prediction that can be made from these equations is that there can exist a wave made of electric and magnetic fields that vary in time. The simplest form is shown in Fig. 6-3. Each field waves at right angles to the other and to the direction of travel. This complex wave is a portion of a plane electromagnetic wave. The speed is predicted to be:

$$c = \frac{1}{\sqrt{\mu_0\epsilon_0}}$$

The constants μ_0 and ϵ_0 are the permeability and permittivity for vacuum, respectively (hence the zeros as subscripts). They can each be measured in the laboratory and the speed of light then calculated. The modern result for c is $2.9979250(10)^8$ m/s, easily remembered as three hundred thousand kilometers per second.

The density of a material that is relevant for light propagation is an optical density, not necessarily related to either the mass density or

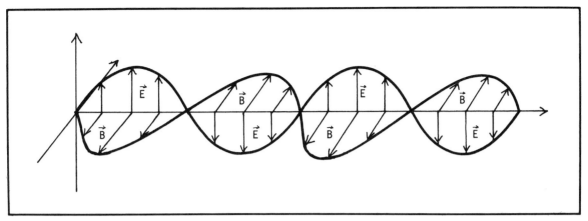

Fig. 6-3. An electromagnetic plane wave propagating through space.

the degree of transparency. The effect of the medium on lightspeed is expressed by means of the *refractive index n*. The actual speed c' in a material with index n can be found from the absolute speed of light by:

$$c' = c/n .$$

As will be seen later, real speeds can never be greater than c, so that n must always be greater than 1. For air it is about 1.0003 and for glass about 1.5.

We might now be able to see how an electromagnetic wave can propagate itself. One of the relevant laws, Equation (c), says that a changing magnetic field makes an electric field. In Fig. 6-3 there is an electric wave accompanying the magnetic wave and resembling it in form. Another law, Equation (d), says that a changing electric field makes a magnetic field. This also appears in the Figure. Each part of the wave continually induces the other part. Not only does the wave propagate in this way, but it appears self-sufficient. The ether is not needed as a medium, but it was several decades after Maxwell's work before this was realized fully.

The wave shown is also polarized. The electric field is confined to one plane, and the magnetic to another. More complex waves are not polarized, or their plane of vibration varies. We are mainly interested in polarized waves for discussing elementary particles. The relation of $\vec{E}$, $\vec{B}$, and the direction of travel is set by—you guessed it!—the right-hand rule. If the thumb points in the direction of $\vec{E}$, then $\vec{B}$ at the same place in the wave is positive along the first finger, and the second finger gives the direction of propagation. The average energy density of the wave is given by either the electric or magnetic parts. These have to be equal since the wave propagates by dumping electric energy into magnetic, then back, and so forth. The result is the same as earlier, $1/2\epsilon_0 E^2$, where E is the maximum strength of the electric field. The relative strengths are that E is c times greater than B.

Maxwell's equations include the means of creating an electromagnetic wave. One need merely move or vibrate charges such as electrons. An electron going in a circle nicely produces an outgoing wave. The energy needed to keep the electron moving becomes energy radiating away. A current of electrons shuttling back and forth in a wire produces radiation also. The frequency of the radiation is determined by the rate at which the charges are moved. Figure 6-4 shows much of the electromagnetic spectrum from long radio waves to very very short, very high frequency gamma rays, emitted only by particles in violent reactions. We shall see later how there is an energy equivalent for each wavelength and frequency.

Heinrich Hertz in 1886 was able to detect electromagnetic waves (radio waves) produced by a spark. An ingredient needed, not understood until later, was that the current had to oscillate very fast to produce sufficient wave energy. A spark represents a very rapid back and forth flow of current, not a one-way flow. Resonance is important, and most radio waves are detected by having an electric circuit of a certain size which will oscillate at one of the frequencies being produced and transmitted. Hertz also showed experimentally that radio waves could be reflected, refracted, diffracted, polarized, and made to interfere, just as light. The effects were bigger because typical radio waves were 1 to 1000 meters long.

At the other end of the spectrum occurred one of the beginnings of modern physics—the discovery of x-rays by Wilhelm Roentgen in 1895. The discovery was accidental, as many are, but it was to Roentgen's credit that he recognized how extremely unusual the new effect was. Like many experiments of the time, the crucial work was done in a glass vacuum tube, using electric discharges in thin gas. Roentgen had the tube completely covered with opaque material, yet some new radiation different from ultraviolet (the most energetic light known then) passed through and was detectable with a fluorescent screen. Later it was found that x-ray wavelengths are so short that they are approximately the size of an atom and that they are produced when fast electrons in a vacuum tube strike a metal target. Roentgen received the first Nobel Prize in physics in 1901 for this work.

6.3 MODERN WAVE OPTICS

The behavior of light waves is summarized here, to establish terminology for use later with particles. Refraction and interference have already been adequately described, except for the effect of diffraction on interference.

Diffraction is the ability of light or other waves to pass around obstacles. Augustin Fresnel demonstrated diffraction about 1814 without knowledge of interference. After some false attempts he decided that wave theory was the best explanation. In the wave model, diffraction occurs because the edge of the obstacle causes new sources of waves to form on the original wavefront. In Fig. 6-5A plane waves are shown (in cross section) approaching a round obstacle. To obtain the effect clearly with light, the obstacle should be very small, less than a pinhead. On the screen S appears a pattern such as in Fig. 6-5B. There is a bright central spot, where one would expect a shadow, and bright and dark rings around it. Far from the central spot the rings become closer together and fade out. To obtain an ordinary shadow, the object would have to be much bigger. For example, Fig. 6-5C shows the bands of light and dark obtained next to the edge of a large obstacle. This is a magnified view, and the bands would be too narrow and faint to notice at the edge of an ordinary shadow. Diffraction can also be done with a pinhole in a barrier, as was shown in Fig. 5-10. When the pinhole is small, the

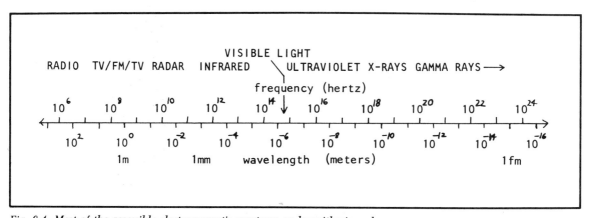

Fig. 6-4. *Most of the accessible electromagnetic spectrum on logarithmic scales.*

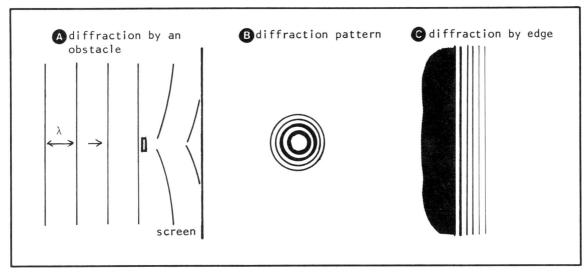

screen

Fig. 6-5. Diffraction of light by obstacles.

same pattern is obtained as for a tiny obstacle.

Diffraction effects are difficult to notice with "white" light, which is a mixture of colors, but quite obvious when laser light is used. The effect depends on the relation of the wavelength to the hole or obstacle size. The smaller the hole or obstacle, the bigger the pattern. Different colors will form their light and dark pattern in different places, so that a mixture of colors will produce an indistinguishable pattern. Laser light is usually so "pure" in color that a small range of wavelengths is present. (It is impossible to obtain exactly one wavelength from any physical system, but the spread in wavelength can be reduced to one part in a billion or trillion.)

When plane waves arrive at an opening that is a very narrow slit, the diffraction pattern from the one slit appears as a wide bright central band and narrower fainter bands on the sides. When another similar slit is located near the first, as shown in cross-section in Fig. 6-6A, the interference pattern obtained on screen S consists of bright and dark bands as indicated in Fig. 6-6C. This is Young's experiment. The interference will appear clearly only in the region where each slit alone would have made a central diffraction band (Fig. 6-6B) that overlaps that of the other. Whereas the diffraction sidebands grow closer and fade rapidly

to the side, the interference bands are regularly spaced.

Interference occurs for two or more distinct slits or other sources of light provided certain conditions hold. The sources must be related in that they have the same wavelength and their *phase* does not vary. Phase describes whether the maxima and minima of the waves are in step or how much they are out of step. They can be out of step as long as their relation remains that way. A plane wave of one color ("monochromatic") impinging on two slits automatically satisfies these requirements. The resulting sources are said to be *coherent*, as laser light is. To detect interference it is also necessary that the wavelength and the slit size and separation be not too dissimilar.

The wavelength of light limits the ability to distinguish two objects with that light. Suppose that two round obstacles such as two particles are side by side and close together, as in Fig. 6-7A. We have seen that if their diffraction patterns overlap, the image of each on the screen should interfere strongly with the other. The image of each cannot be resolved any better than indicated in Fig. 6-7B, where their central spots nearly overlap on the screen. To a useful approximation, the smallest size d between two objects that can be distinguished is given by:

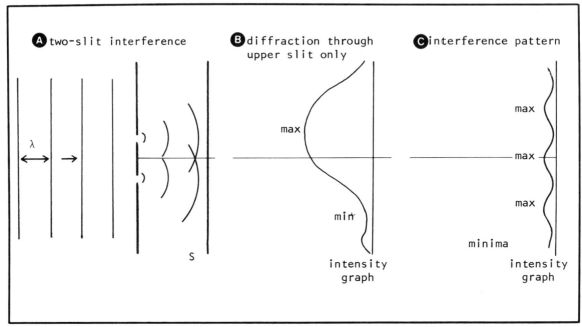

Fig. 6-6. Interference of light for two slits (with graphs of the resulting patterns).

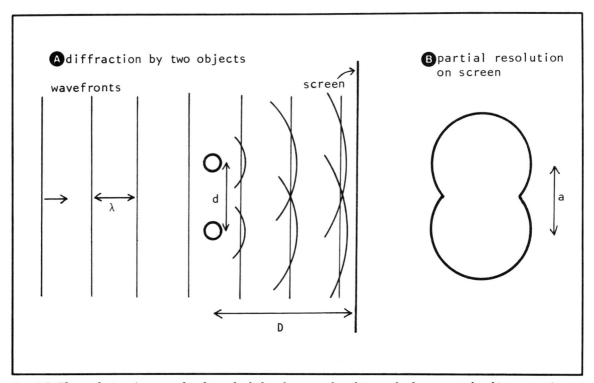

Fig. 6-7. The resolution of two nearby objects by light whose wavelength is nearly the same as the object separation.

$$d = \lambda D/a$$

where a is the measured distance between the central spots on the screen, and D is the distance from the objects to screen. This is the limit of **resolution**. It also can be applied to a single object for purposes of obtaining its size accurately. The first to calculate resolution was Lord Rayleigh in the late 19th century. The ability to measure small distances with waves is an important ingredient of modern physics in both theory and experiment.

An important application of interference is the "diffraction grating." This device is made by ruling a large number of closely-spaced grooves on flat glass. Each groove acts as a slit, and there can be more than 300 per millimeter. The result of the interference of light from thousands of slits is that each wavelength has a maximum at just a few special widely-spaced positions. Any light can be resolved into its component wavelengths and the spectrum seen or photographed. The ability to achieve high accuracy and fine resolution has greatly aided the study of atoms by the light they absorb and emit.

X-ray wavelengths are too small for an ordinary diffraction grating, but nature has provided in crystals a natural way to diffract waves. Atoms are arranged in regular rows in three dimensions in a crystal. An x-ray beam shone through a crystal yields not spectral lines but spectral spots. X-ray diffraction is an important tool for science and technology, not so much for the study of x-rays as for the study of the structure of matter. Max von Laue received the Nobel Prize in 1914 and William H. Bragg and William L. Bragg in 1915 for their work with x-ray diffraction.

Another application of interference enabled the definitive experiment that ruled out the ether. If there is an ether, and lightspeed is measured with respect to it, then a difference in lightspeed should occur depending on whether the measurement is made in a moving frame of reference. The Earth itself moves in the supposed ether at one of the highest speeds accessible to an experimenter, 30 km/s (really, much higher if the motions of the Sun and our galaxy are considered).

Albert Michelson and E. Morley in 1887 compared the speed of light along the direction of the Earth's motion with the speed perpendicular to it. They used the interference of light in an arrangement similar to Fig. 6-8, called an **interferometer**. Light from source S both passes through and is reflected from the partially-silvered "beam splitter" B. It then proceeds in two perpendicular directions to mirrors M and is returned to

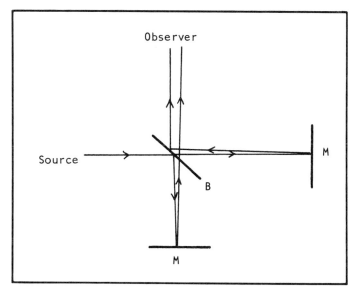

Fig. 6-8. The Michelson-Morley interferometer. (The angles of the beam paths are exaggerated for clarity.)

Observer

Source

M

B

M

B. An observer can see the returned beams superimposed. If the light is in step a bright maximum is seen. If out of phase, a dark minimum is seen. A pattern of light and dark "fringes" is obtained. If there is no difference between the two paths, the pattern will remain steady. If some effect changes the relative travel times, the pattern will move, and the change can be measured quite accurately. The method is sensitive to movement of 1% of a fringe. The effect of the Earth's speed was checked by rotating the apparatus, to compare lightspeeds in two directions. No difference was found, and this and other experiments demolished the notion that light needed a medium to travel. Michelson received the Nobel Prize in 1907.

6.4 LIGHT AS PHOTONS

The modern name for the particle representing light is the *photon*, from the Greek word "phos" for light. The particle model for light was revived at the very beginning of the 20th century and marks the true beginning of modern physics, a physics that has not been substantially changed during the ensuing eight decades of intense research.

The first clue that all was not right with the wave theory of light was the study by Lord Rayleigh in 1900 of the color of light emitted by a hot body. This light energy, called "black-body radiation" whether it is the dull red glow of hot metal or the intense "white" light from the sun, obeys certain laws outside the scope of this book. If the radiation is waves, then the laws require that any hot body emit more radiation at shorter wavelengths, without limit. This has been called the "ultraviolet catastrophe," since an object such as an incandescent light bulb would have to emit huge quantities of light more energetic than visible light, starting with ultraviolet and extending up the spectrum.

Max Planck in 1900 considered the black-body radiation problem in terms of light as particles. In this model a radiator must emit energy in discrete chunks, called *quanta* (from the Latin for "how much"). At this point we will find it more convenient to talk in terms of the frequency of light rather than wavelength. If the energy of a given quantum were proportional to its frequency (and thus inversely proportional to its wavelength), a balance could be found in the spectrum emitted by the radiator. At low frequencies (long wavelengths—microwaves or red light) low energy quanta would be emitted; at high frequencies (short wavelengths—blue light and beyond) high energy quanta would be emitted. An excess emission of short wavelengths would be avoided by the fact that each corresponding photon would, by itself, have to carry much energy, and thus few quanta could be emitted. The theory goes on to predict a certain wavelength at which radiation is maximum, and this result agrees with experiment. Planck received the Nobel Prize in 1918 for this quantum theory of light. He even calculated the relation of energy E to frequency f for light, which is expressed in the Planck equation:

$$E = hf$$

The proportionality constant h is now called Planck's constant and is known to have the value $6.626196(10)^{-34}$ joule second. Its units are sometimes known as "action." This constant is very small, even for phenomena on the scale of elementary particles. We shall see how the size of this constant is vital to modern physics and sets the stage for new models of atoms and new and strange ways of nature.

Albert Einstein in 1905 used the quanta of light to explain the "photoelectric effect." Einstein was a theoretician, and a genius at thought experiments, and his work was either based on previous laboratory experiments or suggested new experiments to be tried. The photoelectric effect was known since 1887, when Heinrich Hertz observed that ultraviolet light somehow enhanced the emission of electric sparks from metals. It was soon demonstrated by Phillip Lenard that the particles emitted were electrons, and he received the Nobel Prize in 1905.

In Fig. 6-9A a light wave, symbolized as a brief wavy line even though we will be consider-

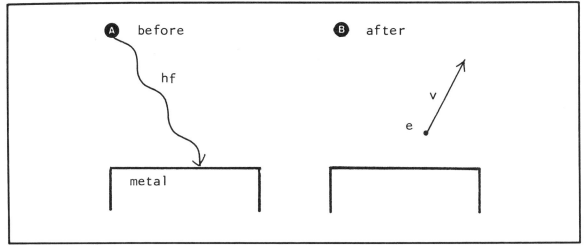

Fig. 6-9. *The photoelectric effect.*

ing it as a particle, approaches a metal surface which has many electrons as part of its atoms. In Fig. 6-9B the light of energy hf has been absorbed, and an electron with a certain speed and energy leaves the surface. Further work established two laws about the effect: that the number of electrons emitted is proportional to the intensity of the light, and that the maximum kinetic energy of an emitted electron depends only on the frequency, not the intensity, of the light.

Einstein saw that these results implied the relation "one quantum, one electron," if the light wave were quantized and acted as a particle. Electrons were emitted as the result of simple collisions; one quantum was required to knock one electron out. The kinetic energy possessed by the electron depended on how much energy the quantum had, which depended only on frequency in the manner first proposed by Planck. The number of electrons depended on how many quanta struck the metal, independent of light frequency as long as the light had the minimum energy necessary to release an electron (a few eV). The quanta of light came to be known as photons. Einstein's work on quanta was done in the same year as his celebrated work on relativity (and major work touching another field of physics). He received the Noble Prize in 1921 particularly for the work on quanta.

If light is quantized as photons, does it act like particles in every situation? For example, will photons interfere or diffract? The answer is that there is no simple way to explain interference in terms of a hail of tiny photons bouncing through slits. In such a model there seems to be no way to explain the pattern of maxima and minima found on a screen on the other side of one or two slits. Particles dropping through an aperture as if they were fine sand will indeed form a pile, but a pile related to the shape of the aperture in the wrong way, and with no small distinct piles beside the main pile.

Does light produce pressure in the way that a hail of small particles would? Yes, and the pressure can be explained with either model. Photons bouncing from a surface cause a pressure on it, which can be measured in the laboratory. It is due to the momentum of the photons being reversed. This pressure will move space vehicles propelled by the light from the sun. The force exerted by the electric and magnetic fields in the wave model of light can also be interpreted as pressure when the electromagnetic wave encounters a surface. Both of these explanations work whether the light reflects from a shiny surface or is absorbed by a dull surface. It will be seen that light carries momentum along with its energy. The energy of light cannot be considered as kinetic energy in the usual way, or as any other previously mentioned energy. It is a unique form.

If we explore the polarization of light waves, we would find that in some cases light waves seem to rotate, a phenomenon known as *circular polarization*. The rotation can be either left handed or right handed. This rotation can be imparted to a body, so that light must somehow carry angular momentum. The planes in which the electric and magnetic fields vibrate can rotate, giving the wave explanation of polarization. If light is photons, where is the angular momentum? It is in the form of "spin" carried by the particles. Each quanta of light carries one unit of spin, so that the spin is also said to be quantized. The size of the unit of spin will be seen later to be Planck's constant itself (divided by 2π). If a beam of light is not carrying measureable angular momentum (unpolarized), then the particle view says that it must be composed of approximately equal numbers of photons with spins in two opposite directions, as in Fig. 6-10. When the beam does carry angular momentum, its circular polarization—right or left—is due to it carrying predominantly right or left handed photons (spins parallel or antiparallel to the motion). Ordinary linearly polarized light waves can be explained by knowing that right and left circular forms can be combined to make the linear form.

6.5 INTERACTION OF LIGHT AND PARTICLES

Collisions between light waves and particles are possible. We need not think of the light itself as made of particles in order to discuss or calculate the results, but the particle model is hard to put out of mind. The classic experiment in this regard is now called Compton scattering. The perhaps surprising result is that light scattered from particles suffers an increase in wavelength. The scattering of interest here is not the scattering from the electrons in atoms and molecules. Such bound electrons may reradiate the light at a different wavelength and scatter different wavelengths with different effectiveness. If light were simply waves, electrons in materials would seem to have no other choice than to re-emit light at the same frequency at which it impinged.

Experiments by Arthur Compton in 1923, shown in simple form in Fig. 6-11, revealed that the scattered light (x-rays in early cases) had lower energy (lower frequency) and longer wavelengths, as in Fig. 6-11B. The change in frequency and wavelength depends on what angles the outgoing light and electron happened to have. A head-on collision, in which the light returns straight back, requires that the electron carry away much of the energy, causing the scattered x-ray to have its

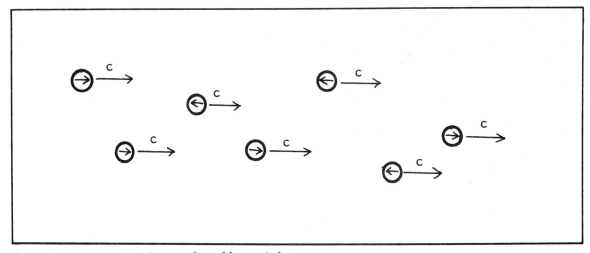

Fig. 6-10. A representation of an unpolarized beam of photons.

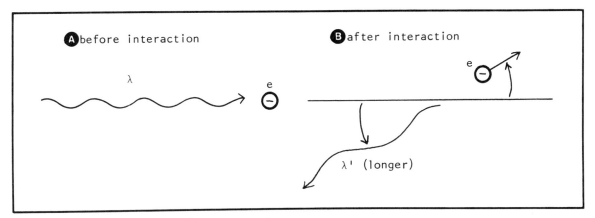

Fig. 6-11. Compton scattering of a photon from an electron.

longest wavelength. A special wavelength called the *Compton wavelength* is defined to be the change in wavelength when the photon is deflected by 90°. Its value is given by three fundamental constants:

$$\lambda_c = h/mc$$

The mass m is the mass of the particle struck, not always an electron.

Chance dominates in such microscopic collisions, and measurements are made on many collisions before consistent results are obtained. But whether viewed as a wave or particle, the outgoing light has a high probability of being detected at a certain position related to the direction of the electron and lower probability elsewhere.

Compton received the Nobel Prize in 1927 for his experiments and his analysis of the collision that explained the wavelength change when x-rays are scattered from electrons. When the particles in a beam of x-ray light were assumed to have a certain kinetic energy and momentum, ingredients necessary for analyzing a collision between two

particles, the analysis of the experiment demonstrated that these assumptions were valid. The photon model of light, including the Planck relation, was utterly convincing. A new feature was needed, the assignment of momentum to the light wave or photon. Compton had to use for the momentum p of light:

$$p = E/c = hf/c = h/\lambda$$

relating the wavelength to momentum with Planck's constant. Light also has a unique relation between its energy and momentum, involving c as one might expect.

Compton's results were immediately put to good use in other developments in modern atomic physics. Later the association of quanta with a field, in analogy to the association of photons with the electromagnetic wave or field, was to prove of immense value in modern particle theory. And the use of two different models, the wave and particle ones, to explain the same phenomenon became an inspiring example of duality in modern physics.

Chapter 7

Relativity

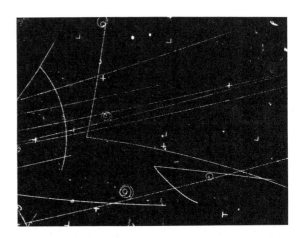

THE WORK THAT LED TO THE MODERN THEORIES of light also started another revolution leading to a modern view of mechanics. The theory of relativity provides, for our purposes, new laws about how small fast particles behave. Einstein's theory has become closely interwoven with other theories about elementary particles. It is a next logical step in our progress toward contemporary theories of the structure of matter. At this point we can move back to a more particle-oriented approach. However, we can never ignore the wave and field aspect of matter and energy.

7.1 FOUNDATIONS OF RELATIVITY

In the late 19th century the physical world was thought to be well understood. Newton's laws expanded to several subfields of physics and seemed to explain everything about motion and energy. Maxwell's equations did the same for light, but a medium (ether) was unnecessarily assumed. Then came the Michelson-Morley experiment (1887) which showed that the motion of the Earth in the ether did not affect light. One logical ex-

planation was that the ether moved along with the Earth. This notion was dashed in 1893 by another experiment which measured the speed of light between moving disks. The speed was not affected, so the disks must not have dragged the ether with them.

Another piece of evidence had been waiting since 1728 when James Bradley accurately measured the position of stars during the Earth's orbit of the sun. The stars seemed to move in a corresponding elliptical manner, since the light from them takes a certain time to reach the Earth. The effect is called the "aberration of starlight." No effect of ether on this starlight was observed.

An interesting and useful but wrong explanation of the null result from Michelson and Morley was advanced by G. F. Fitzgerald in 1892. He proposed that the motion of all earthly materials in the ether caused them to be contracted in length in the direction of motion in the ether. This "Fitzgerald contraction" (sometimes with the name Lorentz added) nicely canceled any ether effects in the Michelson-Morley interferometer.

Experimenters tried other ways to measure the Fitzgerald contraction but failed. The philosophical conclusion was that observers could not measure their own motion in the ether. This being so, it would be impossible to measure a speed in any absolute sense, so that all speeds are relative. We can see the concept of relativity creeping into physics before there was any coherent theory of what was going on.

In 1905 Albert Einstein performed and published his most famous thought experiment which showed what the answer would have to be. Applying Ockham's razor to select the simplest laws, he assumed that any observer must measure the same physical results, regardless of what speed the observer is traveling. In particular, he realized that any experiment should find the same speed for light, and he postulated that lightspeed is a universal constant c. Much has been made of the alleged complexity of Einstein's theory, but in regard to these simple ideas, the calculations are easily set up and carried out. The results have changed the view of the world held by most people in many ways, whether or not they know any physics. The theory is known as the "special theory of relativity," to distinguish it from a later "general theory." We can call it simply *relativity*, and this word is understood now in physics to refer to Einstein's theory unless other context is given.

Certainly Einstein's predictions that moving objects appear shorter, that moving clocks mark time slower, that moving masses are heavier, and other effects struck many physicists as bizarre. It was especially disturbing that the speed of the light emitted from a rapidly moving source is the constant c, regardless of other speeds involved. Attempts continued to find other explanations. At first it was difficult to design experiments that could be done accurately enough to verify Einstein's work. Astronomy provided the fastest light sources, and in 1913 Willem de Sitter appealed to observations of orbiting stars to prove that lightspeed was independent of source motion. The Michelson-Morley experiment was repeated, with changes, by E. J. Kennedy and E. M. Thorndike

in 1932, finding no variation in the speed of light to an accuracy about 2000 times greater.

Work on confirming Einstein's theories continues to this day, but relativity has been considered to be sufficiently correct since the 1920's to build the whole edifice of modern physics on it. Particle accelerators are one of numerous practical applications of the theory. Still, there are a few holdouts in scientific and popular circles who remain unconvinced and try to persuade others. It is possible to raise doubts about relativity because its character is quite different from what we experience ordinarily about the world. In daily life we think we see classical (Newtonian) physics in operation. Relativity is a refinement of Newton's physics, but requires an entirely different means of description. In fact, a world without relativity would be quite a different world, with no magnetism, different materials, different protons, and many other changes. Perhaps there would be no planets and life.

7.2 BASIC RELATIVISTIC EFFECTS

The predictions of Einstein's theory of relativity are, for the most part, observable only when high speeds are involved. We call such speeds "relativistic," and they must be a substantial fraction of c, which is close to $3(10)^8$ meters per second. For reasons that will be seen, the effects at a speed of $0.1c$ are only about 1% different from the predictions of Newtonian physics.

Talking about relativity requires us to be much more careful about the frames of reference we use. We must always have a frame in mind during the discussion of a particular situation. We must know about its motion. Special relativity avoids accelerating frames or objects. It uses frames that are *inertial*, named because there are no apparent forces due to the motion of the frame. An observer in a noninertial rotating frame of reference (Fig. 7-1A) can keep objects fixed in that frame only if an inward force is applied to counteract what seems to be a (fictitious) force pulling outward. Common examples are a rotating merry-go-round and the rotating Earth itself.

An observer in a noninertial frame that is ac-

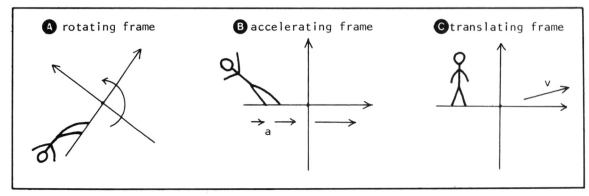

Fig. 7-1. Noninertial and inertial frames of reference.

celerating in a straight line, as shown in Fig. 7-1B feels a (fictitious) force opposite to the acceleration. A common example is the way one feels pushed back into the seat on an accelerating train. Actually, this is one's reaction to the seat pushing forward. The level of liquid in glasses on this train is seen to be tilted, as if gravity were at an angle. In contrast, the inertial frame of Fig. 7-1C is moving at an unvarying velocity in some direction. One can carry on life as usual without any mysterious and fictitious forces to explain.

The Earth would seem to make a poor reference frame for relativity. It has gravity and rotates. We have already abandoned it as an absolute frame, motionless at the center of the universe, but it is also a poor frame for considering relative motion and it is noninertial. Since some humans have now experienced the frame of reference afforded in a satellite in "freefall," orbiting the Earth, let us consider that one (Fig. 7-2A). Although gravity cannot be switched off, the satellite frame seems to have no gravity or

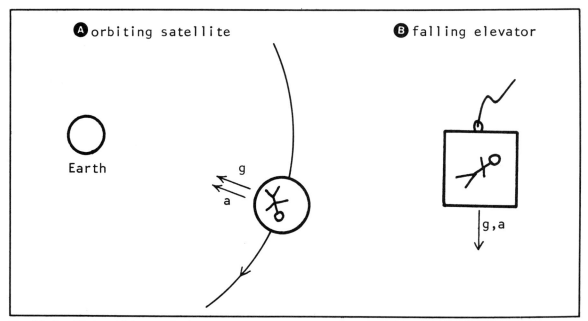

Fig. 7-2. Possible inertial frames of reference.

other outside forces. A person can stay motionless in mid-air. Gravitational force (really, weight) has been canceled by the fact that all parts of the satellite, including person and air, are falling toward the Earth with an acceleration equal to the strength of gravity. This frame would be a good inertial one except for one small difficulty.

A freely-falling elevator makes a good if dangerous inertial frame (Fig. 7-2B). It is accelerating, but the effect is canceled by gravity. A physicist can do relativity experiments inside it and never have difficulty. This and other inertial frames are also called "local" frames because practical reasons prevent their being very big. Another falling elevator used as an inertial frame would be falling in a slightly different direction toward the center of the Earth. There is difficulty comparing experimental results between two local frames.

Relativity has an effect on some basic physical variables but not on others. Lengths, times, and masses are affected; charge is not. In the case of mass, the theory predicts that a moving mass m will be greater by a factor γ (Greek gamma) than the rest mass m_0. The rest mass must be measured in a frame that is at rest with respect to the mass. The unitless factor γ is calculated from the speed of light by:

$$\gamma = \frac{1}{\sqrt{1 - \dfrac{v^2}{c^2}}}$$

Here v is the speed of the object of interest, measured with respect to the frame of the observer. The relativistic factor γ is an important basic quantity for discussing high speed particles, and we should try to become familiar with its behavior. In Fig. 7-3 is shown a graph of γ, a picture of how γ varies as speed v is increased from 0 to the maximum c. At zero speed γ is 1. Thus

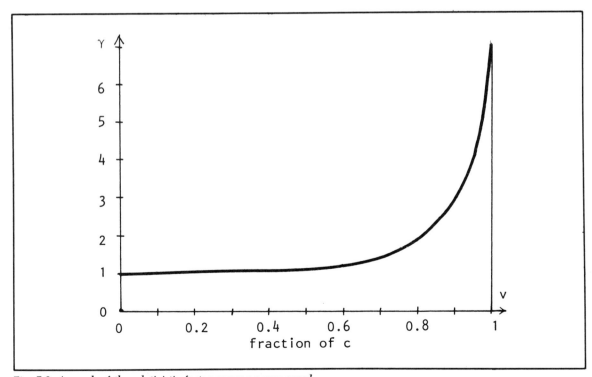

Fig. 7-3. A graph of the relativistic factor gamma versus speed.

when used as a multiplying factor with any other quantity, it has no effect. At one-tenth the speed of light, already a tremendous speed in our experience, γ is only slightly greater, 1.005. At half the speed of light γ is only 1.155, having a modest effect of about 15%. At nine-tenths the speed of light γ is 2.294, definitely a major relativistic effect. At 99% of c γ is a factor of about 7. Some particles move at speeds much much closer to lightspeed and have huge γs. When v approaches c, γ approaches infinity.

Given that moving mass is increased by γ, let us consider the consequences of applying a force to a mass to accelerate it. According to Newton any force can accelerate a given mass to any speed we want, if we wait long enough. Speeds a million times greater than c could be possible. But Einstein found that the mass grows, and more force is required to produce a given acceleration at higher speed. Things get out of hand at speeds near c. A much larger force is needed just to make up for the increased mass. We can see that c is the ultimate speed, never attainable. A nearly infinite force acting on a nearly infinite mass can give it only a slight increase in speed as c is approached. The increase can never put the speed over the top, because the mass continues to increase.

When energy is considered, we see that the kinetic energy can increase without limit. The speed is limited to c, but the mass approaches infinity. The mathematics works out that the old expression for kinetic energy is replaced with the most-widely known formula in the world, $E = mc^2$. The m in this equation of energy and mass is the moving mass. Thus we might rewrite this equation in a more accurate manner as:

$$E = \gamma\, m_0 c^2$$

What about low speeds, such as zero? Then we obtain $E = m_0 c^2$. Instead of no energy at zero speed, we find that a given mass possesses a very definite and large energy, which could be called its "rest mass-energy." This is a necessary consequence of having the kinetic energy behave correctly at speeds near c. One might wonder

what good this result is? A stationary car has no speed and would seem to have no energy (other than being at ordinary temperature, etc.). But Einstein claims that a 1 metric tonne car (1000 kg) has about 10^{20} joules associated with it, enough to create a very large crater if released all at once. (It is no accident that the energies involved are remindful of nuclear weapon energies. We shall see the same principles involved later.) We do not have to worry about this huge energy if we are simply colliding cars at ordinary speeds. But near the speed of light, look out! Nature then provides ways to bring all that "rest energy" into play. It is always necessary to consider a particle's rest mass and energy when studying high speed collisions. Thus relativity is essential to our main topic of particles.

What about light? It moves at c. The photons, if they had any mass at all, would have to have infinite mass, which is ridiculous. The logical deduction is that photons possess no rest mass. There is no difficulty getting them started moving ("creating" them) nor in stopping or absorbing them ("destroying" them). They defy Newton's laws but fit readily into Einstein's rules about motion. We have already seen that the total energy of a photon is $E = hf$. There is no need for a photon mass.

Another relativistic effect is called *length contraction*. Moving objects are shorter in their length in the direction of travel. The length L is reduced by the same factor γ thus:

$$L = \frac{L_0}{\gamma}$$

where L_0 is the rest length. A meter stick moving at $0.99c$ would appear to an observer as about 1/7 meter long. This is not an appearance in the sense of an illusion. It could really be measured to have that length. This stick sliding past a 1/7 meter hole at that speed would drop through (Fig. 7-4).

Since we are interested in elementary particles for the most part, do we care if a tiny particle is shorter when moving fast? What does it matter if it is elliptical instead of round, assuming that

Fig. 7-4. Length contraction: will the meter stick pass through the hole?

elementary particles have such simple shapes? There is little interest (so far) in particle shapes, but we can see the consequences of length contraction in another way. Remembering that the physics is the same regardless of which inertial frame one chooses, an observer who could sit on a high speed particle would see the world contracted as it rushed by. Much larger distances are seen to be covered in the same time, as compared to the situation if length did not contract. Accelerator designers must allow for this and make linear accelerators longer using the factor γ, which increases with speed.

Now the reader can face one of the kinds of apparent paradoxes that arise so often in modern physics. Returning to the story of the fast meter stick sliding over the small hole (Fig. 7-4), consider the point of view of the meter stick. As seen in the frame of the moving stick, the 1/7 meter hole is contracted by $\gamma = 7$ to a size of only 1/49 meter. It cannot go through, or can it? Either it does or it does not. Relativity does not defy the logic of events, and there is a definite answer. One of our points of view is wrong. While it is not in the scope of this book to cover all aspects of relativity, another ingredient of relativity will be provided shortly that aids the answer.

Relativity predicts that moving clocks run more slowly. The effect is called *time dilation.* Seconds, or whatever the unit of time is, are stretched or dilated, to last longer by the factor γ. The notion can be confusing to think about, as we are even less accustomed to unusual changes in the flow of time than we are to shrinking meter sticks.

One of the earliest measurements made on Earth to confirm relativity involved time dilation. A certain particle called a "meson," which we shall meet later, lives a certain length of time and then breaks into other particles. These particles live by their own internal "clocks." When moving fast, their clocks seem slowed, and they live longer, according to an outside observer's point of view. (The particles do not "notice" any difference.) Since they live longer, they can travel farther. Measurements by B. Rossi and D. Hall in 1941 found that mesons produced by cosmic rays high in the atmosphere were living long enough to reach the Earth, which they should not be able to do in Newtonian physics, given their microsecond lifetimes. The increased lifetime was measured and found to agree with relativity within experimental error.

By 1971 the use of atomic time standards had improved to the extent that portable clocks with an accuracy in nanoseconds (billionths of a second) were available. J. C. Hafele and Richard Keating carried several around the world on ordinary passenger jets, going both east and west. Most of the time was spent at speeds of about 250 m/s. Other more sophisticated relativistic effects were expected besides time dilation. The effect on time was in the range of 40 to 270 nanoseconds and was measured to agree with the predictions of Einstein's theories.

7.3 ADVANCED RELATIVISTIC EFFECTS

To explore relativity further, we must define a vital concept, the *event*. This sounds simple, but the ultimate consequences have baffled physicists and philosophers ever since Einstein. An event in relativity must have four coordinates—the three ordinary space coordinates measured with respect to X-, Y-, and Z-axes, and the time T at which it occurs (Fig. 7-5A). An event is not clearly defined

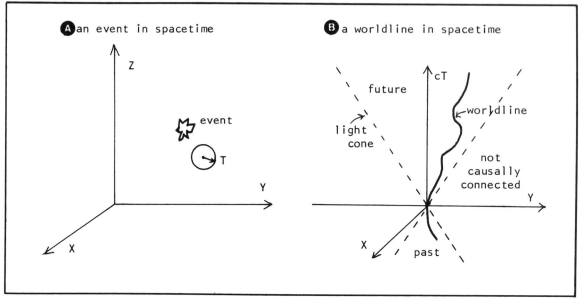

Fig. 7-5. An event and a worldline in spacetime coordinates.

unless all four numbers can be assigned to it. If something takes places over time, then it must be reduced to a series of events, each at a specific time. We will find particles easy to work with because most of the actions they become involved in are identifiable as events—a collision, a disintegration, an emission of light, and so forth.

The four coordinates of an event are called coordinates in *spacetime*, a four dimensional space which physics now uses for all events in both the microworld and in the universe as a whole. Every particle follows a path in spacetime called a *worldline*. A moving particle is a series of events in time. Figure 7-5B depicts a worldline of a particle in two space dimensions and time; the time axis is always shown vertically. We humans have worldlines, the intermingling paths we follow in spacetime during the course of our lives. The fastest particles, photons, have the worldlines shown dashed in Fig. 7-5B. These outline the "light cone," a region of spacetime not *causally* connected or in communication with ordinary matter.

Hermann Minkowski, who happened to be Einstein's former teacher, proposed in 1908 a new invariant in spacetime that corresponds to our no-

tion of distance. As shown in Fig. 7-6A, the points A and B (simple "events") are separated in two dimensions by the distance *d*. This distance should measure to be the same regardless of how the XY frame is placed or moved about. In Fig. 7-6B this idea is extended to three dimensions. Regardless of what XYZ frame is used, A and B are separated by the same distance *d*. By ordinary Pythagorean geometry the square of *d* is given by the sum of the squares of the separate intervals Δx, Δy, and Δz measured along each axis as shown. Minkowski extended this idea to one more dimension. We cannot easily draw a fourth dimension on paper or in space, whether or not it is time, but we can extend the idea of adding squares of intervals to get a new kind of *interval* or separation Δs between two events in spacetime. The only peculiarity introduced is that the time part should be subtracted in the calculation:

$$(\Delta s)^2 \;=\; (\Delta x)^2 \;+\; (\Delta y)^2 \;+\; (\Delta z)^2 \;-\; c2(\Delta t)^2$$

Speed *c* is introduced in order that all terms be in the same units. Since *ct* has units of length, all terms are lengths. Just as distance does not change when coordinate systems are changed, so the spacetime interval remains invariant during changes in coordinate axes or frames. The corresponding symmetry operation is a rotation of the four spacetime axes. In the search for new theories about elementary particles, this sort of invariance is always required. A theory is not universal if it depends on the particular frame that is used.

Einstein found a special way to calculate the coordinates of an event in another inertial frame, given them in one frame. The method is called a *Lorentz transformation* ("transform" for short). The procedure is necessary for purposes such as the study of particle collisions or other events in either the lab frame or the CM frame. Needless to say, the Lorentz transform leaves intervals invariant. It also can be used to find previously mentioned effects such as time dilation, since dilation is a result of changing the point of view from one frame to another. The Lorentz transform is said to be *covariant*, another new term widely used in modern physical theories. The "variant" part refers to the fact that an event has coordinates that *vary* from one frame to another. The "co-" part means that the change in coordinates from one frame to another is *consistent* and depends on only one parameter, the relative speed of the two

frames. The Lorentz transform can also be viewed as a rotation in four-dimensional space from one set of coordinates to another.

Since the Lorentz transform provides for a change in the time of an event as seen by different observers, there arises a difficulty over *simultaneity*. No longer can two observers be sure that they are seeing the same thing occur at the same time. A person on the ground can observe two strokes of lightning strike a rapidly moving train at the same time (provided that the speed of light from each stroke is allowed for). But another person on the train will find that one stroke occurs slightly after the other. These two well-defined events have different space-time coordinates in the train frame than in the ground frame. Therefore they cannot be simultaneous from two points of view. This discovery (initially by thought experiment) has deeply shaken but not destroyed physicists' faith in the idea of causality. Cause and effect cannot be established unless the two events are close enough that a light signal can pass between one event and another. Otherwise situations could be found where different observers must disagree on which of two events came first. The way was prepared in relativity for

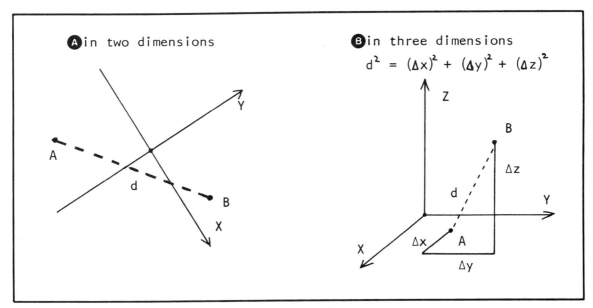

Fig. 7-6. The distance between two points as an invariant interval.

further disturbing developments about causality, to come from the world of the very small.

A relativistic effect observed for sources of light is the Doppler shift. This phenomenon was originally named for C. J. Doppler who observed in 1845 the change in pitch of sound as the source of sound moved toward or away from the observer. Light waves, despite their constant speed, also show a Doppler shift—two kinds, in fact. One kind is due to the usual effect of the stretching of a wave if the source is moving away (lowering the frequency) and the compression of a wave if the source is moving closer (raising the frequency). The other effect, due to relativity, is the time dilation at the source. A moving source, no matter in what direction, has its time pass more slowly and hence emits lower frequency waves than if it were not moving with respect to the observer. The effect is much smaller than the ordinary Doppler shift, except at very high speeds.

The total Doppler shift of starlight (or any light) is often called "redshift" for receding sources and "blueshift" for approaching sources. Laboratory verification of relativistic Doppler shifts was made first by H. E. Ives and G. R. Stillwell in 1938, observing the light emitted by hydrogen atoms moving at 0.003c. After we learn about radiation from atoms, we will see that atoms anywhere in the universe can be identified from their light. The state of their motion can then be ascertained from the shift in frequency (or wavelength) of that light.

If a fast moving source cannot have its speed added to the speed of light to obtain a total speed illegally exceeding c, how does relativity say that speeds should be added? We should be curious, as particle collisions are studied in both moving CM frames of reference and lab frames. The law of addition of speeds is not complex, but only a description of it seems necessary here. When the speeds are low, they add as any ordinary speeds, without considering c. When the speeds are near c, then the resulting speed is greater than the two speeds separately, but only by a small amount. The closer one is to c, the less that is gained by adding speeds. Finally, for a frame at speed c, no additional speed can be added. If somehow a light source could be made to move at c and emit light at speed c, the light would have speed c with respect to the observer. The law of addition of speeds has been confirmed, for example, in the Fizeau experiment.

The three space coordinates and time that define an event are viewed in relativity as four coordinates on equal footing. They may also be viewed as the four components of a vector—an arrow from a reference point to the event in four-dimensional space. This may not be difficult to imagine since one more component is added to our familiar three-dimensional vectors. Vectors in relativity are called *four-vectors*. The space-time four-vector is written in symbols as (x, y, z, ct), or similar forms.

One might ask whether there is a four-vector analogous to velocity. If we consider momentum as the "three-vector" in space, then the analogous four-vector consists of the three momentum components and the total energy. In symbols the momentum-energy four-vector is $(\vec{p}, E)$ or (p_x, p_y, p_z, E). Again energy and time are found to be *conjugate*—that is, to accompany each other. Physicists have found the use of momentum-energy language more convenient in advanced physics than other more familiar concepts such as velocity, acceleration, and force. The square of the momentum-energy four-vector (which gives the magnitude of the vector) is also an invariant. There is another way to calculate the total energy of a moving particle in relativity besides $E = \gamma m_0 c^2$:

$$E = \sqrt{p^2 c^2 + m_0^2 c^4}$$

Although E is related to p, the magnitude of momentum $\vec{p}$, in this way, both E and $\vec{p}$ can still be components of the four-vector. At ultra-relativistic speeds the rest mass-energy is negligible compared to the kinetic energy, and the relation simplifies to the useful form $E = pc$. When fields are present, which they often are, the total E should include consideration of the potential energy for each relevant field. Otherwise the particle is simply a "free" one.

In high speed (relativistic) collisions, the conservation of energy and momentum are restated in the form of an invariance of the momentum-energy four vector. It is often the case that the mass changes during a collision. For example, two particles may collide and react (Fig. 7-7A) and two or more new particles form. Relativistic analysis can only look at input and output (before and after). The total mass of the outgoing particles need not be less than the total input mass. Energy can be converted to mass and vice-versa. The only energy that is available for conversion to mass is that available in the CM frame (Fig. 7-7B). There is a "threshold" input CM energy needed to make a reaction occur. In the lab frame particles 1 and 2 have some kinetic energy in common. This energy is not available for the reaction; particles are "unaware" of any speeds and energies other than that with which they approach each other. This is another argument for the necessity of the CM frame.

Using the invariance of the magnitude of the total momentum-energy four-vector for the incoming and outgoing particles, the collision can be solved in the CM frame at least as well as the earlier low-speed cases. The principle result of interest to us in this book is that the energy available for conversion in the CM frame is proportional to the square root of the incoming energy when speeds near c are involved. When a particle of relativistic energy E (as measured in the lab frame) approaches a stationary particle of mass m, the energy E_{CM} available in the CM frame is approximately given by:

$$E_{CM} \cong \sqrt{2Em}$$

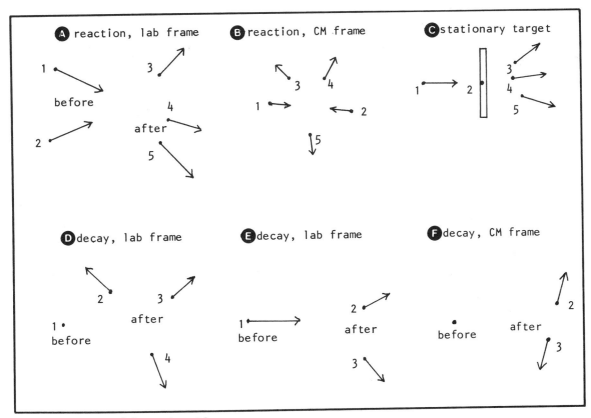

Fig. 7-7. Reactions and decays of particles.

The practical consequences appear when a collision with a stationary target is considered (Fig. 7-7C). Much of the energy of the incoming particle is not available in the CM frame in this case. Any reactions that might occur will only be able to use a fraction of that energy. The fraction gets smaller and smaller as the incoming speed approaches c. The development of colliding beam accelerators has been intended mainly to overcome this relativistic limitation. In such accelerators, both particles are moving at about the same high speed and toward each other. Their CM frame is moving slowly, if at all, with respect to the lab frame. Almost all of the energy generated in the laboratory is available for reactions during collision. To illustrate the improvement, a proton with 500 GeV has only about 32 GeV available when striking a fixed target (of protons), but 1000 GeV is available when two such protons collide head-on. The difference has aided much recent progress in particle physics.

While it may not seem to be a collision, the break-up or *decay* of a particle must follow the same laws. The problem is also easier to solve. If a particle is initially at rest, then the only way decay can result in three new particles is that the original mass be greater than the sum of the outgoing masses. The resulting particles then depart with any "leftover" mass converted to kinetic energies (Fig. 7-7D).

If moving particle 1 decays to only two particles (Fig. 7-7E), then each has a uniquely determined energy and momentum. Particles 2 and 3 must leave in opposite directions along a straight line in the CM frame (Fig. 7-7F), but when we observe them in the lab (Fig. 7-7E), we see the original forward motion of 1 added to the results. Low speeds rarely result from collisions or decays because the amount of energy released from a typical change in mass is very large, thanks to Einstein's mass-energy relation. At least one particle is likely to acquire relativistic speed.

7.4 COVARIANT ELECTROMAGNETISM

The principal concept to be introduced here is the covariance of the electromagnetic field.

"Covariance" means that physical laws such as those of electrodynamics should appear the same in any moving frame of reference. This way of looking at fields (and particles) will be of utmost importance in grasping the accomplishments of the modern revolution in particle physics.

It has been remarked earlier that Maxwell's equations for the electromagnetic field had much symmetry but some peculiarities remained. A new covariant form of the equations was obtained by Lorentz (using his transform) and by Henri Poincare even before Einstein's relativity appeared. The procedure includes changing everything to four-vectors. The current-charge four-vector is constructed to be a quantity such as $(\vec{J}, \varrho)$, where $\vec{J}$ is a current density (current per unit area, in a certain direction) and ϱ is a charge density (charge per unit volume). A four-vector potential is defined as $(\vec{A}, \phi)$, where $\vec{A}$ is the magnetic part, and ϕ is the electric part. This makes the field "gauge invariant" for certain potentials. It is necessary to introduce a new mathematical concept called the *tensor* to provide the covariant equivalent of the electric and magnetic fields. If a vector has three (or four) components, then a tensor has nine (or sixteen) components. A tensor has a set of components—better termed "elements"—for each ordinary component. Unfortunately an illustrative example from ordinary life is difficult to find, although tensors have applications in numerous different areas of theoretical and applied physics. With a tensor, what happens in each of three (or four) independent directions depends on what happens in the others.

For illustration, Fig. 7-8A shows the electromagnetic field strength in its tensor notation as well as the equivalent familiar electric and magnetic field components. This tensor is called "second rank," not because of its inferiority but because rows and columns in two dimensions are needed to write it out in an orderly way. The fourth row and column denotes those variables connected with time. The electromagnetic tensor turns out to be "antisymmetric"; the terms on each side of the principal diagonal (the line of zeros) are the same except for minus signs. The i is a the electromagnetic tensor   covariant Maxwell equations

$$\begin{pmatrix} F_{11} & F_{12} & F_{13} & F_{14} \\ F_{21} & F_{22} & F_{23} & F_{24} \\ F_{31} & F_{32} & F_{33} & F_{34} \\ F_{41} & F_{42} & F_{43} & F_{44} \end{pmatrix} = \begin{pmatrix} 0 & B_z & -B_y & -iE_x \\ -B_z & 0 & B_x & -iE_y \\ B_y & -B_x & 0 & -iE_z \\ iE_x & iE_y & iE_z & 0 \end{pmatrix}$$

$$\partial_\mu F^{\mu\nu} = (4\pi/c)\ J^\nu$$

$$\partial_\mu F^{\mu\nu\dagger} = 0$$

Fig. 7-8. *The electromagnetic field strength tensor in two notations and the equations from which it came.*

mathematical creature called an "imaginary" number, standing for the square root of minus one. It is much needed in mathematical physics but need not concern us here. Every physical result which is measurable in the laboratory is ultimately described with ordinary "real" numbers.

The tensor in Fig. 7-8A is the solution of Maxwell's equations in four-vector language, displayed in Fig. 7-8B. The sub- and super-scripts μ (mu) and ν (nu) are counters with values from 1 through 4, for the four coordinates. The symbol ∂ (partial) is calculus notation for derivatives (changes). The advanced mathematics is needed to express the theory of electromagnetism in covariant form—that is, to have it work at all speeds up to and including c. All this new notation is introduced for the electromagnetic case not just to show how difficult things get but to pave the way for more modern methods of the description of particles. Calculations at this level of mathematics are not given in this book, nor will further display of advanced mathematics be instructive.

The display of the electromagnetic field strength tensor implies that the different elements of it are all on the same footing. In its covariant form there is no distinction between electric and magnetic components. The implication is that electric and magnetic fields might not be distinguishable, despite our ordinary experiences. Moving charges feel generalized electromagnetic forces regardless of their electric or magnetic origin.

The situation is best seen in the example of a charge q moving with speed v parallel to a current-carrying wire. The speed is chosen to be the same as the speed of the electrons in the wire, but in the opposite direction. In the lab frame (Fig. 7-9A) the charge is affected by the usual circular magnetic field $\vec{B}$ around the wire. It feels an inward force toward the wire. We now change to a frame of reference moving along with a current in a wire (Fig. 7-9B), so that q is stationary. The electrons in the wire are traveling twice as fast to the right. The explanation seems completed. The electrons attract the test charge (assumed positive) with an electric force.

But we have overlooked one crucial fact. There is an equal amount of positive charge in the wire, and the net charge of the wire is zero (that is the way wires and all materials are made). We have also forgotten relativity. In Fig. 7-9B the leftward moving positive charges are closer together by virtue of length contraction. The electrons moving at double speed to the left are even more closely contracted together. There are effectively more electrons than positive charges. The wire has a net charge in this frame that is negative, attracting the positive test charge toward it. What was magnetic in one frame is electric in another! Quantitatively, it might seem that since electric currents move rather slowly the relativistic effect would be negligible. But we must not underestimate the huge strength of the electric force. The very tiny net change in charge density in the wire as one moves along it is enough to attract the test charge

95

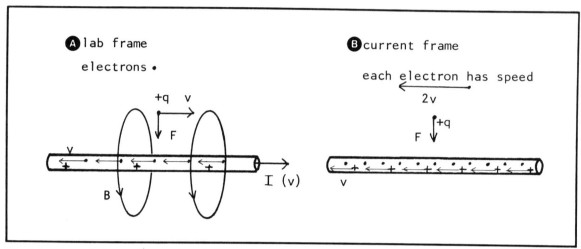

Fig. 7-9. *The equivalence of magnetic and electric forces for a current-carrying wire.*

with exactly the same force that the magnetic field would exert if *q* were moving instead.

The moral of this story goes beyond the fact that we do experience the effects of relativity in our daily lives. Every use of electric current and magnetic fields invokes relativity. Moreover, there is only one basic field from charges, moving or not: the electromagnetic field. Its effect on other charges depends only on their relative motion and no distinction of current, charge, electric, or magnetic is necessary. The fact that electromagnetic theory can be made to work covariantly is an inspiration for the success of other particle-related fields to come.

7.5 FAST CHARGES

The title of this section should be "Relativistic electrodynamics," pertaining to the behavior of charges at high speeds, but this vital section might sound unnecessarily advanced if so-titled. The γ {gamma} for a moving particle can be readily found from its energy using $E = \gamma m_0 c^2$. We will almost always want to work in MeV or GeV, so the calculations will be especially simple. The rest energy of an electron is about 0.5 MeV; this is its rest mass converted to energy and expressed in electron volt energy units. In our chapter on relativity (and elsewhere) it is convenient to think of speeds in terms of γ, the relativistic factor. If

we are told that a particle has energy *E* and rest mass-energy E_0, then its kinetic energy KE is given by:

$$KE = E - E_0 = (\gamma - 1)E_0$$

The relativistic range starts at about $\gamma = 2$. At that speed a particle has KE approximately equal to its rest energy. Its total energy E would be twice its rest energy. The method works regardless of the rest mass of the particle. A proton with rest energy about 1 GeV and a KE of 1 MeV has a γ of about 1.001.

We now explore some other consequences of relativistic electromagnetism. Let us examine closely what happens as a relativistic charged particle *q* passes by another charge *q'*, as in Fig. 7-10. The electric field *E* to the side is briefly felt as a strong pulse of electric force; its strength is enhanced by γ. A weaker accompanying magnetic field (strength $B = vE/c^2$) provides a pulse of magnetic force. Both field strengths are graphed in alignment with the particle position in the figure. Lorentz contraction of the electric field means that most of the field is sideways, as shown, and very little is in the direction of travel. The fields are indistinguishable from a "pulse" of electromagnetic radiation (a wave) at high particle speeds. As *v* approaches *c*, all of the field becomes transverse.

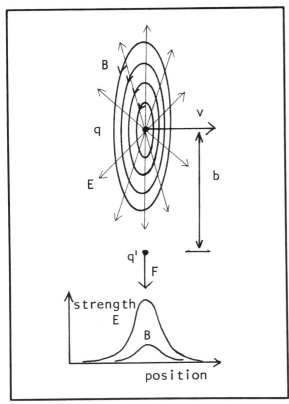

Fig. 7-10. *The electromagnetic impulse delivered by a rapidly passing charge, with a graph of electric and magnetic field strengths.*

The interaction in Fig. 7-10 is sometimes called a "coulomb collision," since the interaction is primarily electrical. The stationary charge *q'* receives a sideways kick, toward or away from *q*, depending on signs. The energy transferred from the incident charge to the disturbed one becomes very large as the impact parameter *b* decreases. Often the charges present when a high-speed particle passes through matter are the electrons that are a part of atoms. Ultrarelativistic charged particles transfer their energy well to the electrons. This bears on the stopping of, or the detection of, high-speed charged particles.

When fast particles are accelerated or detected, they emit radiation. Any accelerating charge emits electromagnetic radiation (light, ultraviolet, x-rays, or worse). A charge, regardless of its speed, that is moving in a straight line at

constant velocity does not radiate. In the linear accelerator previously discussed, the radiation that results from straight-line acceleration would result in a loss of energy by the accelerated particles, but the amount turns out to be negligible. In a circular accelerator, the problematical part of the acceleration is the fact that the charges move in a curved path in a magnetic field. The energy loss can be calculated to be proportional to γ to the fourth power (and other factors). This is an extremely strong dependence on speed, once relativistic speeds are reached. It is difficult to pump energy into electrons rapidly enough to maintain them at more than about 10 GeV because of this radiation.

This is an appropriate place for discussing a major method of obtaining relativistic particles. Other methods will appear when nuclear and particle reactions are discussed. The *synchrotron* is an advance in design that avoids the increasing expense of a magnet with the same diameter as the path of the particles. Particles travel in a narrow evacuated ring with straight sides and curved corners (Fig. 7-11). A magnet is placed at each corner to bend the particles around the curve. The magnetic fields are perpendicular to the plane of this diagram. There is no limit to the size of the ring and therefore the energy of the particles because magnetic fields of limited strength could be used to control small portions of the curve. Electric fields along the straight sections provide the acceleration. To further reduce the cost of the magnets (and their consumption of electric power) as bigger machines are designed, the "alternating gradient" or "strong focusing" version was developed. The magnets have segments which alternate in the way they focus the beam of particles, so that the vacuum ring can be very narrow. The beam intensity can be increased as well.

The radiation from a relativistic charge moving in a curved path (invariably in a magnetic field) comes out as a narrow beam along the direction of motion (Fig. 7-12). The width of the beam is inversely proportional to γ. The charge emits not one frequency but a wide range, up to a maximum frequency approximately γ^3 times the rate

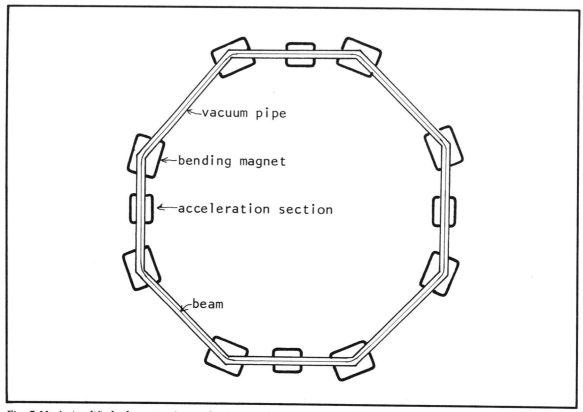

Fig. 7-11. A simplified schematic of a synchrotron accelerator.

at which the particle orbits in a circular path. The radiation is called *synchrotron radiation* after the type of accelerator in which it was first observed in 1948. Until recently this radiation has been an unwanted and hazardous byproduct of particle accelerators. Now it is turning out to be a useful source of intense ultraviolet and x-rays for other research. Synchrotron radiation is also received from interplanetary and interstellar space. If accelerators produce it in space, they are naturally occurring ones. It has several distinguishing features which allow deduction of the strength of the interstellar magnetic field and the energy of the particles emitting it.

Because light travels slower in transparent media than it does in space, it is possible to have particles at speeds which exceed the local speed of light. Such particles will be slowing down as they enter a medium, but for a time they will emit radiation that moves slower than the particles. The radiation is called *Cherenkov radiation* and was discovered by P. A. Cherenkov, Ilya Frank, and Igor Tamm in 1937. They received the Nobel Prize in 1958.

Cherenkov radiation is the result of a large scale effect on the electrons in the medium by the passing electric pulse of the particle. The disturbance of the electrons adds up in a systematic way to produce an electromagnetic "shock wave." We are familiar with the acoustical analogy: a plane traveling faster than sound in air produces a single conical wavefront which we hear as a "sonic boom" when it passes by. For a given particle speed, the wavefront of the radiation travels in a certain direction (see Fig. 7-13), and measurement of the angle enables calculation of the velocity of the fast particle which produced it. The radiation is usually visible light of various colors. Since emission of this radiation represents an energy loss by the particle, it must soon be slowed below the speed necessary for producing the radiation.

A particle passing near atoms and among electrons in matter suffers various collisions. Each interaction results in an abrupt loss of speed. This constitutes a deceleration and therefore radiation must be emitted. It is called *bremsstrahlung* after the German for "braking radiation." Ultrarelativistic particles suffer most of their energy loss by this mechanism. Since each particle that is collided with then accelerates, it also must emit radiation. The lighter of the two particles suffers the greater change in speed and is the dominant emitter. A broad range of frequencies are emitted, up to a limiting frequency based on the time during which the collision occurs. The radiation is emitted in a narrow forward beam. When the particle is relativistic, its path is changed only slightly during each collision with ordinary matter and it loses very little energy each time. The significant collisions for energy loss are those with the nuclei of atoms—that is, near encounters with the electric field of the nucleus—not those with electrons.

Another phenomenon should be mentioned which prepares us for later discussion of nuclear and weak interactions. It is observed that nuclei occasionally emit electrons at high speed. The electron is somehow accelerated from near rest to an energy as high as 60 MeV. Radiation should accompany the acceleration, and it has been observed. The amount, however, turns out to be a very small fraction of the other energies involved.

When a force is applied to a very small object, even a particle, relativity guarantees that the particle may not accelerate all at once. The force reaches different parts of the particle at different times, propagating at a speed less than or equal to c. The differential force across it should temporarily "squash" it. We can deduce from this simple application of relativity that the electron, if it is a fundamental particle, must be a point particle without size. Otherwise large enough accelerations, easily obtainable in modern collisions, would break it into smaller parts—a phenomenon never observed. The proton gives a different result, as will be seen.

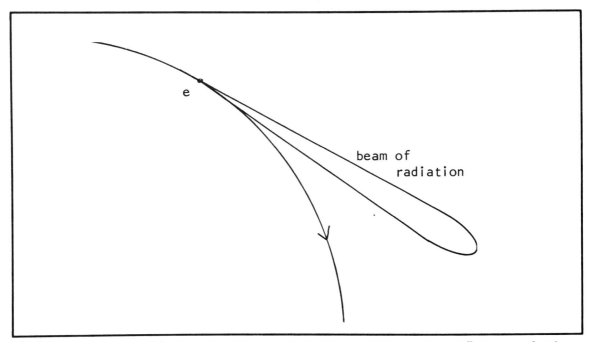

Fig. 7-12. A representation of the beam of synchrotron radiation from an electron moving rapidly in a curved path.

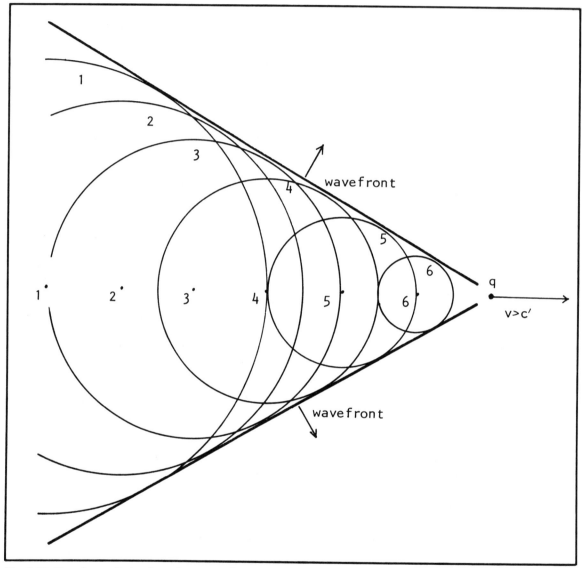

Fig. 7-13. Cerenkov radiation by a charge traveling faster than light in a medium.

7.6 TACHYONS

Nothing in relativity prohibits the existence of particles which travel faster than light. What is not permitted is the acceleration of a particle from low speeds to lightspeed and higher, as that would require infinite force. *Tachyons*, named for the Greek for "swift," would have a number of properties opposite to those of "tardyons" (ordinary slow particles, named for the Latin for "slow"). Tachyons prefer to move very fast, and force must be applied to slow them down to nearly lightspeed. They cannot be slowed farther, as that would require infinite force. If they have zero mass, then they can travel at infinite speed.

It would be useless to mention tachyons unless there were schemes to detect them and use them, say for interstellar faster-than-light communication. Many proposals have been made for detecting them, based on supposed interactions with ordinary matter (tardyons), but no positive results have been obtained and the search is all but abandoned at present. Tachyons could be produced in collisions without requiring an unusual amount of energy, but they would appear only as missing energy and momentum. They would supposedly be created already moving faster than light and no impossible acceleration is required. Trying to count tachyons could be a problem. Different observers moving differently would disagree on whether tachyons were emitted or absorbed, and how many. Their faster-than-light travel allows them to violate our notions of causality. Also, since their speed exceeds lightspeed in vacuum, they should continuously emit Cerenkov radiation, quickly reaching low energy (and high speed). If tachyons were to exist, we would expect a whole zoo of different kinds as is the case with tardyons. There may be no way of detecting them. Another limit on their existence is imposed by the fact that light can travel for billions of years with no apparent change. If tachyons exist, photons should decay to them, at some low rate.

Chapter 8

The Quantized Microworld

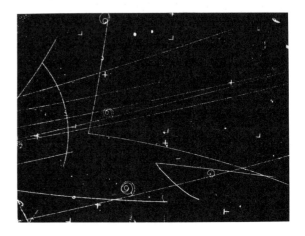

S OME OF THE FOUNDATIONS OF THE MODERN quantum view of nature were established with the discovery that light could act as particles (photons). We have seen that the proton and electron were established as elementary particles before the 20th century. The electron was discovered by J. J. Thomson in 1897. The proton was known as the hydrogen atom (without its electron) although its identity as a particle was not clear until Rutherford established the mass and charge of nuclei in terms of hydrogen atoms in 1911. We now turn our attention to major developments in regard to some peculiar behaviors of the three particles—photon, electron, and proton. Much physical and philosophical complexity will appear. Later we will assemble the results to form the modern view of the atom, which is more complex physically but not so perplexing philosophically. Many of originators of the new ideas to be referred to are pictured in the group portrait given as Plate 2.

8.1 DIFFRACTION OF MATTER

We have seen that light, despite its acting sometimes as particles (photons), can be diffracted and interfered, that is, to be made to act as waves. If light has this dualistic aspect, what about matter, which is supposedly different from light? In the case of matter, we want to consider objects which are normally thought of as particles, whether stars, baseballs, grains of sand, or electrons. In 1925 Louis de Broglie made the analogous consideration, that matter ought to act as waves, at least under certain conditions. Everyone knows that the path of a baseball does not bend just because it passes near a post. But could electrons, protons, and possibly grains of sand somehow "diffract" as they passed by an obstacle or through an aperture?

Looking at the Planck and Compton relations for obtaining the energy and momentum from frequency and wavelength, respectively, for a photon, de Broglie supposed that matter would be treated the same way and wrote these relations:

$$f = E/h$$
$$\lambda = h/p$$

Here f and λ are values for the proposed "matter wave" that would correspond to a particle of mass m moving at speed v, thus having momentum $p = mv$ and total energy $E = mc^2$. E is not the kinetic energy but the total energy required by Einstein's relativity, and m is the moving mass. Planck's constant h is included. De Broglie had knowledge of relativity and was guided by the fact that both the wave variables (λ, f) and the particle variables ($\vec{p}$, E) formed covariant four-vectors (if λ and p are defined as vectors). He received the Nobel Prize in 1929 for this work.

Let us make a few order-of-magnitude calculations to see what might be expected from these de Broglie relations for matter. We will round h off to 10^{-33} J s. For a baseball of mass 0.1 kg thrown at a typical speed of 10 m/s, its momentum is 1 kg m/s in SI units. Its alleged wavelength would be about 10^{-33} meters, extremely small. To observe diffraction effects one must have a wavelength approximately the same size as the obstacle size. Clearly any diffraction of a baseball would be far below the limits of measurement. The baseball lands where it is thrown, and there is no significant possibility that it will end up anywhere else.

It will be instructive to try to find the "frequency" of the baseball. Suppose the only energy we know about is its kinetic energy, which is calculated to be about 10 J, ignoring the 1/2. Then the f would seem to be this quantity divided by h, giving the ridiculously large number 10^{34} Hz. Whatever this means, it is far beyond the usual frequencies we can measure. Suppose that we must believe Einstein and use mc^2 as the total energy. Then we get about 10^{16} J for E and about 10^{17} for f, high but more plausible. Soft x-rays have such a frequency. At this point it seems that Einstein's total energy would give more realistic calculations for matter waves.

Nevertheless, the baseball's wavelength seemed ridiculously small. What sort of particle might have a big enough wavelength that we could see the effect? By playing with the values for the masses and speeds of some known objects, we would eventually discover that things make sense only if we look at electrons and protons. An electron (mass about 10^{-30} kg) moving fast (10^6 m/s, about 1 % of lightspeed) has a matter wavelength of about 10^{-9} m. The frequency is 10^{20} Hz, high but in accord with what we will learn later about such high energy particles and their interactions.

An electron wavelength of 10^{-9} m still sounds small, but it is almost ten times bigger than an atom. We could use atoms as the obstacles and we would expect to be able to detect the diffraction of electrons. If the electron has interacted with a set of atoms and been somehow deflected from its course, we can try to observe it from a long distance back. The deflected electron will arrive at our electron detector at a position far from its original path. Things are simpler if we arrange for orderly diffraction by using atoms arranged regularly as in a crystal (Fig. 8-1A). Electrons leaving the crystal in almost any direction will land on a screen (or photographic plate or other suitable means of detection). If there is diffraction some bright spots, bands, or rings are expected at places where the supposed "electron wave" interferes constructively with itself and the electron strikes. The diffraction occurs on a very small scale, and only ordinary electrons as particles will emerge from the interaction region and be detected as particles. We do not expect to receive literally an "electron wave," nor would we know how to recognize it. At each position on the screen, either a whole electron lands or no electron lands.

This experiment was first done by Clinton Davisson in the 1920s; electrons were fired at a nickel crystal. In 1927, working with Lester Germer, he found a pattern of spots sufficiently distinct (see Fig. 8-1B) as to be identified with the recently developed theory of matter diffraction. Electrons could be waves, a result that astounded the scientific world and set much of the course of

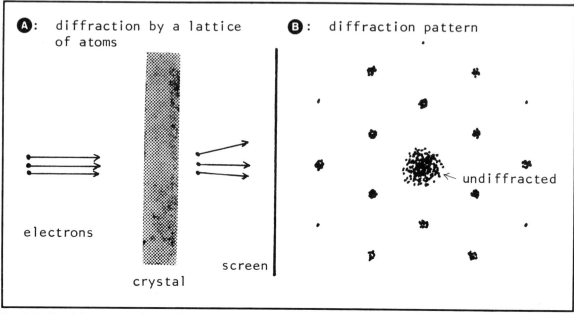

A: diffraction by a lattice
of atoms

B: diffraction pattern

electrons

crystal

screen

← undiffracted

Fig. 8-1. *The diffraction of electrons by a crystal lattice (cross sectional view).*

modern physics! The analogous experiment with light waves (x-rays) at similar energy and with similar material had already been done, and the expected diffraction pattern was similar. (At this point it does not matter whether we call the process interference or diffraction. It is the result of waves interacting with matter.) George Thomson, son of the electron discoverer, obtained pictures of electron diffraction for a metal foil. Such a material is composed of many randomly oriented crystals and yields a set of rings around the beam axis for the pattern, resembling closely the result of using x-rays on a powder sample. Davisson and Thomson received the Nobel Prize in 1937 for this work.

The philosophical implications of matter diffraction were soon realized, and these astonished the world of physical scientists. To have a crystal transform a tiny beam of electrons into a regular but complex pattern of spots all over a photographic plate implies that there are many places the electrons avoid and a few places they congregate. The spot positions could be exactly predicted by simple theory using the de Broglie wavelength for the given electron speed. More-

over, a probabilistic character of the electron behavior was revealed that had been previously unrealized. If x-rays were interfering, it would be understood that wave intensity would be greater at certain places on the photographic plate, producing spots there. But when particles are involved, it must be inferred that the spots are located where there are high probabilities of receiving particles.

The experiment works even if the electron beam intensity is turned down so that one electron at a time passes through the apparatus. Most of the time the electron lands on the screen where it should according to wave diffraction. Rarely does it land in a place where waves would cancel. Within this constraint, each electron arrives randomly. One electron might contribute to a spot in one area, and the next electron contribute to another spot somewhere else. No one can predict where the next electron will go, but one can be almost absolutely certain where an aggregate of billions of electrons will land—on the places predicted by diffraction theory! Each electron, regardless of other electrons, acts as if it were a wave when it interacts with matter on a scale approximately the size of that wave.

The truth will be seen to be stranger still. The next complication is that the electrons interfere with themselves. Electrons aimed one at a time at a small enough slit will form a diffraction pattern behind the slit, as if each electron were a wave. But electrons aimed at two slits which are sufficiently narrow but not too close together will interfere and form the usual interference pattern on the screen (Fig. 8-2). A photographic plate left long enough to capture the results from a flow of individual electrons will show an aggregate interference pattern. There will be a set of bands where the electrons underwent constructive interference, giving maxima on the plate. It does not matter whether one electron at a time or many pass through the experimental apparatus. Interference occurs for each one. If the many electrons are not all identical in speed (therefore wavelength), then there will be some blurring of the pattern just as occurs for light with a spread of colors.

Matter waves are not peculiar to electrons. Helium atoms were diffracted by a crystal in 1931. As other elementary particles were identified and could be controlled in the laboratory, matter diffraction was found. Neutrons, almost 2000 times heavier than electrons, form the same sort of diffraction patterns, allowing for a different wavelength because of the different mass and momentum. Physicists carried out many thought experiments about diffraction, too. They became quite adept at making wild predictions based on this new view of matter, then eventually finding a way to confirm them in the laboratory.

That a matter wave is not real can be seen by finding its wave speed. The usual rule is to multiply f by λ to find the speed of a wave. When this is done for matter, the result is the speed c^2/v, where v is the speed of the particle. Its speed v must be less than c, so the wave speed is always greater than c. If the particle is moving slowly, then the wave is at its fastest. A stationary particle would have an infinitely fast wave. An ultrarelativistic particle would have a wave speed just over lightspeed, another reason why fast particles act much like light. The general result is that it would be impossible to detect a matter wave as such because it moves faster than light and has no reality as we know it.

8.2 HEISENBERG UNCERTAINTY

Those who considered the interference of single particles acting as waves had to wonder about the logic of the experiment. When a moving electron faces two openings as in Fig. 8-2, reason tells us it will go through one or the other. Does it, or has our human-sized reason and logic failed in the microworld? We cannot tell from where the electron lands where it went. It can and does land almost anywhere on screen S. If the electron does indeed go through one slit, then how can it "know" or detect the existence of the other? It does *somehow* take into account both slits since it arrives at the screen in accord with the interference pattern characteristic of the particular two slits.

The question of what logic applies when experiments are done on a small scale is still a matter of great controversy. One of our mistakes would be to assume the electron must act as a wave and a particle at the same time. But another mistake would be to assume it switches roles according to some sort of built-in wisdom about its environment. What the electron may really do is follow another model not yet simply described by

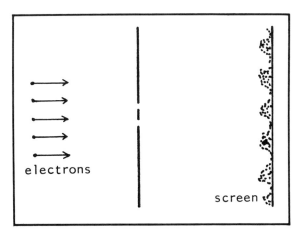

Fig. 8-2. The interference of electrons through two slits (cross sectional view).

scientists, but one which reduces to waves in certain situations and to particles in other (overlapping) situations. What is certain is that Planck's constant determines the scale on which things will behave differently from what our human experience can predict. There is also an in-between range where we can directly perceive some rare effects of the wave aspects of matter. But for the most part, the microworld ruled by Planck's constant is far removed conceptually from our own macroworld.

Pondering the question of where the electron goes led to the principle of *uncertainty*. It was realized that there must be a fundamental uncertainty in tracking tiny particles during their motions. Indeed, one could never be certain whether a particle is sitting still. Precedent had been set in the case of the microscope. Optical technology had proceeded to the point where the object to be seen was hardly bigger than the wavelength of light used. Thus the smallest bacteria could barely be resolved with visible light, and viruses remained invisible (and unknown) until technology using smaller wavelengths could be developed. X-rays provide smaller resolution, and the reader may be familiar with the use of electrons at high speed in the electron microscope for working at sizes approaching that of atoms. It should be clarified, however, that the usual electron microscope focuses the electrons as particles, after they have interacted with the specimen as waves. There is no simple "lens" for matter waves.

The Heisenberg uncertainty principle, first stated by Werner Heisenberg in 1927, tells us that the attempt to measure some of the properties of a tiny particle automatically introduces a certain amount of uncertainty into the measurement process. The uncertainty refers to random errors that are found in measured data after all known errors are eliminated. In the microworld one cannot *ever* measure anything exactly. Worse, there is a limitation on how accurately certain quantities can be measured simultaneously.

In Fig. 8-3 is shown an attempt to measure a particle's position and speed (momentum). This experiment is a thought experiment, but parallels real and more complex ones. The position we are interested in is the y-position—where it is with respect to the y-axis shown. We will attempt the measurement by locating the small hole through which the particle must go, as shown, but trouble develops. If we shine a light on the particle so we can make a photograph of it in the hole, the light must have a small wavelength in order to resolve the image of the particle and hole. Short wavelength light has high energy and momentum and will give the particle quite a kick. The light must come at least partly from the side, in order to see the y-position, and the kick will knock it sideways. We no longer know its y-position very well, as it acquires a momentum component Δp_y in the y-direction, as shown.

Since this experiment does not seem to work well, let us abandon the photography and just try to find an aperture it goes through, then detect it later. If the aperture is small, to pinpoint the particle's y-position, its matter wave will diffract substantially in the aperture. After this encounter it again acts like a particle but is going in a new direction. Its position and momentum in the y-direction are both uncertain. We are not any better at knowing its position and momentum in the x-direction, for the same reasons. Any use of light or passage through an aperture will change the

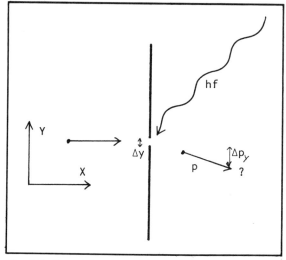

Fig. 8-3. Uncertainty in position and momentum.

x-component of its momentum. These difficulties are at the heart of the uncertainty problem.

Position and momentum in any given direction are called *conjugate* variables. If we know one accurately, the other must be inaccurately known. There is no way around it. Countless real and thought experiments have confirmed this initially disturbing result. The relation of the uncertainty in two conjugate variables can be estimated mathematically. Using Δ to denote uncertainty, we can write:

$$\Delta x \, \Delta p > h/2$$

The variable x stands for any position component, and p is the corresponding component of momentum in that direction. The symbol ">" denotes "greater than." The product of the two uncertainties must be greater than a certain number, about 10^{-34} in SI units. It should be no surprise that Planck's constant enters into this fundamental statement about our knowledge of the microworld.

Another pair of conjugate variables are energy and time. The uncertainty principle states mathematically:

$$\Delta E \, \Delta t > h/2$$

This relation says that if we know the energy of a particle quite accurately, we cannot know accurately how long it possessed that energy, and vice-versa. Again, the effect is only seen in the microworld, where events can occur in very short times. Another way to look at it is that this relation tells over what length of time the law of conservation of energy can be violated. For example, if we wish to see conservation violated by the amount of 1 joule, a perceptible quantity of energy, this can only happen for a time less than 10^{-34} second. Conjugate variables just happen— as if there were any coincidence at the foundations of physics—to be those useful with the principle of invariance. Invariance with respect to translation in space leads to conservation of momentum, as discussed earlier.

In the human-sized world, the uncertainty principle does not set uncomfortable limits to what we can know. For example, if a baseball passes through a hole of diameter 0.1 m, which establishes the accuracy with which we know its position, then its momentum can still be measured to an accuracy of 10^{-33} in SI units, hardly a challenge to our instrumentation.

The boundary between the microworld, where some things are absolutely unknowable, and the macroworld, where all is knowable to great accuracy is crossed by another principle. Neils Bohr first formulated the *correspondence principle* in 1923 in recognition that a new physics called *quantum physics* was conflicting with the older classical physics. No one doubted the accuracy of classical theories on the large scale, so that results in the microworld had to *correspond* to classical results as the scale was increased from small to large. The diffraction of an electron through a tiny hole had to reduce to the normal passage of a particle through a hole as the hole was scaled up. Planck's constant sets an absolute scale in the microworld, but at larger scales it is irrelevant. The effect of Planck's constant must fade out as sizes are increased (or times lengthened). Consequences of the correspondence principle will be pointed out as they are encountered.

8.3 WAVE MECHANICS

Most physicists in the 1920s tentatively accepted that a wave-particle duality existed for all radiation and all matter. A few kept their disbelief for long periods, most notably Einstein. All accounts of this period by its participants report great excitement at the revolution in physics. The bulk of the work started in Europe but gradually moved to the U.S. as political turmoil in Germany and its neighbors distressed intellectuals there.

Neils Bohr stated the *principle of complementarity* in 1927. The principle says that either the wave or the particle view can be used for radiation and matter, but not both simultaneously. The search began for the possibility of describing the behavior of matter completely in terms of waves. This was partially an intellectual exercise, to see if it could be done and to see what new practical ideas might come of it. Physicists already had a perfectly good set of theories—classical physics—which seemed to explain nearly everything about particles. At first only a few unusual phenomena such as electron diffraction signaled the need for a totally new kind of physics. But as the new theory developed and experiments were attempted, many new realms of nature opened up, leading ultimately to the applications that are modern technology.

The first major step in a *wave mechanics* or *quantum mechanics* to replace or complement the existing classical mechanics was to ascertain the nature of the wave. There was no known medium for the electron and other matter waves. They propagated in free space, and the speed of a matter wave was greater than light. Their frequency depended on the new total energy of a particle as defined by Einstein, an unusual development but in parallel with the quantization of electromagnetic radiation.

Since what was observed in microworld experiments were the statistical results of what seemed to be the interference of waves, the notion of *probability wave* was born (an inappropriate pun since Max Born proposed it in 1926). A theory was sought which would tell everything about the particle but in wave language. It could have speed, momentum, energy, position, and so forth, but also wavelength and frequency. Erwin Schrodinger formed such a theory in 1926 using mathematics. His procedure for setting up the theory was philosphically weak and founded primarily in the applied mathematics of classical physics, but it worked. He used the correspondence principle to create a wave equation from the classical equation for the corresponding particle. The statement of the Schrodinger equation is deceptively simple:

$$H\psi = E_n \psi$$

where H is an *operator*, not just a multiplying symbol, that acts on the symbol ψ (Greek psi) for the wave. H and ψ can be very complicated, and a new era of applied mathematics faced researchers and students in physics. E_n is one of the allowed energies for the wave in the given system. The wave function ψ is also called an *eigenfunction* (from the German "eigen", for "characteristic" or "self") and E_n is an *eigenvalue*.

Special wave functions and energies were found to fit each particular physical situation described succinctly by H. Physicists proceeded to find the actual form of the probability wave in various simple physical problems. It became possible to "solve" the hydrogen atom to a high degree of accuracy and the agreement with experiment was near perfect, as will be seen in the next chapter. As often happens, some theoretical mathematics that had been developed with no application in sight now found application in quantum mechanics. Schrodinger received the Nobel Prize in 1933. For many early quantum physicists, their work was recognized by the Nobel committee when they applied the new theories to atoms. Some quantum pioneers will be so-noted in the next chapter.

At first Schrodinger thought that his wave function specified that the electron or other particle is smeared out in space. However, this does not make sense in collisions, when the electron hits all at once. Born received the Nobel Prize in 1954, partly for early work such as interpreting wave functions as probability waves. The probability wave cannot be seen nor detected in any direct manner. It is a mathematical construct that works, leading to predictions that can be observed when certain rules are followed. The new quantum mechanics contains within it not only the correspondence principle but also the uncertainty principle. The matter wave, and therefore the particle, are limited to a set of energies (eigenvalues) more restrictive than hitherto suspected. These are also called energy *states*. The analogy between waves on a string and the probability wave for a particle is partly appropriate. Shorter wavelengths represent higher energy. However, a larger probability wave amplitude tells not how energetic the particle is, nor how big it is, but how likely it is to be found in a given place. (A difficulty over negative amplitudes is removed mathematically.)

Simultaneously with Schrodinger another physicist was wrestling with the same problems and finding a different approach in 1925. Werner Heisenberg pondered the extensive knowledge available about the light emitted by atoms. He found that the amounts of the different frequencies of light had to be recorded not as a list but as a table, a two-dimensional arrangement. He knew that operators had been found to make the new wave mechanics. He wished to avoid the wave aspect and get to the heart of some new mechanism in nature. He found an analogy between certain operators and a new branch of mathematics that had been invented by Arthur Cayley in 1858, now called by various names such as "matrix algebra." A matrix is a two-dimensional array of numbers which relates one set of variables, say x_1 and x_2, to another set, say y_1 and y_2. If, for example, the variables are related by the simple set of equations:

$$y_1 = a_{11} x_1 + a_{12} x_2$$
$$y_2 = a_{21}x_1 + a_{22}x_2$$

then the four values of *a* form a square array or two-by-two matrix:

$$\mathbf{A} = \begin{matrix} a_{11} & a_{12} \\ a_{21} & a_{22} \end{matrix}$$

A matrix can be bigger, such as three-by-three, four-by-four, and so on. It also need not be square. It is much like the tensor mentioned earlier. In comparison, a vector can be written simply as a column of numbers, one number for each component. The matrix elements tell how, in the above example, the components (x_1, x_2) of $\vec{x}$ each contribute to the components of $\vec{y}$. In a symbolic shorthand, the matrix equation above can be written: $\vec{y} = \mathbf{A} \, \vec{x}$. $\mathbf{A}$ can be viewed as an operator which converts vector $\vec{x}$ to $\vec{y}$.

An unusual fact from matrix algebra is that the result of using matrices $\mathbf{A}$ and $\mathbf{B}$ in succession as operators (which can be done by multiplying them with special rules to get the operator $\mathbf{AB}$) is not the same result as using $\mathbf{A}$. The order of use or multiplication matters. Heisenberg realized that this agreed with his data, which seemed to call for operations which did not *commute*—that is, when the order of operating is reversed, the result is different. He was especially concerned with using momentum $\mathbf{p}$ and position $\mathbf{x}$ as operators. He soon found that $\mathbf{px}$ was not equal to $\mathbf{xp}$ and that their difference is:

$$\mathbf{px} - \mathbf{xp} = h/2\pi i$$

Again h and *i* appear in the mathematics of the microworld. This noncommutativity occurs for conjugate operators, corresponding to variables for which simultaneous accurate measurement cannot be made. It would seem that variables for which the uncertainty principle holds are also variables which do not commute when used as operators. Heisenberg was able to form equations describing physical problems, which could be solved. He received the Nobel Prize in 1932, partly for his version of the new quantum mechanics.

Paul Dirac, who knew of this work in 1925, immediately saw the connection between this quantum mechanical result and an older method in classical mechanics. He provided the *correspondence* between classical and quantum theory. He created a matrix-based version of quantum theory which had noncommutativity at its core. He was able to convert any classical variable into a quantum operator and carry out many kinds of calculations which ultimately agreed with experiment.

8.4 FREE AND BOUND PARTICLES

Let us look briefly at two important applications of the new quantum mechanics. We will use the Schrodinger and Heisenberg approaches interchangeably. We have already mentioned a connection between a moving particle and a traveling probability wave. It is not possible for a single infinitely long smooth sinusoidal wave to represent a particle (Fig. 8-4A). Such a wave would have the particle be almost anywhere in space. It has a single wavelength, implying a single momentum. The infinite uncertainty in position and the zero uncertainty in momentum can be in accord with the uncertainty principle, but the object described is not useful. We have no idea where it is. The wave has no distinguishing features. It can carry no information about the particle.

What is needed is a useful way to describe a "free" particle with a wave. Mathematical results from wave theory provide a convenient tool. A combination of sinusoidal waves can be added together to make almost any form of wave. A certain combination of sine waves forms a *wave packet* (Fig. 8-4B), a wave which "waves" but is confined to a small region in space as it travels. A range $\Delta\lambda$ of wavelengths centered around λ were used to make it. The corresponding particle would have an uncertainty Δp computed from λ. The wave occupies a region in space of width Δx. The Schrodinger theory can be used to find that this uncertainty in position is consistent with the uncertainty in momentum. As time goes on, a wave packet naturally spreads out in time. This agrees with the increasing uncertainty of position as a particle moves. The dashed curve in Fig. 8-4B

shows the probability of finding the particle. It has a maximum where the particle is most likely to be, and it moves to the right with the speed v of the particle. The sine waves from which this wave packet were constructed continue to move at greater than lightspeed. But the constructive interference between these waves forms a group, or wave packet, which has a "group velocity" always less than c.

The free particle can have any energy. It must necessarily have its total energy E greater than any local potential energy PE. This is so that it can have a positive KE and indeed be free to move. In regions where PE is greater than E, the particle cannot exist and there is no corresponding traveling wave. But, as we shall see, there is a small probability of the particle being there. A PE that varies over space enables the creation of a "potential well." A "square" one with steep sides is shown in Fig. 8-5A. In the center portion PE is less than E and the particle can move freely. Its motion is restricted by the boundaries formed where PE rises to a high value. Classically, the particle can move back and forth in one-dimension (the X-direction) and bounce from each rigid wall.

In terms of wave mechanics some new options are open. The situation of Fig. 8-5A could remind us of a string stretched between two points and

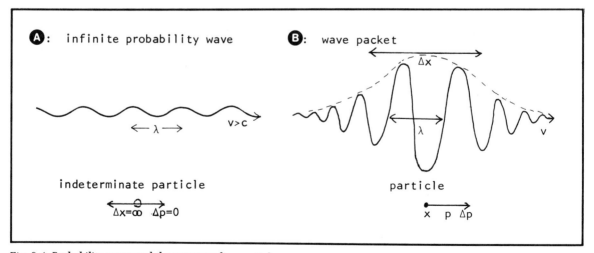

Fig. 8-4. Probability waves and the corresponding particles.

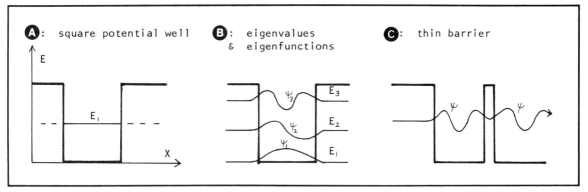

Fig. 8-5. Square potential wells.

vibrating. Fig. 8-5B shows several of the lowest modes of vibration, graphed at certain positions. With minor changes, sine-like probability waves of certain wavelengths can occur between the two boundaries. A probability wave can travel in the "well," reflecting from each wall, and the result is a standing wave. This represents a "bound" particle. Calculation with Schrodinger's theory gives exact permitted values of wavelength, momentum, and energy for this particle in the well. Three permitted energy levels (eigenvalues) E_1, E_2, and E_3 are shown in Fig. 8-5B. The eigenfunction for each is graphed at the corresponding energy level for convenience only.

The astonishing result is that a particle with any old energy E thrown into this well might not stay in. It will remain only if it fits one of the special energy states. Once the particle is in the well, there are some places where it can never be found. These are at the nodes where the probability wave is zero. A particle with energy E_2, for example, will never be found in the center of the well. In this regard we might also observe the symmetry of the wave functions. Function ψ_1 is symmetric about the center and called "even." Function ψ_2 is "antisymmetric": one side must be flopped over to resemble the other side. It is called "odd." The next one is symmetric and even, and so on, in alternating fashion. This general symmetry behavior is called *parity.*

Close attention to detail in Fig. 8-5B reveals another astonishing prediction. There is a non-zero probability of finding the particle outside the well, the more so at higher energies. Each probability wave tapers off beyond the boundary, falling rapidly to zero. Indeed, we shall learn that nature forms some wells with boundaries so narrow (Fig. 8-5c) that the particle can "leak" out. The tail of its wave function extends through the barrier and enough remains when the outside is reached that it again is a wave. On a probabilistic basis, occurring at a time no one can predict for an individual case, the particle will suddenly be outside, reconstituted with wave function and all, and proceed on its way according to its KE. Statistically, when many such cases are watched, one can predict how long one must wait to see such a thing happen. The wait is not very long when the well is shallow, E is large (but less than the barrier height), and the barrier is thin. J. R. Oppenheimer realized the possibility of quantum leaks or "tunneling" in 1928. Later Ivar Giaever first found tunneling experimentally between two narrowly-separated metals in 1960 (for which he shared the Nobel Prize in 1973 with Leo Esaki and Brian Josephson for related work).

Both Schrodinger and Heisenberg-Dirac theories can be used to calculate exactly what energy states are possible in a given well or other situation where a particle is constrained. For the simple straight-sided well, the states depend on h, particle mass, well width, and nothing else. The energy values increase steadily according to the squares of the series of integers used to label them.

It is also not possible to have a constrained particle with $E = 0$. This would imply it is motionless and therefore extremely uncertain in position. The well has limited width, implying certain knowledge of its position, and a paradox is created. Thus the minimum energy of the particle must be $E_1 > 0$. This of course gives the particle just the right uncertainty in momentum (speed) and position (width of well) to satisfy the uncertainty relation.

Let us consider briefly another way to bind a particle, very common in nature. In Fig. 8-6 is a harmonic potential well, one where the PE increases with the square of the distance from center. We should recall that this PE pertained to a stretched spring, whose force law is sometimes called the "harmonic" force. As before, the bound particle is described by certain wave functions matched to the boundaries, somewhat different from sinusoidal curves. Each wave function ψ is labeled with an integer and is associated with an energy state. At higher energies, the functions "leak" further into the forbidden region outside the well. The energy values are determined by half-integers of the form $n + 1/2$. The first level is E_0, for $n = 0$. It is not zero energy because $1/2$ is used in its calculation. This is the minimum or ground state of the harmonic well and satisfies the

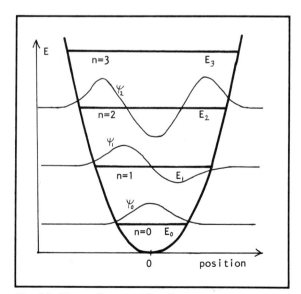

Fig. 8-6. Harmonic potential well.

uncertainty principle. Higher states are uniformly spaced in energy above it.

8.5 SPIN AND RELATIVISTIC QUANTUM MECHANICS

That the electron has *spin* was suspected since 1925 when S. Goudsmit and G. E. Uhlenbeck reasoned that orbiting electrons might spin just as planets rotated. Wolfgang Pauli agreed, in order to explain other frustrating data about the light from atoms. Magnetic fields were known to affect electrons in certain situations as if they were tiny magnetic dipoles with their charge rotating to create a magnetic moment. The amount of the spin was found to be in units of—you guessed it—h (actually $h/2\pi$, which has the same units as angular momentum). Observations indicated that electrons had a spin of $1/2$ unit and that it is quantized. Physically spin is the same as angular momentum, but spin is intrinsic to a particle, and angular momentum refers to motion, usually orbital. In considering conservation, spin may be counted separately or with other angular momentum, depending on whether some field such as a magnetic one couples them together.

If the electron and other particles can be represented by a matter or probability wave, where is the spin? The dilemma was particularly bad because electrons had spin $1/2$, unlike photons with spin 1. Polarization of matter waves, similar to that for light waves, did turn out to be a plausible way to carry a spin $1/2$. Wolfgang Pauli undertook to find a matrix method to represent spin, along the line of work of Heisenberg. C. G. Darwin searched for a wave representation. Paul Dirac made the breakthrough in 1928 while introducing relativity into quantum mechanics.

Relativity must be considered, not only because free particles were known to achieve high speeds but also because some bound particles were relativistic, for example, some electrons in heavy atoms. The rest energy of the particle, the new momentum-energy relation, length contraction, time dilation—all the paraphernalia of relativity—would have to enter quantum theory. Dirac produced a four-vector covariant relativistic quantum mechanical theory. Its form is as much more complex as its verbal description would imply, and we would gain little by displaying the basic equation here. It resembles a Schrodinger theory with vectors and matrices added, but it is much more difficult to use. Not only did it contain matrices that represented electron spin (called Pauli matrices and measuring two-by-two), but it also predicted a new particle. Dirac shared in the Nobel Prize for 1933.

At first there was puzzlement over the prediction of a particle with electron mass, spin, and charge but negative energy. Then it was realized that the new particle simply had positive charge, opposite to the electron. It was called a *positron*. Since it was only hypothetical, physicists agonized over where they had gone wrong in formulating such an advanced theory that explained everything known about particles and atoms but required the existence of a new particle which had never been observed. In 1932 Carl Anderson found the positron in cosmic rays—that is, in the results of collisions of high-speed particles from space with matter on Earth. The new particle had a curved path just as an electron has in a magnetic field, but it curved in the other direction, indicating an opposite charge. For this vindication of Dirac theory and for related work Anderson received the Nobel Prize in 1936. Dirac went on to search for a quantum mechanics for protons, hoping to wrap up the full story of the microworld, but the discovery of nuclear physics again changed the whole game, as we shall see later.

With the appearance of the positron and its peculiar relation to the electron, another new concept was born in particle physics, the notion of negative energy states. It was realized that an electron or any particle could materialize from empty space (the vacuum) if the corresponding "hole" were also counted. It was sufficient that the hole have the opposite charge in order to avoid violating the conservation of energy and charge. The positron is not simply a hole. The issue that arose is whether "empty" space is empty and continuous if particles can be created from it. The uncertainty principle also raises the issue, because we cannot examine space at an arbitrarily small scale. Huge energies would be required. Space itself may be quantized, exhibiting a granularity or discontinuous structure, like a tiny grid. Much more recent work in particle theory seems to require this, as will be seen.

8.6 PHILOSOPHICAL ISSUES IN QUANTUM MECHANICS

We close this story of the theoretical foundations of quantum mechanics with some remarks on the ongoing attempt to obtain philosophically and experimentally satisfactory proof of validity. After the initial revolution in the 1920s, efforts continued to find better explanations of what was happening deep within the theories. The concept of determinism was continually at stake. It seemed that particles, and their wave functions, knew more about the world than they should; they even seemed to be able to outguess an experimenter's intentions. The true nature of reality was up for investigation in deeper ways than ever before realized.

A world containing many particles, each obeying Newton's laws exactly, is still too complex to keep track of. An experiment can never have a perfectly certain result, so that we must endure a form of classical indeterminacy (to be discussed further in a later chapter). Quantum indeterminacy occurs at a deeper level. Particle motion has a fundamental uncertainty so that any given measurement does not have a definite result. Some have said that there must be Newtonian mechanisms hidden deep within particles (and waves) which determine exactly what will happen. If physicists could persevere enough, they might uncover new rules of nature that might allow a return to a deterministic picture. Such a new level of theory would be much more complex than what we have now, thus raising the need to use Ockham's razor to retain the most simple view possible. Moreover, basic experiments have been done which show fairly conclusively that hidden mechanisms ("hidden variables") are not at the heart of the problem but rather that nature is intrinsically, stubbornly, profoundly indeterministic about what happens.

The Einstein-Podolsky-Rosen (EPR) experiment was conceived by Albert Einstein, Boris Podolsky, and Nathan Rosen as a thought experiment in 1935 but could not be tested until techniques had much improved. It was proposed as a paradox, because the predicted results were so contrary to views of reality that its proponents did not want to believe them. Einstein particularly hoped that any real experiment along these lines would cast doubt on the foundation of quantum mechanics and its uncertainty.

The statement of the EPR experiment has many forms. In Fig. 8-7 two particles A and B, perhaps photons, are shown emitted from another particle (or atom). There is supposed to be a correlation between a convenient property such as spin possessed by both particles. It is the nature of the emitter that it has no angular momentum. Therefore the spin of A should be oriented opposite to that of B in order for the conservation of spin to yield zero for oppositely moving particles. The spins of photons can be detected with ordinary polarization detectors and the process is well understood. The crux of the experiment is that once detector D_1 receives A and establishes the spin of A, the spin of B can be predicted with no further measurements needed. Detectors D_1 and D_2 can be placed so far apart that this information cannot be transmitted from one to the other, even at the speed of light, in time to actually determine or affect the result that will be found at D_2.

It is desired to preserve "local causality;" events in one place should not determine instantaneously events at another location. The apparent paradox is either that local causality is violated, and relativity with it, or that inner mechanisms ("hidden variables") established, at the time of emission, exactly the facts about each photon for all later time. But it is also possible that neither case holds and quantum theory remains what it is, stating that nature does not determine beforehand what will happen in this experiment. This point of view is that there is no "objective reality," only quantum uncertainty. The source of logical difficulty could be in the assumption that the spin of B has objective reality even if it is not measured, only predicted. Philosophically there exists what is called the "Copenhagen interpretation" of quantum mechanics, so-named because the creators of the theory worked in Copenhagen in the 1920s. This interpretation states that there is no objective reality besides what is measured. The sound of a tree falling in the forest is said not to exist unless some record was made of its occurrence.

The argument did not approach resolution until 1965 when John Bell found a mathematical relation, called "Bell's inequality," which could be checked by measurement. The inequality was a comparison of correlations between results, calculated differently for quantum mechanics and for hidden variable theories. The possibility that probability waves can interfere affects the quantum result. If Bell's inequality is found to hold in an experiment, local causality and objective reality are correct. If not, then Copenhagen quantum uncertainty reigns. Much ingenuity is needed to test nature on this particular philosophical issue, and even the thought experiment is difficult to understand.

In the 1970s and early 1980s a series of actual experiments were devised to test Bell's in-

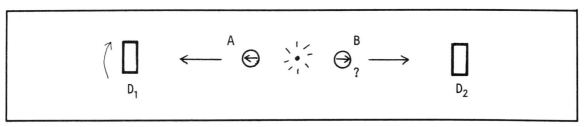

Fig. 8-7. The Einstein-Podolsky-Rosen apparent paradox.

equality. In some experiments a source of pairs of photons is used. Certain atoms will emit photons going in opposite directions. The photons have correlated spins: either both are in the direction of photon motion, or both are opposite, as discussed earlier for the EPR thought experiment (Fig. 8-7). The experiment cannot be done for a single instance; instead numerous events must be observed and the results statistically analyzed. The orientations of the pairs of photons will vary randomly but the polarization detectors each have a set orientation. A photon will be detected, or not detected, in both. The correlation of the spin-states of both photons can be found.

To test the EPR issue one detector (D_1) can be rotated, reducing the number of correlated counts made of the randomly oriented photon spins at each detector—that is, the number of times photons with opposite spins are detected by each detector simultaneously. The correlation can only be measured by comparing results from both detectors. No unusual effect has occurred thus far. The crux of the experiment occurs in a comparison with counts made at another orientation of D_1. The correlated count decreases, but not enough. Photon B seems to "know" what result is recorded at D_1 for photon A. Although D_1 was rotated to obtain less correlation with D_2, B seemed to rotate its spin in an attempt to agree with what is measured for A. Bell's inequality is violated in some experiments, most definitively in 1982 in ones by Alain Aspect.

Has local causality then been violated? Must we believe that information is transmitted instantaneously between natural particles? What happens at detector D_1 does seem to affect what happens at D_2. Further reflection shows that violation of local causality is not a required conclusion. The results obtained at each detector are truly random, due to the nature of the source. There is nothing wrong with a correlation between two sets of random measurements. Two die "loaded" in the same way will show a correlation between the results of otherwise random tosses. The fact that the results are recorded also alters the nature of the experiment. Measuring the spin

of one photon randomly alters it. One cannot know what it really was and therefore what spin the other had. It is the old story of the observer affecting the experiment to cause a fundamental uncertainty.

Efforts to test local causality continue. In case it holds firm, some theoretical work is being done on nonlocal hidden variables theories, which so far are experimentally identical to quantum mechanics. A hidden mechanism shared by some or all particles but which provides no new predictions which can be tested is a theory still at the rudimentary stages. Most physicists have accepted quantum mechanics and are not troubled by the notion that observers seem necessary to have reality. A crucial turning point occurs when one accepts that reality in the microworld is fundamentally different from reality at our scale, just as reality at high speed is different from what we experience at low speeds. Not only does nature play with dice but all sides can be seen at once at relativistic speeds.

A different viewpoint about quantum mechanics is the *many worlds* interpretation. It was developed by Hugh Everett, Bryce DeWitt, and John Wheeler for quantum theory, although thought along this line is traceable back to Gottfried Leibniz, a 17th century philosopher. The probabilistic picture of a particle being potentially at many possible points until observed at one is extended to many different possible worlds (or universes). When a measurement is made, a single world is singled out. We would seem to be living in a kind of infinite dimensional space with every possibility occurring at once. We only observe one possibility at a time, however.

It would seem that the nature of reality is such that there is a subtle unity in nature. Particles cannot be isolated from the systems in which they are studied, including the human experimenters. Particles in this sense could not be fundamental building blocks of nature. How an individual particle behaves—if such is definable—may be too linked with what other particles are doing. But the indeterminism of quantum mechanics remains basically unchallenged. The issues are too involved to explore further here. The reader should at least be aware that the microworld is also the frontier of the capabilities of human consciousness to understand nature. All our prejudices and assumptions must be identified and examined before progress is possible. Quantum mechanics has altered, and will continue to alter, the way everyone views the world.

Plate 1. Ernest O. Lawrence, inventor of the cyclotron, and the Bevatron building at Lawrence Berkeley Laboratory (courtesy of LBL).

Plate 2. Many of the founders of quantum mechanics, modern atomic physics, and nuclear physics in a group portrait at the 1933 Solvay Conference. From left to right in bold type for those mentioned in this book, front row: **Erwin Schrodinger, Irene Joliot, Neils Bohr,** A. Joffe, **Marie Curie,** P. Langevin, O. Richardson, second row: E. Henriot, F. Perrin, **Frederick Joliot, Werner Heisenberg, Enrico Fermi, Ernest Walton,** P. Debye, B. Cabrera, M. Rosenblum, W. Bothe; third row: H. Kramers, E. Stahel, **Paul Dirac,** N. Mott, **George Gamow,** P. Blackett, (courtesy of LBL).

*From left to right with boldface type for those mentioned in this book, front row: **Ernst Rutherford**, T. De Donder, M. de Broglie, **Louis de Broglie, Lisa Meitner, James Chadwick**; second row: E. Bauer, **Wolfgang Pauli**, J. Verschaffelt, E. Herzen, **John Cockcroft**, R. Peierls, L. Rosenfeld; third row: J. Errera, M. Cosyns, C. Ellis, A. Piccard, **Ernest Lawrence**, (courtesy of LBL).*

Plate 3. The Bevatron at LBL (Bevatron on the left, Alvarez linear accelerator in the center, and Cockcroft-Walton accelerator on the right (courtesy of LBL).

HERA

PETRA

DEUTSCHES ELEKTRONEN-SYNCHROTRON DESY
2000 Hamburg 52 Notkestraße

Plate 4. Aerial view of Deutsches Elektronen Synchrotron, Hamburg, West Germany, accelerator rings, including HERA under construction (courtesy of DESY).

Plate 5. *Aerial view of Stanford Linear Accelerator Center, with LINAC in background, PEP storage and collision rings curving from the end of LINAC approximately where the road is, LINAC beams fanning out to experimental areas in the center, and SLC planned as shown with dashed line (courtesy of SLAC).*

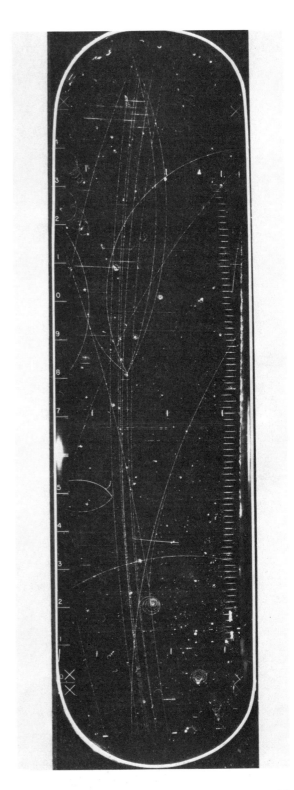

Plate 6. Photograph of tracks in the 72-inch bubble chamber at LBL (courtesy of LBL).

Plate 7. The Crystal Ball detector with thousands of photodetectors wired (courtesy of DESY).

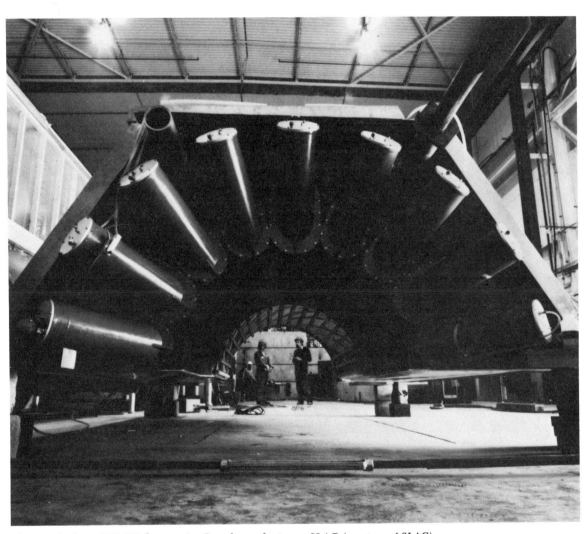

Plate 8. The huge DELCO detector for Cerenkov radiation at SLAC (courtesy of SLAC).

Plate 9. The six-sided MAC detector for PEP at SLAC, with 200 tons of iron plates to measure particle energy (courtesy of SLAC).

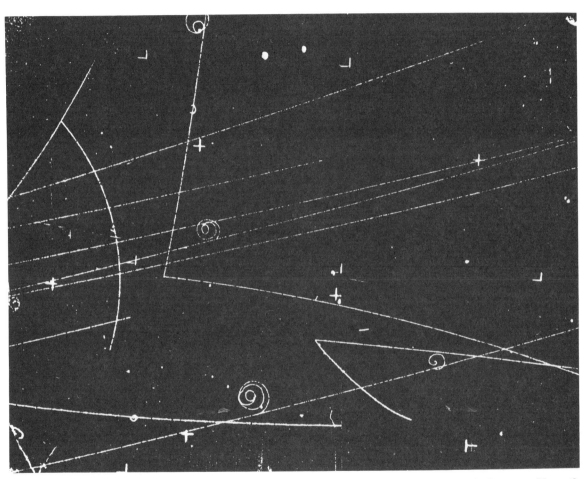

Plate 10. A pion strikes a proton in a bubble chamber and produces a lambda and a neutral kaon, which are invisible until they decay to charged particles (courtesy of LBL).

Plate 11. Physicists at a 1966 high energy conference (from left to right with boldface for those mentioned in this book: **Edwin McMillan, Val Fitch, Murray Gell-Mann,** *Victor Weisskopf,* **Geoffrey Chew,** *and* **Sidney Drell** *(courtesy of LBL).*

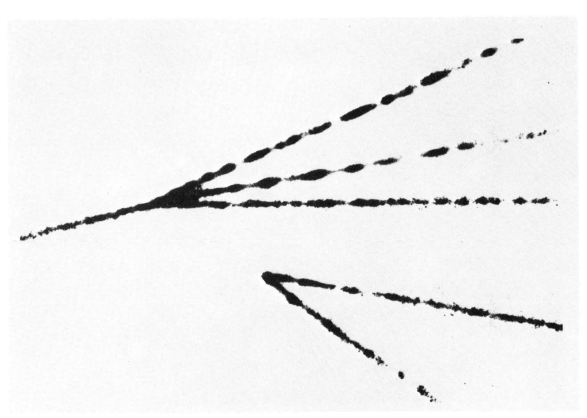

Plate 12. Photon-proton collision producing a charged charmed baryon and a neutral charmed meson, each of which decays into charged uncharmed particles (courtesy of SLAC).

Plate 13. Burton Richter, one of the discoverers of the J/ψ meson and a Nobel Prize winner (courtesy of SLAC).

Plate 14. Carlc Rubbia and Simon van der Meer receive accolades upon winning the Nobel Prize in 1984 for finding the W and Z intermediate vector bosons in weak interactions at CERN (courtesy of Photo CERN).

Chapter 9
Atoms

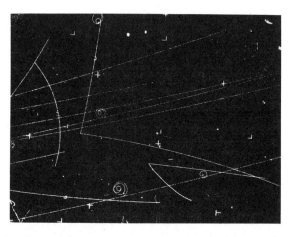

THE CONCEPT OF AN ATOM AS A BUILDING block of matter has been traced to early Greece and to other early civilizations. We resume the story, starting two centuries ago when chemistry developed from alchemy, and carry it to the present, utilizing the modern quantum mechanical view. The focus is on the hydrogen atom, in which a fascinating array of different phenomena can be found. The specialized language of quantum mechanics is gradually introduced and extended to many atomic applications.

9.1 THE BEGINNINGS OF THE CHEMISTRY OF ATOMS

In the late 18th century some modern chemical results were obtained by scientists still seeking to use the phlogiston theory based on fire, air, water, and earth. In 1754, for example, Joseph Black demonstrated the existence of a gas distinct from air—carbon dioxide (he called it "fixed air"). For the first time a known substance was shown to be transformed to another known substance by

heating, without reference to the imaginary phlogiston. In 1766 Henry Cavendish identified the most basic element hydrogen, as a gas. More gases followed, all of them compounds, although this was not known at the time. It is curious that invisible gases predominate among the first substances to be identified without a need for phlogiston.

Metals followed later even though substances such as gold and mercury had been noted for their "purity" for millenia. Antoine Lavosier attacked the alchemical theories directly with numerous experiments and found by 1780, together with Joseph Priestly, that air was about 4 parts nitrogen gas and 1 part oxygen gas. He established oxygen as that which combined with many other substances upon heating. The emphasis was now on explaining chemical reactions in terms of the weights of reactants and products. In 1781 Priestly found that water was made of hydrogen and oxygen and therefore not an element. Chemistry then moved rapidly forward.

Laws of equivalent proportions and constant

compositions were established late in the 18th century by Joseph Gay-Lussac and others. By experiment it was found in what proportions various substances would combine and that these were always fixed for a given substance. Progress toward the recognition of chemical elements accelerated, and the atomic theory was brought back from exile. Again the study of gases, principally by John Dalton, was a major aid. By 1804 he had shown that there were elements differing in size, weight, and chemical behavior. Taking hydrogen as unit weight, Dalton made the first table of elements, necessarily incomplete and also containing many compounds. Until the 1860s the idea that two atoms of the same element could combine was rejected. It is now known that all the gases identified at the time did have their atoms bonded in pairs to form molecules.

The weights of the elements were known sufficiently well to realize that they seemed to be multiples of some unit weight, the weight of hydrogen. Thus, without knowing it, William Prout in 1815 formed the first nuclear theory of matter. Faraday found with electrolysis (in 1833) that a certain amount of electric current caused a certain number of atoms to react. Valency was discovered, the preference of certain atoms for combining with certain numbers of other atoms. Looking at valencies and weights, many scientists, culminating in Lothar Meyer and Dimitri Mendeleef in 1869, formed a *periodic table* of the known elements. All elements with similar properties were classified in columns with weight increasing down the column. Gaps appeared in the table, and unknown elements were predicted on the basis of expected properties. Some were found.

A structure for atoms that foretold of the modern view was proposed by J. J. Thomson in 1904. Already having discovered the electron, he placed a large number of these negative particles around a positive center (Fig. 9-1A). At first it was not known that the positive center had most of the mass of the atom. The discovery of radioactive elements showed that some massive body must be at the center since the mass of an entire hydrogen or helium atom, positively charged, would emerge in radioactive decay (discussed later), leaving the charge in the atom decreased by one or two units. Each atom seemed to prefer to stay neutral, so that a change in positive charge was accompanied by the same change in electrons. The atom remaining after a radioactive decay has the chemical properties of a different atom, one or two steps earlier in the periodic table.

Rutherford's work on scattering helium nuclei (also called alpha particles) from metal foils showed that the center of an atom was indeed positively charged but very small and contained most of the mass. Rutherford's model of the atom was like a tiny solar system (Fig. 9-1B), with a positive nucleus in place of the sun and electrons orbiting like planets. The solar system model had a grave defect. Charges moving in a curved path such as an orbit were known to emit radiation, so that the electrons in an atom must lose their energy steadily. They should be slowing down and spiraling toward the nucleus (Fig. 9-1C), assuming that somehow they were losing their angular momentum as well. A major breakthrough was needed in atomic theory, or else atoms were quite unstable, contrary to all observations about nature.

9.2 ATOMIC SPECTRA

Meanwhile, evidence about the behavior of atoms as they absorb and emit light was accumulating despite the lack of suitable theoretical explanations. In 1859 Robert Bunsen and Gustav Kirchhoff developed the spectroscope, which could identify suitably prepared atoms by means of the light they gave off when heated. For a long time spectroscopes used prisms to disperse the light into its component colors. Sets of *spectral lines*—distinct wavelengths of different colors—were found for elements. Their wavelengths could be measured, and each element had its own special set. Not until 1895 was the inert gas helium found on Earth by William Ramsay; it is the second simplest element and exists as isolated atoms. Its spectrum had already been identified in the sun, and it is the second most common element in the universe. Henry Moseley first used x-ray spec-

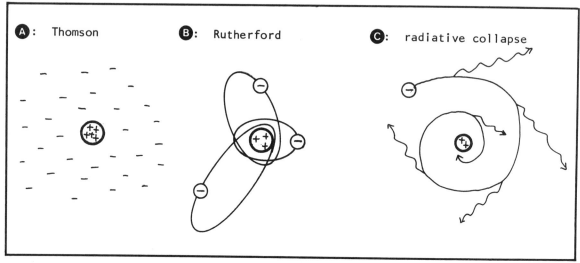

Fig. 9-1. Early 20th century models of the atom.

troscopy to identify elements and determined that there should exist 92 elements (through uranium) by 1914, although seven had not yet actually been discovered physically. The periodic table has since been extended beyond uranium, as discussed later.

Almost all that we know about atoms has come to us by virtue of *spectroscopy*, the science of measurement and interpretation of the different wavelengths or frequencies of light emitted or absorbed by atoms. The set of colored lines emitted by a particular element is called its *spectrum*. In 1885 Johann Balmer analyzed the relationships of the various wavelengths of light emitted by hydrogen. In schematic form his data from photographs taken through a spectroscope appeared as in Fig. 9-2E. This series of lines in primarily visible colors is called the Balmer series, and the lines are labeled traditionally with Greek letters. Starting with the longest wavelength (on the left), he was able to calculate the wavelengths of the others with a simple formula involving an integer assigned to each. Whatever explanation of the atom was to come, it had to provide such a formula.

Other series of lines for hydrogen were found, both a series in the ultraviolet (by Lyman) and many series at longer wavelengths in the infrared. All wavelengths followed the same simple rules based on integers. While the closest agreement was found with hydrogen, the same simple relationships were found for the spectra of many elements. Spectral lines often come in groups or *multiplets*, and the spacing between members of a multiplet is a simple constant. Major work was by Hartley (1883) and Rydberg (1890). Numerous other regularities in spectra were discovered, all awaiting explanation. Rather important was the failure of atoms to display every possible line in a set. Some lines seemed to be "forbidden," and a set of *selection rules* was found, specifying or selecting which lines could occur. Of course, these rules do not constitute any sort of explanation but are just an expression of empirical laws of nature.

The x-ray spectra from elements were also measured. Charles Barkla discovered that heavier elements emitted this energetic radiation as spectral lines and received the Nobel Prize in 1917. Further work by Karl Siegbahn in this area earned him one in 1924.

The effect of strong electric and magnetic fields on atomic spectra was noticed, even before the electrical nature of the atom was understood. Both kinds of fields cause splitting or shifting of spectral lines from the wavelengths they would have without the disturbance. The Stark effect is caused by the strong electric fields of ionized

134

atoms acting on each other. It is a shift in wavelength, different for each of many different randomly moving atoms. The net result is a broadening of each line. The Zeeman effect provided the first proof that electrons played a role in the emission of light from atoms. Pieter Zeeman, together with Hendrik Lorentz, in 1896 found that a magnetic field split emission lines into three or more lines, and moreover, that the light was polarized in certain ways. They were even able to deduce that the electron was negative from the evidence. Both received the second Nobel Prize in physics in 1902 for this work.

In 1914 James Franck and Gustav Hertz showed that other agents besides light could evoke spectral lines from elements. They bombarded atoms with electrons and obtained the same spectra, establishing energy levels as inherent in atoms. For this they received the Nobel Prize in 1925.

9.3 ATOMIC MODELS

In 1913 Neils Bohr was acquainted with the Rutherford atomic model and suggested a solution to its apparent instability. Noting that radiation emitted from atoms had to be in discrete amounts (photons), he supposed that the electrons had discrete orbits. The orbits of the electrons had to be quantized, reminiscent of the Planck quantization of light. Only certain orbits were possible. They were numbered with the integer n from the innermost outward. This was the first use of a *quantum number* to designate a quantized atomic state. A quantum of light was emitted whenever an electron jumped to a closer and lower energy orbit (Fig. 9-2A). The conservation of energy was preserved by having the difference in orbital energy appear as a photon of a certain frequency and energy. For the converse process, the long-observed absorption of light by the atom, exactly the right frequency of light had to come along and boost an electron to a higher permitted orbit. This theory beautifully explained the fact that spectral lines have rather specific wavelengths and frequencies, and Bohr deservedly received the Nobel Prize in 1922 for this work. He and later contributors to the modern theory of the atom appear in the famous group photograph from the Solvay conference of 1933, given as Plate 2.

The Bohr model takes a classical approach to calculating the forces in the atom but the results are relevant even today. The approach is simple but powerful and worth showing briefly in mathematical form for the hydrogen atom. An equation, based on Newton's law of motion, is formed, using the electric force on the electron due to the single proton at the center (Fig. 9-2B). The inward acceleration of the electron is that of a particle moving in a circular orbit with speed v. The equation states:

$$F = e^2/4\pi\epsilon_0 r^2 = mv^2/r$$

where r is the radius of orbit, m the electron mass, and e the unit of charge. The total energy of the atom consists of $PE = -e^2/4\pi\epsilon_0 r$ and $KE = 1/2mv^2$. The angular momentum L of the electron is assumed to be quantized in units of $h/2\pi$, giving:

$$L = mvr = nh/2\pi$$

These three facts can be combined to find the total energy E of the electron:

$$E_n = \frac{-me^4}{8\epsilon_0^2 n^2 h^2}$$

and the radius of its orbit:

$$r_n = \frac{\epsilon_0 n^2 h^2}{\pi m e^2}$$

Several features of these classical results should be noted, since they will be correct even in the modern Schrodinger theory to follow. First, it is most remarkable that the size of an atom can be calculated from fundamental constants such as electron mass and charge and Planck's constant. By virtue of the correspondence principle, the average orbit of the electron is indeed the same as calculated classically, and similarly for its energy.

It was sufficient to introduce Planck's con-

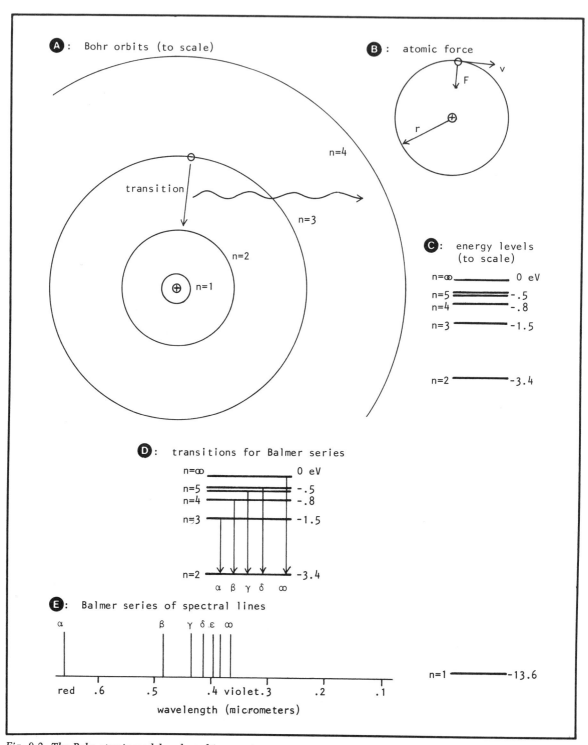

Fig. 9-2. The Bohr atomic model and resulting spectroscopy.

stant in quantizing the angular momentum in order to achieve quantized results for both energy and orbital size. The energy is negative, as it should be for a bound electron which cannot leave the atom. There are an infinite number of discrete energy levels available, labeled with n, the principle or energy quantum number. Higher energy levels have larger orbits. The energy levels or states are neatly arranged in an energy level diagram as in Fig. 9-2C. The lowest or *ground state* ($n = 1$) is at the bottom and has the value -13.6 eV for hydrogen. The states are closer together in energy as n increases and reach a limit at very large n. Beyond that, the atom is ionized; the electron has too much KE to remain bound and is free to depart. Light is emitted when the electron—somehow—jumps from one state to a lower one. The Bohr theory almost exactly predicts the wavelengths of the Balmer series in the hydrogen spectrum. Fig. 9-2D shows which jumps are needed to produce which spectral lines shown in Fig. 9-2E.

The "Bohr radius," the radius of the ground state orbit, is calculated to be $5.29(10)^{-11}$ m. The Bohr theory works well if certain assumptions are relaxed. In an isolated atom the heavy nucleus (the proton) makes small orbits around the center of mass while the electron makes large orbits about the same point. The theory can be extended with high accuracy to any atom so ionized that only one electron remains bound to it. An atom with three protons in the nucleus and one electron orbiting would have nine times more electric energy, and all spectral lines would be nine times more energetic, having 1/9 the hydrogen wavelengths. The electron would be three times closer to the nucleus. But Bohr theory does not explain why the electron is restricted to quantized energy levels or certain angular momenta. The assumption of quantization is at odds with the rest of the classical theory.

Spectral lines had been observed to have a "fine structure," with each principle line being resolved into a pair or multiple of lines. Further explanation was needed. Arnold Sommerfeld in 1915 conjectured that there must be closely-spaced suborbitals of slightly different energies and depicted them as ellipses of varying eccentricity (Fig. 9-3A). The ellipses would not remain stationary but rotate slowly or "precess" around the nucleus. The resulting electron motion might be as shown in Fig. 9-3B. This pictorial model remains today a symbol of the atom despite several inaccuracies.

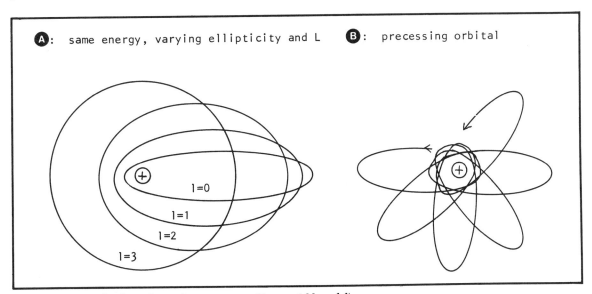

Fig. 9-3. Atomic orbitals with angular momentum (Sommerfeld model).

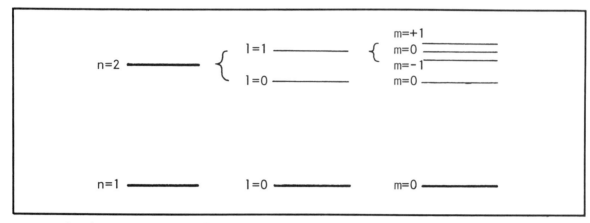

Fig. 9-4. The splitting of atomic energy levels by spin-orbit coupling or external magnetic field.

The different possible elliptical orbits associated with a given principle orbit (quantum number n) differ with respect to the angular momentum L of the orbiting electron. Each is numbered with a new orbital quantum number—now called l. Thus began the proliferation of quantum numbers in atomic and particle physics. The most elliptical orbital has $l = 0$. For a given principal orbit, l could take on values from 0 to $n - 1$. The corresponding energy states are indistinguishable unless some force besides the electric one acts differently on each state. In the absence of such a force, the n different L states for a given n orbit are said to be *degenerate*, in this case n-fold. They are only potentially different.

To explain the Zeeman splitting of lines by a magnetic field, each l orbital has to be further spread or split into several sub-suborbitals when a magnetic field is present. A third quantum number—now called m for "magnetic"—was introduced to label the different orbitals. It can have $2l + 1$ values, ranging from $-l$ to $+l$, both negative and positive. The effect on $n = 1$ and $n = 2$ states is shown in Fig. 9-4. It should be emphasized that this is a state diagram, not a display of spectral lines. Spectral line energies would be obtained by considering an electron to jump between two of the states shown. The existence of different suborbitals is not observed unless a magnetic field is present, removing the degeneracy.

Then each energy level is split into states distinguished by both l and m. Angular momentum states are not split by themselves, without magnetic splitting. We shall see that magnetic fields exist within the atom which cause a small amount of splitting without the need for an external field.

Experimental verification for the quantization of angular momentum in atoms was obtained by Otto Stern and Walther Gerlach in 1922. A beam of neutral silver atoms was sent through a nonuniform magnetic field as shown in Fig. 9-5. Because of its orbiting electrons each atom not only has angular momentum but also acts as a tiny magnet—a magnetic dipole. The field applies a force to the magnetic dipole that depends on its orientation. The atoms were found to spread into two narrow bunches rather than to be dispersed into a continuous smear. This indicated that only two kinds of dipole were present, up and down (in the magnetic field), despite the otherwise random orientations of the atoms. The quantized angular momentum of each atom is thus limited to two possible states, up and down, each with the same magnitude. This "space quantization" is expected for any quantum mechanical variable which is a vector, as angular momentum is.

9.4 THE QUANTIZED HYDROGEN ATOM

When the Schrodinger quantum mechanics is applied to the hydrogen atom, the overall results

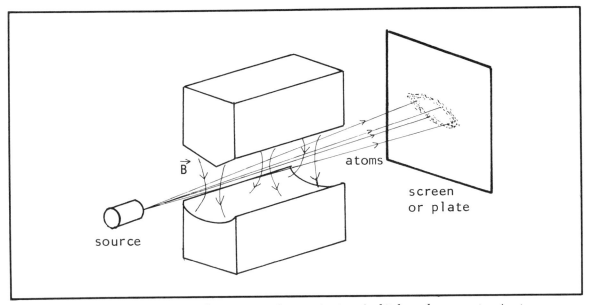

Fig. 9-5. The Stern-Gerlach experiment demonstrating the quantization of orbital angular momentum in atoms.

are the same as those achieved by Bohr. However, the electrons are no longer treated as orbiting particles but as wave functions confined in the potential well formed by the electric attraction between positive nucleus and electron. The mathematical details that occur in the apparently simple Schrodinger equation $H\psi_n = E_n\psi_n$ are complex because of the multiplicity of three-dimensional shapes possible in an atom. The remarkable fact is that the atom can be almost completely solved at this relatively simple level. Almost every aspect of its behavior can be predicted and verified by experiment.

The probability wave for the electron turns out to be a rather complex distribution in three dimensions. All three quantum numbers, n, l, and m, follow naturally from the analysis. For the ground state ($n = 1$, $l = 0$, $m = 0$), at the same energy predicted by Bohr, the electron has its simplest distribution, a spherically symmetric shape with the greatest concentration at the Bohr radius r_1 as shown at the upper left in Fig. 9-6. This corresponds to the classical precession of the most elongated elliptical orbit, as shown in Fig. 9-3. If it were a classical particle, the electron would spend the most time at the outer ends of

the complex elliptical path. The interpretation of the quantum mechanical results was philosophically unpleasant at first, despite all that was known about electron waves. The electron could no longer be located as a "dot" orbiting around the nucleus but instead was somehow smeared out. The smearing was soon realized not to refer literally to a pulverization and spreading of the electron but rather to the probability of finding it.

Any instrument which could measure one electron would find it randomly somewhere in the region denoted by the smeared probability distributions of Fig. 9-6. Whenever the electron is found, it appears as a whole electron. In a hydrogen atom it is more often found in the vicinity of the Bohr radius than anywhere else. There is a 25 % chance that the electron will be found at a distance greater than conservation of energy would permit. If enough energy is put into the atom by shining strong ultraviolet light on it, a single electron will come out and act as a free particle (a wave packet), leaving a naked proton to fend for itself. The proton would then have no electron probability wave around it.

The electron cannot be localized in the atom

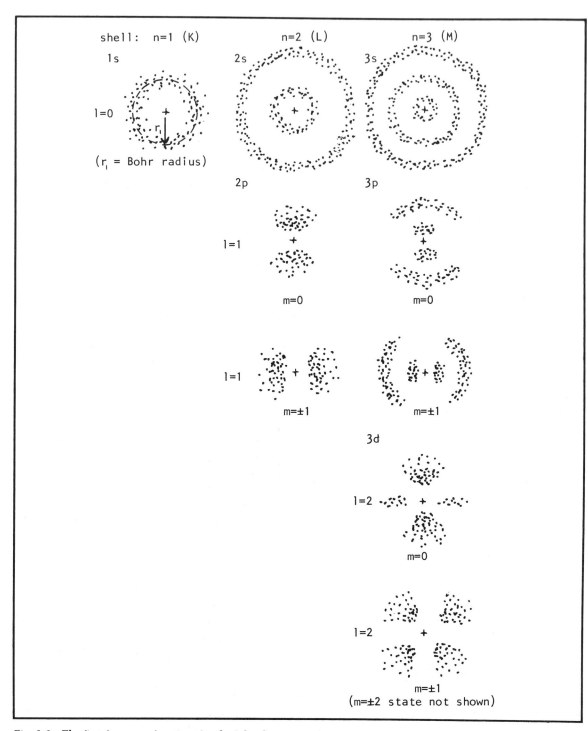

Fig. 9-6. *The first few wave functions for the Schrodinger one-electron quantum mechanical atom (represented in cross section as probability distributions for the electron).*

because it is moving too fast. If the uncertainty principle is used to find its momentum when confined to a region about 10^{-10} m (twice the Bohr radius), its speed must be at least 10^6 m/s, almost 1 % the speed of light. What is worse, the electron distribution is smeared out spherically, which means it cannot be found in an orbit in a plane. Whereas a planar orbit would obviously represent angular momentum, an orbit which can wander all around the atom in all directions, smearing out to form a thick spherical shell, must represent an average of zero angular momentum. This fits well with the theoretical requirement that the ground state be permitted only $l = 0$ for zero angular momentum. Thus the electron has speed and KE, and orbits in some way, yet possesses no net energy or momentum associated with the circular aspect of its travels.

Just as a wave on a ring can have many different wavelengths that fit compatibly, there are many other ways the electron wave can fit around the proton of the hydrogen atom. If the electron is somehow given energy E_2, it wants to move faster. It will tend to reside farther from the nucleus and may possess some net angular momentum. The three different probability distributions possible for $n = 2$ are sketched in Fig. 9-6 in cross section. The value of l could be zero, in which case the electron is again spherically distributed in two spherical shells. It is more likely to be found in the bigger outer one. There are some places where the electron would never be found, corresponding to the nodes of the wave function.

If the electron has acquired angular momentum ($l = 1$), it will be found in two regions (lobes) "above" and "below" the atom (the $m = 0$ state) or in "side" lobes ($m = +1$ or $m = -1$). Unless there is some other influence, the energy of the electron when $l = 0$ or $l = 1$ is the same and it is equally likely to be in either state. Then the state $n = 2$ is doubly degenerate. Ordinarily we have no idea of the orientation of the atom. If a field such as a magnetic field is present that might influence the orientation, then the electron wave will have reason to distinguish the m-states and will be found preferentially in one. The atom then ac-

quires a definite axis of symmetry that it otherwise would not have. Magnetic fields intrinsic to the atom cause the energies of different angular momentum states to differ slightly, with a lower l value usually having lower energy.

There is no limit to the number of states the atom can have. Figure 9-6 shows those for the next energy level, $n = 3$. Also shown is another common way of labeling the states. An integer for the n "shell" is followed by a small letter denoting the l "subshell." The letter "s" denotes $l = 0$, the symmetric state; "p" denotes $l = 1$ states (two lobes); "d" denotes $l = 2$ states (four lobes); "f" denotes $l = 3$ states (six lobes); and so forth. [The reader is not expected to memorize all this nomenclature in order to proceed.] Table 9-1 enumerates the possible electron states up to $n = 4$. The shells are commonly named with upper case letters K, L, M, etc. (with L not to be confused with angular momentum L). The number of possible states increases rapidly as n increases. Some details of the Table will be explained in the next section.

9.5 THE MODERN PERIODIC TABLE

In 1925 Uhlenbeck and Goudsmit proposed another quantum number s which could have two values, in order to explain further details previously observed in spectra. It was needed that the electron exist in one of two states, for each of the other available (n,l,m) states already identified. Electrons in an atom seemed to act as if a magnetic field were already present in the atom, originating from the other moving electrons and possibly from a spinning charged nucleus. The electron has an "intrinsic" angular momentum, called *spin*. Because of its charge it would act as a magnetic dipole and interact with any magnetic fields. In Fig. 9-7 a particle with spin is represented in a magnetic field. For a field pointing up, the particle prefers to have its spin up to have lower energy. The spin down state has a slightly higher energy. Thus a fine splitting of spectral lines could be explained with electron spin. Moreover, to fit both theoretical and experimental constraints, the electron had to be assigned the value of 1/2 of a unit of angular

Table 9-1. Atomic States for the Light Elements.

Quantum numbers				Shell	Subshell	Number of possible states		Element	
n	l	m	s			in shell	in subshell	name	number
1	0	0	1/2	K	$1s^1$	2	2	hydrogen	1
1	0	0	-1/2		$1s^2$			helium	2
2	0	0	1/2	L	$2s^1$	8	2	lithium	3
2	0	0	-1/2		$2s^2$			beryllium	4
2	1	1	1/2		$2p^1$		6	boron	5
2	1	0	1/2		$2p^2$			carbon	6
2	1	-1	1/2		$2p^3$			nitrogen	7
2	1	1	-1/2		$2p^4$			oxygen	8
2	1	0	-1/2		$2p^5$			fluorine	9
2	1	-1	-1/2		$2p^6$			neon	10
3	0	0	1/2	M	$3s^1$	18	2	sodium	11
3	0	0	-1/2		$3s^2$			magnesium	12
3	1	1	1/2		$3p^1$		6	aluminum	13
3	1	0	1/2		$3p^2$			silicon	14
3	1	-1	1/2		$3p^3$			phosphorus	15
3	1	1	-1/2		$3p^4$			sulfur	16
3	1	0	-1/2		$3p^5$			chlorine	17
3	1	-1	-1/2		$3p^6$			argon	18
4	0	0	1/2	N	$4s^1$	32	2	potassium	19
4	0	0	-1/2		$4s^2$			calcium	20
3	2	2	1/2	M	$3d^1$		10	scandium	21

Notes: The order of the filling of states is shown and is the same as the element order for light elements. It is determined by what arrangement gives the lowest energy atom. Electrons prefer having their spins parallel when possible. m = 1 states are lower energy than m = 0 and m = − 1 states. The 4s subshell starts filling before the M shell (3d subshell) is filled, because the electrons have lower energy in the former. The subshell named gives the state in which the current electron is being placed. The element named is the resulting one if the corresponding number of electrons up to that point has been placed in the atom (with corresponding positive charge in nucleus).

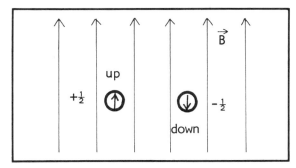

Fig. 9-7. *The spin of a particle (half-integer spin).*

$\frac{1}{2}\frac{h}{2\pi}$. Its two spin states are $s = +1/2$ and $s = -1/2$.

With four quantum numbers available to assign to an electron in an atom, not only could many spectra be explained but also a correspondence with the old chemical periodic table was possible. To make the correspondence, Pauli in 1925 formulated his *exclusion principle* that stated: in any atom, each electron must have a unique set of the four quantum numbers. No other electron could have exactly the same values for all four numbers. For this work Pauli received the Nobel Prize in 1945, somewhat belatedly for such insight.

Atoms could now be built (metaphorically) by taking the appropriate number of electrons and throwing them at a nucleus with an equal number of positive charges (protons). The first electron would drop to the K shell and into the 1s subshell ($n = 1$, $l = 0$, $m = 0$, $s = 1/2$), emitting a flash of light. This would form not hydrogen but a highly-ionized larger atom. The second electron would also emit a flash as it settled into the 1s subshell ($n = 1$, $l = 0$, $m = 0$, $s = -1/2$). If there are only two protons in the nucleus, then helium is formed. Helium is so unreactive that its electrons must be especially happy with their $1s^2$ state (the superscript tells how many electrons are in the subshell).

The third electron would choose the lowest $n = 2$ state in the L shell, the state which has zero angular momentum ($l = 0$ again). Table 9-1 illustrates the early stages of the process. The 2s and 2p subshells (and therefore the L shell) are completely filled when a total of 8 electrons have been added. With 2 electrons also in the K shell and 10 protons in the nucleus, neon is formed. Higher states are filled first by putting an electron with spin up in each possible state, starting with $m = 1$. Then the remaining electrons are filled in to pair the spins, one up and one down. It is energetically more favorable to fill the 4s subshell before filling the 3d subshell (which has 10 states). The number for each element is its *atomic number*, giving the number of positive charges and electrons in the atom.

There is a connection between the completion of a shell and the inertness of the element so formed. There are some "magic numbers" with which nature is particularly satisfied. For filling atomic shells, these numbers are 2, 8, 18, 32, 50, and so on. These numbers automatically follow from the basic principles of quantum mechanics. When a shell is incomplete in a certain way, the elements so-formed have related reactivities (the same valence). Every group of elements so-related forms a column in the periodic table. Table 9-2 gives the beginning of the periodic table and corresponds to the elements built up in Table 9-1.

The first electron placed in a new shell (say, 3s) is very loosely bound and the element (sodium) is very reactive. Most atoms react with each other to form molecules, borrowing or donating enough electrons so that each "thinks" it has a completed shell. Thus do we have a complete modern quantum mechanical theory of atoms as the building blocks of matter. The atoms are made of protons and electrons in an orderly but complex arrange-

Table 9-2. *Periodic Table (Abbreviated, to Correspond with Table 9-1).*

1. hydrogen						2. helium
3. lithium	4. beryllium	5. boron	6. carbon	7. nitrogen	8. oxygen 9. flourine	10. neon
11. sodium	12. magnesium	13. aluminum	14. silicon	15. phosphorus	16. sulfur 17. chlorine	18. argon
19. potassium	20. calcium	21. scandium	etc.			

ment. At this point the weights of the atoms so-formed are totally wrong as compared with measurement, and we leave it to a later chapter to resolve this problem. The mathematics needed to "solve" atoms such as helium and beyond has remained difficult or impossible; however, fast computers are being used to solve the Schrodinger equation numerically for increasingly complex atoms and molecules.

9.6 ANGULAR MOMENTUM AND SPIN IN QUANTUM MECHANICS

As variables in quantum mechanics, angular momentum L and angular position θ are conjugate. They are analogous to linear momentum p and position x. Since there is no known absolute rotational orientation for a body, by the invariance principle angular momentum is a conserved quantity. With regard to uncertainty, L and θ cannot both be measured simultaneously. The product of their uncertainties must be greater than Planck's constant. As operators, $\mathbf{L}$ and θ do not commute.

Of particular interest are the components of $\vec{L}$ as a vector. The vector angular momentum is the quantity conserved, both in magnitude and direction. We have seen that quantization of angular momentum is vital in quantum mechanics, especially when a spherical system such as an atom is studied. Quantization of a vector implies "space quantization;" that is, $\vec{L}$ can point only in certain directions in space. Unless we place a frame of reference on an atom—a real frame, such as a magnetic field, not a fictitious one—we will be unable to observe any such quantization. The reference axis is usually the Z-axis, and the component of $\vec{L}$ measured with respect it is L_z (Fig. 9-8A).

In quantum mechanics L_z is quantized in length and permitted m units of $h/2\pi$. L_z can be positive or negative. Fig. 9-8B, C, and D show the various possible values of L_z and m when the magnitude of $\vec{L}$ is $1/2$, 1, and 2 units of $h/2\pi$. Spin can be considered as $\vec{L}$, although the symbol S is usually used for it. Since we cannot know where $\vec{L}$ points, any size component of it might be observed, within the restriction of quantization. Thus for $L = 2$ one could detect 0, 1, or 2 units of $h/2\pi$, either positive or negative, along the Z-axis. No other component of $\vec{L}$ can be found simultaneously with L_z.

Each atom has a total angular momentum

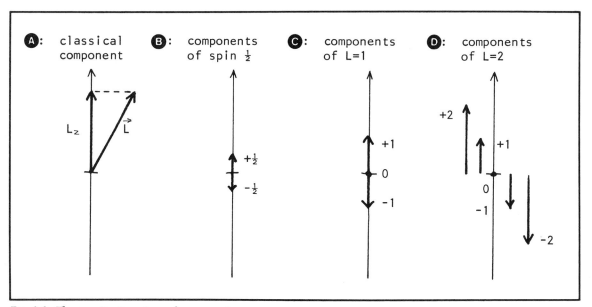

Fig. 9-8. The quantization of angular momentum.

(symbol J) composed of an orbital part for each electron and its spin. If each part is represented by a little arrow up or down, there is usually an almost equal number of ups and downs. The net total J is 0, 1, or perhaps 2 units. Each electron likes to pair up-down with another in its subshell, and each orbital part likewise. The effect of the magnetic field of the orbiting electrons on their energy states is a "fine structure" splitting of spectral lines. The electron spins feel this magnetic field, and are said to be *spin-orbit* coupled. When atoms possessing a net total angular momentum are combined to form solids, some useful properties are found such as large scale ferromagnetism due to unpaired electrons (beyond the scope of this book).

Electron spin has a further peculiar quantum property. It requires two turns to preserve its symmetry. In ordinary life we expect that if we rotate an object such as a book 360°, we see the book exactly the same as before. Not so with some elementary particles. The electron has to be rotated 720° to return to its original configuration. Needless to say, this is not observed directly but rather is predicted theoretically and verified by rather indirect experiments. Behavior such as this should convince the reader not to think of the electron as a rotating ball despite the loose language often used in both popular and technical sources.

Because of the Pauli exclusion principle, electrons behave oddly toward one another in confined regions. Each has to have its own state. There is a fictitious force between them that repels them apart, in addition to normal electric repulsion between negative charges. Electrons cannot be brought close to each other unless a separate state is somehow provided for each. Pauli exclusion "forces" are the main reason atoms are built in an orderly fashion, with each electron—really its wave function—keeping its proper place. In a way difficult to explain, the repulsion between two identical electron waves is related to their 720° symmetry.

Relativity cannot be neglected in the case of the electron spin. The effects appear—you guessed it!—as more splitting of spectral lines. The splitting from the spin-orbit effect is altered by "Thomas precession." Because of its high orbital speed (and regardless of our difficulties in observing said orbit), the electron suffers from time dilation. The effect is not large in hydrogen since the electron at best moves at $0.01c$, but it is substantial in heavy elements. Speaking loosely and classically, the spin of the electron (1/2 unit) is slightly out of step with its orbital angular momentum (1 or more units), as seen in the rest frame of the nucleus. Instead of "spinning," say, exactly twice per orbit, it might spin 1.999 times.

9.7 RADIATION BY ATOMS

At an elementary level some of the interactions with atoms—treated as particles—will be summarized. Of principle interest are radiation effects such as absorption, emission, and excitation. When light of appropriate frequency strikes an atom, it will be absorbed, raising one electron to a higher energy state corresponding to the energy absorbed. If the frequency is not sufficiently correct, the photon is absorbed briefly and re-emitted without a change in frequency. This is the way light is scattered. The electron state can also be boosted during a collision, and it may be put into a *metastable* state, one for which no emission is ordinarily allowed. Thus the atom may remain "excited" for a time. In the process of collision KE has been trapped as internal energy in the atom, so the collisions of atoms are often inelastic.

If the incident light is sufficiently energetic, or the collision is sufficiently rapid, an electron may receive enough energy to leave the atom, forming an ion. Above a certain minimum, any amount of energy can result in ionization. The ion can recapture a stray electron and a certain frequency of light will be emitted as the electron drops into an available state (one not occupied by other electrons). The electron may cascade through a series of states, with several photons emitted. The emission of frequencies different from (and less than) the exciting frequency is called *fluorescence*. Mercury atoms in fluorescent lights are excited by high speed ions, then emit a set of seven principal spectral lines from ultraviolet through red. We observe predominantly "white" light because a coating inside the tube converts the ultraviolet to a broad band of visible wavelengths.

When an atom emits a photon, it "recoils" (to conserve momentum). The motion of the atom Doppler shifts the frequency of the photon, so that a collection of atoms moving about randomly cannot emit a sharp spectral line but rather a broad line. Uncertainty itself also leads to line broadening. An excited state can persist only a short time; therefore the uncertainty principle tells us that the emitted energy from a number of such atoms must be spread in its value. Each photon that is emitted represents an electromagnetic wave of limited length, usually about a million cycles.

Angular momentum ought to be conserved when an atom emits radiation. The photon carries away spin angular momentum of 1 unit. When an electron jumps to a lower state, certain restrictions apply to its initial and final orbital angular momentum. In their simplest form, these "selection rules" state that l, and possibly m, must change by 1 unit. Since the photon can have its spin oriented either way, l either increases or decreases. It is beyond the scope of this book to explain how an orbiting and jumping electron can result in an electromagnetic wave (a photon), but the spin of the photon is obtained from the angular momentum of the atom. An atom radiates as if it were a classical electrical oscillator (an electron vibrating sinusoidally), but the size of the oscillating electric field is much larger than the atom.

Heisenberg's matrices are especially useful in the calculation of the amount of radiation. The rows and columns of a matrix describing an atom correspond to the possible eigenstates. These may mix together during radiation, with each eigenfunction contributing a certain amount. Before radiation the atom is considered to be in an initial state, described by its electrons' eigenfunctions and quantum numbers. After radiation it makes a transition to a final state, described by a somewhat different set of eigenfunctions and quantum numbers. The initial and final states are said to be connected by a "matrix element." When the calculations are done carefully, not only are the allowed transitions found, but the predicted intensity of each spectral line agrees with experiment.

9.8 FURTHER ADVANCES

Despite the tremendous success of the Schrodinger theory of the atom, there remained experimental and theoretical shortcomings. In 1928 Dirac started on the equivalent relativistic theory and quickly found the theoretical justification for the spin of the electron. Heretofore, spin had been arbitrarily included to achieve success. The matrix approach pioneered by Heisenberg was found to be essential in the Dirac theory and thus the competing quantum mechanical theories were amalgamated to one. For their work Dirac and Schrodinger received the Nobel Prize in 1933. Dirac's theory went on to deal with how light interacted with atoms. It also explained the parts of the "fine structure" of spectra that were not explained by spin and orbital angular momentum. Inevitably, improved measurement of spectra has found another flaw, the "Lamb shift." In 1947 Willis Lamb found that the ground state of hydrogen was split into two closely-spaced states, where Dirac and earlier theories had predicted a single one. We will meet the superseding theory which explains this later. For finding this flaw Lamb received the Nobel Prize in 1955.

The finer details of atoms are also being explored by the formation of oversize atoms, also known as Rydberg atoms after Johannes Rydberg, an early spectroscopist. Since there is no limit to the principle energy state, n can be 100 or 1000. Such an electron is very loosely bound and very far from the nucleus, up to 10^5 times the Bohr radius, or about 0.01 mm—big indeed! It does not particularly matter what element one starts with once the electron is so far from the nucleus and other electrons. One goal of working with such inflated atoms is to check the correspondence principle. The electron should act more and more like a particle in a Bohr atom as it recedes from the nucleus. The frequency of the radiation the electron would emit if it jumped should become the same as the frequency of its orbital motion. (Closer in the electron radiates much more slowly than it orbits.) The huge atom is fragile and a very low energy separates it from being ionized, losing the electron altogether. Unusual values of angular momentum can also be explored. When n is large and $l = 1$, the orbit is extremely elliptical. A peculiar atom such as the Rydberg atom is useful for studying many details of the quantum theory of atoms, particularly the effects of magnetic fields.

Atoms need not be made just of electrons and protons. Any unlike charged particles will form a bound quantum mechanical atom with energy states, a Bohr radius, and all the rest of the above paraphernalia. Most curious perhaps is the positronium atom, an electron orbiting a positron, or a positron orbiting an electron—it hardly matters since each orbits the center of mass in a symmetric way because they are the same mass. H. C. Corben and S. DeBenedetti discovered positronium in 1954. As already prophecied by Dirac, electron and positron are anti-particles: they can combine to eliminate each other. The atom is readily formed by spraying positrons at electrons so that some are mutually captured in bound states. There is a finite probability that one's wave function overlaps the other's, and they suddenly annihilate in a burst of radiation, after just 140 nanoseconds on the average. In the ordinary hydrogen atom the electron rarely interacts with the tiny proton, and it is moving too fast to make permanent "contact." When it ever does, nothing unusual happens. Positronium is a useful research tool for testing more recent theories as well as the quantum structure of simple atoms (while it lasts).

Another useful if bizarre atom is muonium, combining an electron with a positive muon (which we shall meet later). This was first done by Vernon Hughes and coworkers in 1960. The muon is about 207 times heavier, and a nearly normal atom is formed which does not "vapor-ize" unexpectedly. It does not live long for another reason, however: the muon breaks up and the atom flies apart. The ground state of muonium is −13.5 eV, similar to hydrogen. Besides exploring effects on spectral lines, this odd atom is used for studying spins and other aspects of elementary particles to come later.

It may sound simple, but counting atoms is as much of a challenge as "seeing" them. Since 1949 it has been possible to count slow electrons lost from atoms. Other methods of identifying single atoms have used radioactive decay processes. Recently the laser has aided in direct counting. By one of several methods a pulsed tunable laser applies a precise frequency matched to the particular element of interest to excite its atoms, and its alone. An electron is removed from each atom. Then the ionized atom is collected with an electric field. The process is called "resonance ionization spectroscopy"—for the matching of laser frequency to a particular atom, the ionization of it, and the identification of an atom and/or the sorting of types of atoms.

Field emission and other forms of electron microscopy have provided pictures of the arrangement of certain atoms in materials. In the field emission or field ion microscope, invented by Edwin Muller in the 1950s, a high electric field causes electrons to spray from a tiny crystalline sample at the end of a conducting needle. The pattern of the atoms in the crystal is much magnified when the electrons land on a distant screen. Recently the scanning tunneling microscope was developed by Gerd Binnig and Heinrich Rohrer. It uses a fine probe to scan very near the surface of a material. Electrons tunnel through space from the surface to the probe and are detected as current which is then used to construct an image on a screen. The accuracy is as good as one-hundredth the size of an atom, although this does not mean the inner structure of individual atoms is made visible, just their locations and general shapes.

Single atoms are treated as particles for acceleration, too. Initially the atom must be ionized so that it has a charge which the accelerator's electric and magnetic fields can act upon. After it is brought up to high speed, an electron can be returned to the atom, resulting in a high speed neutral particle unaffected by any electromagnetic fields and not inclined to emit any radiation. Particularly heavy fast atoms are useful for research. The trend in technology is to construct materials with special surface properties atom by atom, firing atoms at the surface. A beam of ionized atoms is a current and can be precisely controlled.

The hydrogen atom to this day remains a testing ground for many aspects of particle physics. In some sense the whole universe can be found in it. Many details of the properties of electron and proton are revealed in how they interact in the hydrogen atom. Almost all that can be measured is the complex pattern of spectral lines that results. A recent and important tool is the tunable laser, so that precise radiation can be provided to excite the atom. Spectral studies of atoms with the laser earned Nicolaas Bloembergen and Arthur Schawlow the Nobel Prize in 1981. This new field is sometimes called "particle chemistry."

Nevertheless physicists have yet to fully test some predictions of the Dirac theory of the atom, and have not been able to observe the naturally narrow line width of certain hydrogen lines. It may be possible to test the most advanced theories of particle physics by their subtle effects on the hydrogen spectrum. Checking nature to just one more decimal place can yield either a new revolution in physics or confirm that a theory seems to be remarkably accurate.

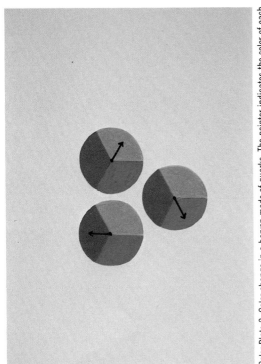

a: a baryon (a proton)　　b: a meson (a pion)

Color Plate 1. Baryon and meson made of quarks. The colors shown vary periodically. The quarks are actually point size.

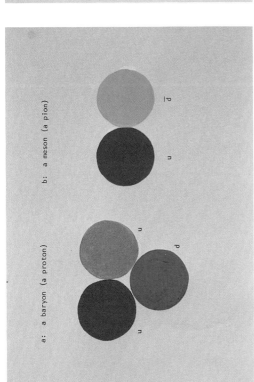

Color Plate 2. Color change in a baryon made of quarks. The pointer indicates the color of each quark at the moment.

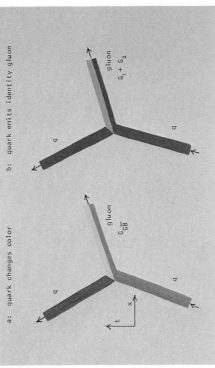

Color Plate 3. The eight possible gluons which carry the color force and charge. The gluon that changes red to green, for example, has the symbol $G_{R\bar{G}}$ and carries the colors red and antigreen (magenta). There are only two identity gluons which leave color unchanged; they are shown grey and have the symbols G_1 and G_2. Gluons are actually point size.

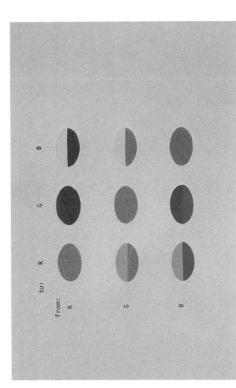

a: quark changes color　　b: quark emits identity gluon

Color Plate 4. Quarks emitting gluons in a Feynman diagram. Particles or colors traveling downward (backward in time) are antiparticles and have anticolor.

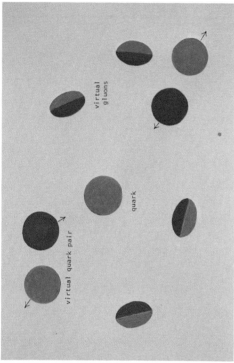

Color Plate 5. Quark-gluon details of the nuclear force when a pion is exchanged between proton and neutron. Note that the continual color change of quarks is accomplished by exchanges of gluons. A Feynman diagram.

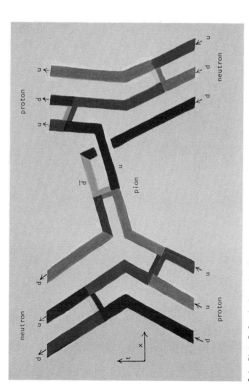

Color Plate 7. An antineutrino emits an X to change into a blue d quark, in the Grand Unified Field Theory. A Feynman diagram.

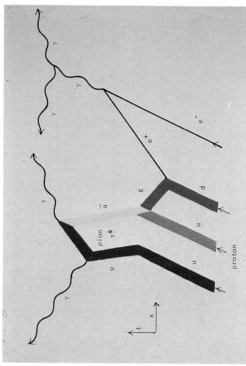

Color Plate 6. Enhancement of the color charge of a quark by virtual gluons and quark-antiquark pairs. Any unlike colors attract; like colors repel.

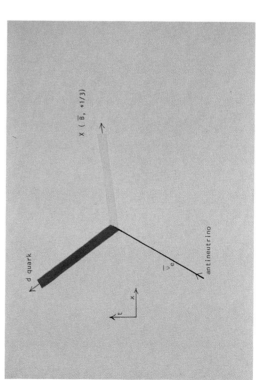

Color Plate 8. Proton decay by means of an X in the Grand Unified Field Theory. Even the electron of the hydrogen atom is annihilated. A Feynman diagram.

Color Plate 9. Aerial view of Fermi National Accelerator Laboratory, showing the main ring and the Central Laboratory on the left (courtesy of Fermilab Photo).

Color Plate 10. Conventional (red and blue) and superconducting (yellow) magnets in the main ring tunnel at Fermilab (courtesy of Fermilab Photo).

Color Plate 11. Experimental area at the end of the proton beam line at Fermilab (courtesy of Fermilab Photo).

Color Plate 12. Experimental area at the end of the neutrino beam line at Fermilab (courtesy of Fermilab Photo).

Color Plate 13. Experimental area at the end of the meson beam line at Fermilab (courtesy of Fermilab Photo).

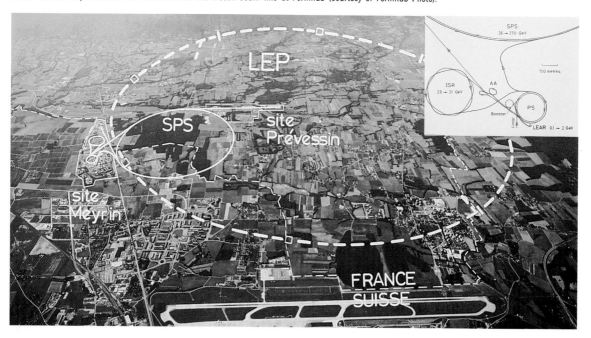

Color Plate 14. Aerial view of European Laboratory for Particle Physics (CERN), showing present rings and one under construction across the boundary of France and Switzerland (courtesy of Photo CERN).
(Inset Color Plate 15.) Diagram of the 28 GeV Proton Synchrotron (PS) and 450 GeV Super Proton Synchrotron (SPS) at CERN, to which a 50 GeV Large Electron Positron colliding ring (LEP) is being added (courtesy of Photo CERN).

Color Plate 16. Part of the PS ring at CERN (courtesy of Photo CERN).

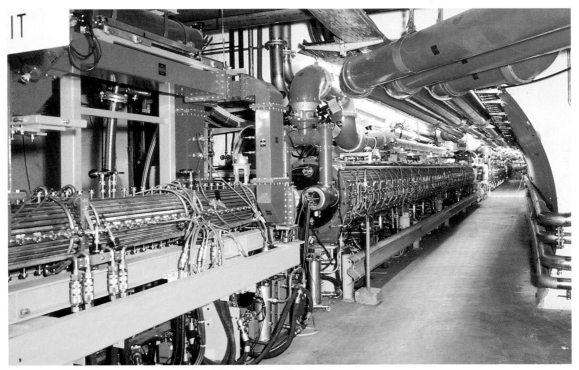

Color Plate 17. Magnets in the tunnel of the SPS at CERN (courtesy of Photo CERN).

Color Plate 18. The Antiproton Accumulator (AA) at CERN (courtesy of Photo CERN).

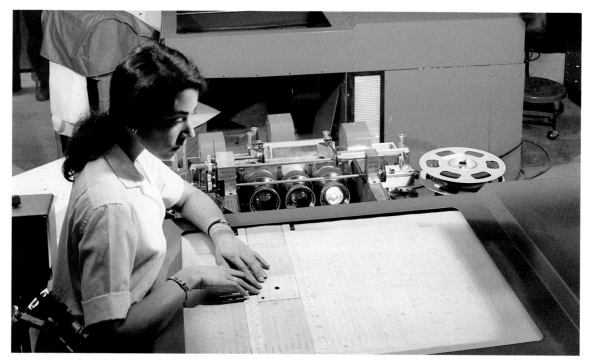

Color Plate 19. An operator measuring particle tracks on a bubble chamber photograph with the Alvarez computer-assisted scanner at Lawrence Berkeley Laboratory (courtesy of LBL).

Color Plate 20. Spark chamber detector at Deutsches Elektronen Synchrotron (courtesy of DESY).

Color Plate 21. Assembling the Plastic Ball detector (815 scintillators and photodetectors) at Lawrence Berkeley Laboratory (courtesy of LBL).

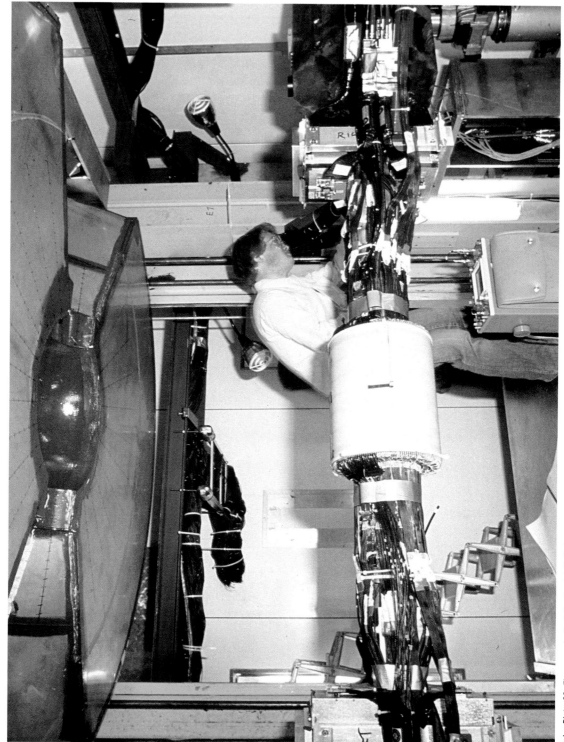

Color Plate 22. The upper half of the Crystal Ball detector, to be lowered onto the beam tube at the interaction region at DESY (courtesy of DESY).

Color Plate 23. Many photomultipliers surrounding the beam at DESY (courtesy of DESY).

Color Plate 24. Six-sided Time Projection Chamber built at LBL for the positron-electron collider at Stanford Linear Accelerator Center (courtesy of LBL).

Color Plate 25. Eight-sided iron plates surrounding magnets, scintillators, and spark chambers in the Mark III detector at SPEAR, Stanford Linear Accelerator Center (courtesy of SLAC).

Color Plate 26. Wiring a detector at DESY (courtesy of DESY).

Color Plate 27. The discoverers of the upsilon meson posing with their equipment at Fermilab (courtesy of Fermilab Photo).

Color Plate 28. The UA2 detector for *W* and *Z* bosons from proton-antiproton collisions at CERN (courtesy of Photo CERN).

Color Plate 29. The UA1 detector for *W* and *Z* bosons before assembly at CERN (courtesy of Photo CERN).

Color Plate 30. Bevatron (in building in foreground) and Heavy Ion Linear Accelerator (HILAC, in building in background) linked (dashed lines) for obtaining relativistic uranium nuclei to create Big Bang conditions upon collision (courtesy of LBL).

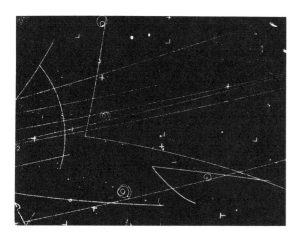

Chapter 10

Many Particles

CONSIDERATION OF THE PROPERTIES OF A group of many particles arises naturally when single particles are being discussed. Besides showing some details of how classical particles act together, we shall see the unexpected quantum mechanical effects of many particles and what is implied for single particles. This chapter covers the elements of thermodynamics, the statistical properties of particles, some further effects for atoms with many electrons, and other aspects. Part of this material is necessary for the modern theories of elementary particles that come later.

10.1 THE DEVELOPMENT OF THE CLASSICAL THEORY OF GASES

A *gas* is a collection of similar if not identical particles in such great quantity that their average behavior is predictable even if the individual motions of the particles cannot be followed. The particles can be atoms such as helium, molecules such as the O_2 and N_2 in air, electrons, or other more exotic ones. Usually the particles are rushing about

at high speed, several kilometers per second for gases at ordinary temperatures. Various terms pertaining to heat and gases will be used casually at first, drawing upon the common experiences of the reader, until groundwork has been prepared for rigorous definitions.

In early times, when a gas such as air was recognized as existing at all, it was presumed to be a thin but continuous form of matter. Robert Boyle in the 17th century found one of the first experimentally-based laws in physics, the inverse proportionality between the pressure and volume of a gas (now called "Boyle's law"). He went on to apply the early atomic theory and supposed a gas to be made of tiny spheres in motion. The study of gases continued to be a source of breakthroughs in science after that. Jacque Charles found in 1787 that the pressure of a gas increases in proportion to its temperature ("Charles' law"). John Dalton stated a "law of partial pressures" and attempted to explain it. It was known that the pressure of each kind of gas (which we now know as different molecules or atoms) behaved independently of the other gases in a mixture such as air. He also found the chemical implications of the behavior of gases, as discussed earlier. Amadeo Avogadro, using known laws about gas reactions, hypothesized in 1811 that there are the same number of discrete particles in a given volume of gas, regardless of the kind of gas and as long as other conditions are kept the same.

The science of thermodyamics was born in studies that led to the law of conservation of energy. In the 17th century Newton, Boyle, and others supposed that heat represented rapidly moving particles in solid matter and suspected that higher temperatures corresponded to more rapid motion. Heat was thought for a while to be a separate substance ("caloric") released or absorbed during mechanical or chemical actions; this was the caloric theory of heat. Benjamin Thompson (Count Rumford) in 1798 measured the approximate amount of heat produced during the cutting of metal. (Again physics was related to war, this time the boring of cannons.) He measured heat by means of the rise in temperature of water and proposed that heat was nothing but mechanical motion, not caloric. This was called the "mechanical theory of heat" and received little support for some decades, although vibration of particles or waves was proposed as a mechanism. The relation of heat in materials to infra-red light was identified in 1807 by Thomas Young.

Jean Fourier created in 1822 the mathematical theory of heat flow or conduction through solids and viewed heat as something distinct from mechanical motion. Among many studying the steam (heat) engine, Sadi Carnot showed in 1824 the relation between the heat in a gas (steam) and the mechanical work or energy obtained. He realized that a fall in temperature was required to extract work from steam and saw temperature as a new form of potential but failed to see for some years that heat was actually converted into mechanical motion. He did correctly show that perpetual motion was impossible. No engine could be made that could move heat to a higher temperature without needing a greater heat

and temperature somewhere else to run it. Robert Mayer in 1842 was able to calculate the relation between the heat in a gas and the mechanical work done by the gas. Mayer's pursuit of a general law of conservation of energy was firmed up by Hermann Helmholtz soon after, and proven experimentally (in part) by James Joule about 1840. Joule measured the relation between heat and electric energy and the relation of work to heat.

William Thomson (Lord Kelvin) proposed an absolute scale of temperature. At a temperature of zero degrees (now called 0 Kelvin or 0 K) no heat should be available, and all gases should shrink to zero volume (if they did not solidify). Further studies of theoretical heat engines by Rudolf Clausius resulted in his definition of *entropy* in 1865. The word comes from the Greek for "inward turning." Entropy is the heat energy divided by the temperature, a quantity which remains constant during the heat engine cycle even though more heat enters an engine than returns at a lower temperature. In the case of real engines or other processes involving heat, entropy was found always to increase. In other words, less heat could be converted to work than was theoretically possible. Clausius stated the notorious *second law of thermodynamics*: that the size of the entropy could only increase in all processes.

Clausius also began the new *kinetic theory of gases*, calculating the heat energy, pressure, and other features of gases from the motion of molecules. Maxwell (as did many ambidextrous physicists of the time) worked on kinetic theory as well as electromagnetism. He recognized that a gas was an ensemble of fast particles with a range or distribution of speeds spread about some average value. At any time a few molecules would have zero speed and a few would have extremely high speeds. Ludwig Boltzmann showed in 1877 the relation of entropy to the random behavior of the molecules. In particular, a bunch of moving and colliding molecules tends to arrive at the Maxwellian distribution of random speeds, no matter what their original order or disorder was. The assumption of a huge number of particles was all that was needed to guarantee that these calculations could be done and would result in simple predictable results for the energy, pressure, volume, and other properties of the gas. Challenges to classical determinism also occurred. Although, particles were not known to do anything but follow Newton's precise law of motion, the huge number of particles made it all but certain that individual particles could not be tracked and that systems tended to disorder or randomness.

Evidence had been accumulating for the existence of atoms (usually as molecules) in solid and liquid matter. The possibility of random restless motion of atoms was suggested by the Greek Lucretius from observation of dust in sunbeams. In 1827 Robert Brown noticed with his microscope that pollen grains seemed to be kicked about when suspended in water. Named "Brownian motion," this phenomenon was pondered by Einstein. Having hardly exhausted his inspiration with the publication of landmark papers in relativity and quantum physics in 1905, Einstein also published in that magic year an explanation of Brownian motion that allowed the calculation of the number of molecules in a given amount of matter, and therefore their size. We shall learn this number later as Avogadro's number. Einstein recognized that statistical variations in the speed and direction of molecules in a liquid would cause larger observable specks to move erratically as collisions with molecules occurred. His work clinched the atomic/molecular theory of matter, including the kinetic theory of heat, and paved the way for an understanding of the diffusion of particles through any kind of matter.

10.2 SUMMARY OF THE CLASSICAL KINETIC GAS THEORY

The reader must be briefly acquainted with the absolute temperature scale. The processes of nature that depend on temperature occur in accord with the absolute or Kelvin temperature. Our ordinary centrigrade or celsius temperature scale is connected to it as follows: 0 °C, the freezing point of water, is about 273 K (formerly written as 273 °K). The boiling point of water, 100 °C is at 373 K. Therefore 100 degrees is the same span

of temperature in both scales, and both use the same size degree (now called a "kelvin"). Figure 10-1 shows the two scales side by side with more accuracy. The current reference point for temperature is defined to be 273.16 K, at which temperature pure water is in equilibrium with ice and vapor at a certain pressure.

Very low temperatures are a frontier of physics. It is very difficult to achieve temperatures below a few degrees Kelvin, although research efforts have gone as far down as a millionth of a degree. Each additional decimal place attained in low temperatures requires much greater ingenuity, so that absolute zero is probably an unattainable goal. We shall see that working close to zero

Fig. 10-1. *The absolute and centigrade temperature scales (shown to scale).*

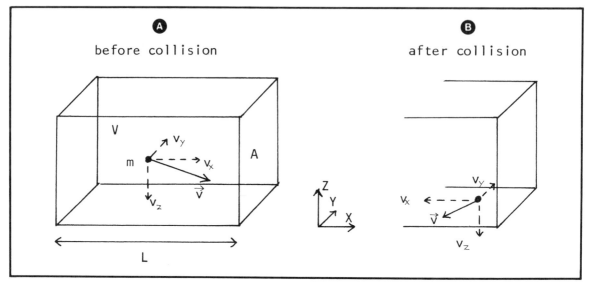

Fig. 10-2. *A gas particle colliding with the wall of a box.*

temperature is important in extending the understanding of quantum properties of matter. The quantum behavior of matter sometimes manifests itself on a large scale near 0 K—for example, superconductivity.

At any temperature the behavior of large numbers of particles, whether as a gas, a liquid, or a solid, can be viewed on two different scales. Individual particles can be studied on the microlevel, with or without quantum effects included. The piece of matter, or container of gas, can also be studied on our ordinary human-sized macroscale. Then the effects of individual particles cannot be detected (usually). Instead the overall average effect of trillions of trillions of particles is felt, for example, as heat or pressure. Theory at the microlevel is called the "kinetic theory" and at the macrolevel "thermodynamics." We shall confine ourselves mainly to gases.

The classical kinetic theory of gases can be found by considering first a single molecule in a box. The box has a rigid wall of area A and a volume V as shown in Fig. 10-2A, and the particle is moving toward the wall with velocity $\vec{v}$. Because we are interested mainly in the particle's progress toward one wall, we will consider the component v_x of velocity toward that wall. After the colli-

sion with the wall, the component v_x has exactly reversed and the other components are unchanged. The molecule moves as in Fig. 10-2B. To undergo the change in speed and momentum that it must have suffered, the particle must have had a force act on it. The force needed to reflect the molecule from the wall is given by Newton's law and has acted in some average amount of time Δt. This time can also be found from the length L of the box and the particle speed by $\Delta t = L/v_x$.

These mathematical ingredients can be assembled to arrive at a relation between molecular speed and average pressure on the box walls. Pressure P is defined as force per unit area thus: $P = F/A$. It is now measured in pascals (Pa), the same units as newtons per square meter. When many molecules are considered but not so many that they collide significantly with each other, the gas is called *ideal*. The result for an ideal gas is:

$$1/3 \ Nmv^2 \ = \ PV$$

where N is the total number of molecules in the box and m is the mass of one molecule. The special average velocity v is called the "root mean square" velocity (rms, for short). During the averaging of the calculation over many molecules, the square root of the average (mean) of the squares of the speeds was taken. The rms is often the best way to express an average for any quantity that can randomly have a variety of positive and negative values.

In modern thermodynamics Boyle's and Charles' laws are combined to give the *equation of state* for a gas at the macrolevel:

$$PV = NkT$$

where T is the temperature and k is the Boltzmann constant for gases. This law has been found from experiment and the constant has been evaluated. The mass or other properties of a specific gas are not needed in order to have a relationship for its pressure, volume, and temperature. We do need a count of the number of particles. The equation of state is illustrated in Fig. 10-3, where all the quantities are in balance or equilibrium. (Lest the reader think we are dwelling on gases excessively, it should be noted that most points made in the

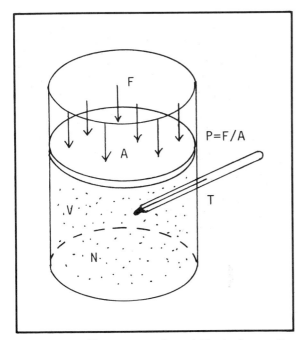

Fig. 10-3. Equilibrium among the variables in the equation of state for an ideal gas.

discussion are of general significance. More exotic particles, for example, have equations of state relating their pressures and temperatures.)

Since the microscopic kinetic theory and the macroscopic thermodynamics of gases above share *PV* in common, a relation between particle speed, kinetic energy, and temperature can be found. Anytime we see a term such as mv^2, we can somehow identify this as KE. The final result is:

$$KE \text{ (per molecule)} = 1/2 \ mv^2 = 3/2 \ kT$$

This is the classical connection between molecular motion and a gross effect, the temperature. The Boltzmann constant k connects temperature to kinetic energy and has the value $1.380622(10)^{-23}$ joules per kelvin. If we wish to know the energy of a gas with N molecules (its total heat energy), we simply multiply the result for one molecule by N. When the temperature is absolutely zero, the molecules should be stationary.

We would seem to know more than enough to find out how many molecules are in a gas. If we know pressure, volume, total mass, temperature, and the constant (which can be found by another procedure) then we can find N. Standard conditions have been defined for a gas. Without getting into the determination of "gram molecular weight," let us say that 16 grams of oxygen, for example, at 273 K, at one atmosphere pressure (about 10^5 Pa), in 22.4 liters of volume (0.0224 m^3) has about $6.022169(10)^{23}$ molecules. This number is called Avogadro's number or one *mole* (a basic SI unit) of particles. The estimate 10^{24} gives a convenient estimate of the number of atoms or molecules in a few grams of any substance. One mole of water occupies about 0.018 liters (0.000018 m^3) of volume, but the same amount of a typical gas is a thousand times less dense and occupies about a thousand times more volume.

When left alone, the enormous number of molecules in a gas continually collide with the walls of the container, and occasionally with each other. In the classical theory the molecules collide elastically like tiny marbles, not losing any energy in reacting with or sticking to each other. Some molecules gain speed after a collision, some lose it. But the gas stays in equilibrium by virtue of the enormous number of collisions. The speeds retain their random distribution, and the pressure and temperature are unchanged. If the temperature is somehow increased, then there are more molecules with higher speed and less with lower. There are always a few with as high a speed as one wishes to consider, although ones near lightspeed are extremely rare.

Of possible interest is the distribution of energy in the gas into its different *degrees of freedom* (df) or modes of motion. When left to itself the gas follows the *equipartition law*: an equal amount of energy appears in each possible degree of freedom. The particles keep hitting each other until the energy is as randomized as possible. Particles moving in three dimensions possess three df, one for each dimension or component of velocity. Each df represents $1/2 \, kT$ of energy. If the particle is not a tiny sphere but dumbbell-shaped (two atoms joined as a molecule) it can rotate in two different independent ways. Together with its linear motion, this molecule has 5 df and can carry $5/2 \, kT$ of energy.

There are more complex theories for gases, formulated by making more realistic assumptions about the properties of the molecules. Especially important is the consideration of molecules with certain sizes, which collide with each other. The particles need not be molecules but any small particles. They will always tend to equilibrium with a certain distribution of speeds and therefore energy among the molecules. The distribution is called the *Maxwell-Boltzmann* distribution and arises principally from the fact that the particles are in principal distinguishable. Their paths could be followed if the experimenter were sufficiently ingenious. This is in contrast to quantum particles to be considered shortly.

10.3 ENTROPY, DISORDER, REVERSIBILITY, AND INDETERMINISM

Let us suppose that we have a two molecule gas and that each molecule could have one of three possible values of energy. Then energy state diagrams such as Fig. 10-4 can be drawn (borrowing terminology from quantum mechanics), and the energy levels can be "populated" in various ways by the two molecules. It is important to decide whether our molecules are distinguishable or not. Although they are identical molecules, classically they are in principal distinguishable from each other as discussed earlier. If one molecule is labeled A and the other B, we could not tell if the configuration b has A above B or B above A. So configuration b is counted as two possible cases. There is only one way to obtain configuration a with molecules A and B in the same energy level. There are fairly simple mathematical procedures for obtaining a count of the total number of ways certain configurations can occur. The re-

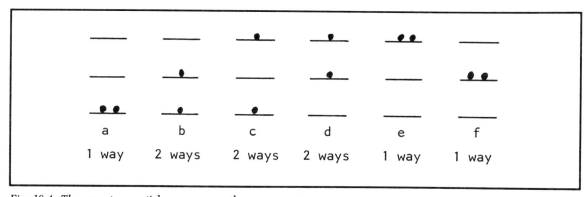

Fig. 10-4. The ways two particles can occupy three energy states.

sult in our case is 9 ways for the 6 distinct configurations shown. As the number of particles is increased, the number of configurations increases much more rapidly.

If this simple example were to resemble a Maxwellian distribution of kinetic energies, configuration f, where both molecules have an intermediate energy, might be the most likely configuration, followed by b and d. When the number of particles and states is increased, an arrangement such as f where the middle energy states are the most populated is the most likely. When there are many more molecules, there are many more ways they will bunch around an average energy level than ways in which they will settle in unusually low or high levels. If our molecules were not distinguishable, an average state such as f would be no more popular than the other 5 states shown. The assumptions made about the particles have a strong effect on the total count of possible arrangements; these are whether the particles are distinguishable, whether more than one can occupy a state (remember Pauli exclusion), whether the states are discrete (quantized), and so forth.

Generalizing this approach to obtain a count w of the total number of possible energy configurations, the entropy S of a gas can be calculated by:

$$S = k \ln w = Q/T$$

Boltzmann worked out the calculation of entropy from the order (or disorder) of particles in this way, and his constant k appears. w is some enormous number, much much greater than N, the number of molecules. The logarithm function using the base e is symbolized ln. If w is a huge number expressed in a power of 10, the logarithm ln extracts that power, except for a multiplying constant. On the right side appears the equivalent expression in macroscopic quantities. Q is the heat content of the gas and T is the temperature. S is likely to be increasing over time and has its largest value for a random distribution of speeds.

The fact that entropy increases provides a direction for the flow of time as well as a form of classical indeterminism. If entropy can only increase as time goes along, then it should be impossible to reverse the flow of time and see entropy decrease. A simple example is shown in two frames in Fig. 10-5. In frame A a particle is shown near a cluster of other particles. One might envision snapshots of billiards "before" and "after" a breakup. In frame B the same six particles are caught instantaneously in another arrangement, more dispersed. There are no arrows placed on the particles showing which ones are moving which way, yet if we are presented with the two pictures and asked which came first, we would name frame A. We suspect that there is negligible chance that particles in frame B, however suitably they might be moving, will later coalesce, ejecting a single one as in frame A.

If the number of particles is made much larger, the tendency of disorder to increase will virtually guarantee that a picture such as Fig. 10-5A preceded one such as Fig. 10-5B. Thus is the direction of time, sometimes called "time's arrow," related to the increase in entropy. Otherwise one might occasionally run short of breath in a room because the randomly moving air molecules happened to cluster in the other side of the room. No such unfortunate behavior of molecules has ever been observed, because the number of configurations available to the molecules in which they are more or less uniformly spread throughout the room outweighs the number of configurations in which they are in half the room by a number too ridiculously large to state. In another context, the same restrictions on order prevent us from making an engine which will convert all the available heat (disorderly motion) into work, which is orderly motion. The second law of thermodynamics thus sets a limit on the efficiency of our devices as they change one form of energy into another.

The order of events and the increase of entropy cannot be considered without initial conditions. If the particles in Fig. 10-5B are each given a very special velocity, then the arrangement of Fig. 10-5A will result at a certain time later.

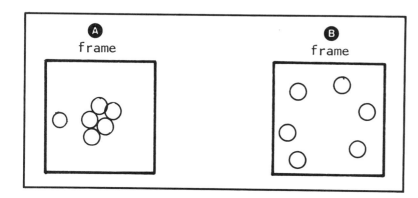

Fig. 10-5. Which frame comes first?

Where philosophers especially like to argue is how our universe started with such special initial conditions that the Earth, life, and us (you and me, now, reading this) are the result. Could we be the results of chance? Life, including human beings, is a special ordering of particles that is far from equilibrium. We process extra amounts of energy in order to retain, for a time, that special condition, but the signs of dissipation of that order are constantly nagging at us. Ilya Prigogine received the Nobel Prize (in chemistry) in 1977 for clarifying the entropic behavior of ordinary matter versus that of living systems. He showed that we all follow the same thermodynamic laws but that we humans live farther from physical equilibrium than inert matter does.

Entropy is also related to information. To a physicist or engineer, information is a much more definite quantity than simply words on a page or dots in a picture. Information involves order arising from disorder. There is no law against creating order; writers, artists, builders, scientists, and so forth create order everyday. They also process energy in doing so, either by breaking down food or by consuming the materials with which they work. The overall effect is an increase in disorder, with most work ending up as heat, even though the creative product stands out for its uniqueness (its order in a disorderly universe). In the computer age we know that information can be broken down to bits, formulated by Claude Shannon in 1946. One bit is the choice between two possibilities, such as 0 or 1, up or down. A minimum energy is needed to encode each bit and

more is needed every time the bit is preserved during processing. The minimum energy can be calculated by the same principles already given, but that would lead us astray. There is a minimum necessary amount of information needed to describe any given situation, whether natural or human-designed. Entropy and information are opposites, and in fact their sum is conserved. The higher the entropy, the less the information, and vice-versa, as long as a *closed* system is being considered that has no contact with the rest of the universe.

What might be greater interest in a book about nature is how nature handles information, independent of human affairs. We have already puzzled over how an electron can know there are two slits to go through or how a photon can know the state of another photon far away. The truth is that these particles cannot "know" these things unless two conditions are met: that information, a piece of order among the disorder, was transmitted from one place to another, and that energy was enlisted in encoding the information. We can wave a flashlight at the night sky and make it appear as if something passed almost instantaneously from one star to another. But unless energy, as waves or particles, actually makes the trip and is "modulated" or made to vary in an orderly way, nothing real passed from one star to the other. Aliens out there would not marvel at our violation of relativity.

In the microworld we have learned that energy comes in relatively big packets, so that when information is passed around, it "accidentally" changes the situation. The information may become irrelevant. We have also learned that a sine wave carries no information. It must be turned on and off; more than that, it must be left on a certain length of time in order to be sent and received as a piece of a sine wave. An uncertainty principle holds in the world of information in analogy to the uncertainty principle for elementary particles. There are limits to the rate at which information can be sent, and this is not unlike the limits on how much a wave function can "know."

Another way indeterminism sneaks into the classical world is through the difficulty of making perfect measurements. It has been mentioned that the molecules of a gas cannot be tracked in any practical way. Even a system as simple as a rotating wheel with a marked spoke illustrates indeterminism. After the wheel has turned a large number of times, the mark could be at any position. No matter how accurately the initial position of the mark was measured, a sufficient number of rotations will amplify the initial error into an amount so large as to exceed the circumference of the wheel.

Consideration of entropy, order, and informa-

tion tends to lead us from classical into quantum matters, where indeterminism reigns. The interrelationship of entropy, information, structure, and even cosmology is a vaster field than can be covered adequately here, but it all stems from a consideration of particles.

10.4 QUANTUM STATISTICS

The consideration of the number of ways a collection of classical particles can distribute their energy is a part of *statistical thermodynamics* or "Maxwell-Boltzmann statistics." We now turn our attention to the particles when they are allowed their quantized behavior and thus will sketch a bit of *quantum statistics*. The first hint of new phenomena occurs in the multi-electron atom. The electrons are indistinguishable, yet each must occupy a unique state by Pauli exclusion. Classically, the electrons might prefer to bunch down near the positive nucleus, but their special quantized character spreads them out in an orderly way. The result is large rather "solid" atoms even though they are mostly "empty" space. No electron carries a label. If one is thrown into an atom and later

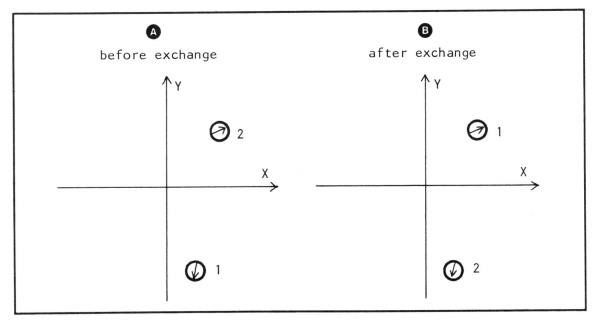

Fig. 10-6. The exchange of position and spin for two particles.

one is knocked out, there is no way to say if it is the same electron.

A new application of the invariance principle is possible. The physical behavior of an atom or other system should remain unchanged if, somehow, some electrons are switched about or *exchanged*. In Fig. 10-6A two electrons with spin are shown (in a simplified manner) labeled and placed on a coordinate system. In Fig. 10-6B the coordinates and spin of 1 and 2 have been exchanged. This is not the same as simply trading particles. Electron 2 has been given the spin as well as the position of 1. The reader may object that this is merely a piece of mental magic and means nothing physically. But it will be argued that very real "exchange forces" with pervasive consequences result from this apparently banal property of electrons.

If a system is invariant under exchange, what must be conserved? The symmetry of the system is conserved. The symmetry of an atom is determined by the symmetry of its wave functions, and those must be constructed with a symmetry that permits electrons (and therefore their wave functions) to be exchanged without effect. The actual symmetry is called "antisymmetry" since experimental evidence indicates that only antisymmetric wave functions describe systems of electrons. When any two electrons are exchanged mathematically, a minus sign is introduced in the overall wave function of the system.

While the probability of finding the electron is unchanged by an exchange, there are various other real consequences. One is that the electrons tend to avoid each other. Another is that a gas of electrons behaves differently in regard to its energy and pressure; its equation of state is different from ordinary gases. And Pauli exclusion follows: no two electrons will occupy the same state. When these consequences are translated into what can be observed, the effects are found in the periodic table, in chemical bonding, in magnetism, in less familiar substances such as liquid hydrogen and helium, and in collisions. Atoms bound together as molecules are continuously exchanging electrons. In the CM frame for a collision, the sym-

metry includes one of exchange. We cannot tell which electron took which exit path, and this affects the quantum mechanical interpretation of collisions.

Electrons with spin 1/2 are part of a more general class of particles which, as a group, act in similar ways. They are called *fermions* and follow statistical laws called *Fermi-Dirac statistics*. The particles must be assumed identical, indistinguishable, and rarely interacting, so that they can be in equilibrium. Only one particle can be in each energy state, so the number of available states must be at least equal to the number of particles. In contrast to an ordinary gas, a fermion gas has all energies represented up to a certain maximum (called the "Fermi level") even at zero temperature. Some electrons are forced to have high energies because there is a shortage of states at low energies, since each electron must have a unique energy. Only one electron could have zero energy and uncertainty forbids even that. At higher temperatures, higher states are available and the spread in energies is greater.

The electrons in an atom are fermions, and in its lowest (ground) state, the atom has a variety of energy levels, each filled by one electron. This is true no matter how "cold" the atom is. This behavior of fermions extends to solids and provides us with the very useful behaviors of metals and semiconductors so needed in modern technology.

There is another class of particles which behaves differently in a group. We have met one member, the photon, with spin 1. These *bosons* (named for the physicist Satyenda Bose) have spin 0, 1, or other whole numbers and follow *Bose-Einstein statistics*. They are also identical and indistinguishable, but their wave functions are symmetric under exchange. They do not mind being all in the same energy state and attract rather than avoid one another. As a gas bosons have their own unique equation of state. More low energy levels are available to bosons than to classical particles. Bosons also cannot be counted. A collection may increase or decrease in number as time goes on and they interact with atoms. We shall be emphasizing how bosons are the equivalent of waves;

therefore bosons must indiscriminately work together on a large scale to act as, for example, a light wave. There are no ordinary cases of materials acting like bosons. Those materials which do are known as "quantum materials" because they exhibit quantized behavior on a large scale.

We shall occasionally be interested in whether matter, or a group of particles, is in equilibrium. Three different kinds of equilibrium have been mentioned, depending on whether the particles are distinguishable and their spin. Many processes in nature, especially those involving a few particles, are nonequilibrium situations. Equilibrium requires time for all particles to participate equally and randomly in a process. It makes little sense to discuss the temperature of a single particle, or the temperature of a collection of atoms all excited to the same higher energy. Temperature requires a certain distribution of speeds and energies, according to one of the three statistics mentioned.

10.5 QUANTUM GASES AND FLUIDS

Photons make a peculiar gas, since they have zero rest mass. Nevertheless, a gas of photons in a box can be found to have exactly the same energy and pressure as if light waves were confined to the box instead. The particle and wave theories of light are again equivalent. The equation of state for photons in equilibrium is:

$$P = \frac{1}{3} aT^4$$

where P is the pressure and a is a special constant (Stefan's).

The superconduction of electric current, a manifestation of quantum behavior on a large-scale at very low temperature, involves electrons paired as bosons (net spin 0) in certain materials. Any set of particles, including atoms, will act as bosons if the total angular momentum is 0, 1, or another integer. About half the chemical elements are bosons because there are no unpaired electrons.

A strange collection of bosons is liquid helium. The helium nucleus, about which we will learn more later, is a boson with total zero spin. (Helium's two electrons do not change its boson character.) Below about 2 K (2 degrees above absolute zero) many of the bosons "condense" into the bottom energy level for bosons. In this state all bosons have the same low energy and constitute a very orderly "fluid" called a *quantum fluid*. It exhibits many bizarre properties. Just to describe a few, it has no viscosity, flows uphill, conducts heat (what little there is) infinitely well, and has zero entropy. It takes on the most extreme properties of any form of matter accessible to us. It exhibits literally a collection of the properties of the quantum world on a scale large enough for us to see. Heike Onnes received the Nobel Prize in 1913 for work with liquid helium before its boson character was known, and Lev Landau received it in 1962 when quantum phenomena were becoming apparent.

Recently (1979) hydrogen atom gas has been cooled so low (about 0.1 K) as to form a quantum gas that can be condensed, not to form a liquid but to form the lowest boson state as a dense gas. It was not easy for the experimenters Isaac Silvera and Jook Walraven to find a way to keep the atoms from reacting to form molecules. The ordinary random orientation of the angular momentum of the atoms had to be ordered in a magnetic field. Hydrogen atoms are bosons because both the electron and proton have spin 1/2. However these add in the atom, a spin 0 or 1 boson is formed. It is not likely that much heavier atoms will be condensed as a boson gas or fluid because the atoms are more eager to interact with each other and form a solid. Although this book does not cover solids, it should be mentioned that their properties can also be described with waves and quantized particles, a sort of solid quantum mechanics known as the *solid state*. The cold side of nature has been emphasized here, in order to find large scale quantum behavior. We shall soon move to the other extreme, very hot fast particles.

Chapter 11
Nuclei

IN OUR PURSUIT OF THE ULTIMATE BUILDING blocks of nature we now make a major jump down in size and a huge jump up in the level of energy. To strain what has become a popular metaphor, nuclear physics is a "quantum leap" from the physics of the atom. Because of the small sizes involved, the quantum mechanical approach is vital for the subject. We have already encountered one particle, the proton, which is a major constituent of nuclei. Later we shall make another major jump up in energy and learn more about the proton. Since the early history of nuclear physics includes a search for certain elementary particles, some early methods of detecting new particles are introduced here.

Nuclei have already been mentioned as being the location of the heavy positive charge that holds an atom together. If we strike a nail with a hammer, it is the electrons in the metal atoms of the hammer and nail which interact electromagnetically (and through Pauli exclusion). But the inertia of the hammer, nail, and wood is contained mainly in the nuclei of those atoms. The disturbed electron shells transmit the force to the nuclei, which then move along with the electrons. At this point it is important to distinguish "atomic" from "nuclear." Atomic refers primarily to the behavior of the electrons of an atom; nuclear refers to the behavior of the matter in the nucleus of the atom. Atomic properly refers to chemistry, and the best interpretation of "splitting the atom" is in terms of chemical reactions. When a nucleus is split, by far more energetic means, then the atom also breaks apart.

11.1 HISTORICAL BACK-GROUND OF NUCLEAR PHYSICS

That a new, perhaps deeper, aspect of nature existed was first shown by Antoine Becquerel in 1896. He found *radioactivity*, so-named for the strangely powerful "radiation" coming from minerals containing uranium. This radiation would ionize gases, pass through solids, and affect photographic plates. It should not be thought of as ordinary electromagnetic radiation since it was not x-rays. It was emitted regardless of the chemical state of the minerals. In 1898 Marie and Pierre Curie found the element radium to be emitting the radiation. They also found other radioactive elements, thorium and polonium, by processing uranium ore in large quantities. For this work all three received the Nobel Prize in 1903. Marie Curie also received the Nobel Prize in chemistry in 1911 for her discovery of radium. Curie and other early contributors to nuclear physics are shown in the Solvay Conference group photography Plate 2.

Rutherford found in 1899 that there were two kinds of radiation, a weakly penetrating kind which strongly ionized gases, called "alpha," and a strongly penetrating but weakly ionizing kind called "beta." Beta radiation, loosely called beta "rays," was found identical to fast electrons from the way magnetic fields bent its path. The name beta particle has remained to this day to refer to electrons emitted by radioactive nuclei. Alpha radiation was also recognized as being particles. Villard found in 1900 a third kind of radiation, named "gamma" radiation (or "gamma rays"). It is also emitted by radium, is very penetrating, and is unaffected by magnetic fields.

Rutherford and Frederick Soddy found by 1903 that one heavy element transforms into another more or less down the periodic table, emitting the various radiations in the process, and they formulated the law of radioactive decay. Soddy later (1921) received the Nobel Prize in chemistry for related work. Alpha decay was found to cause an element to lose four units of weight and two of charge. Beta decay resulted in a gain of one unit of positive charge. Gamma decay seemed to have no effect on an element despite the large chunk of energy that was emitted. The heat from decaying radium was measured by Curie and Laborde in 1903. Because the number of alphas could be counted (they made flashes on a fluorescent screen), the energy per particle could be estimated, and it was high (in the MeV range, to use later units). In the next year William H. Bragg found

how deeply alphas would penetrate matter and estimated their energy at 4 to 9 MeV.

Three different series of radioactive disintegrations were found, two starting with uranium (element 92) and one with thorium (element 90). All ended with the element lead (element 82), but in each case the lead had a different weight! Many specialized names were given to the products of radioactivity because of their unfamiliar weights, but eventually they were all identified in the periodic table and given familiar (or new) element names. Along the way Villard established (1903) the existence of forms (*isotopes*) of the elements that had different masses. Isotope means "same place" in Greek, referring to the periodic table. Details of the radioactive series will be given later.

An instrument designed to detect and count electrically the radiation from nuclei was made by Rutherford and Hans Geiger in 1908. It consists of thin gas in a glass or metal chamber, with a positive potential (voltage) connected through a resistor to an electrode in the center (Fig. 11-1). When radiation passes through and interacts with the gas, ions are formed. Since only a few eV of energy are needed to produce an ion, a particle with several MeV can produce many ions without losing much energy. The ions move toward the negative electrode and are measured as a proportional current through the external circuit. This ionization detector will detect any radiation, such as beta particles (electrons), that ionizes the gas (argon and alcohol are now used). Alphas usually will not penetrate the tube. Gamma radiation ionizes the gas much less readily. The more modern form of this detector is called the Geiger-Muller counter. It uses a voltage so large that amplification of the number of ions occurs and so is sensitive to the production of a single ion in the gas.

One step toward the discovery of the nucleus was made after 1906 by Rutherford, Geiger, and Ernest Marsden, when they found gold atoms to contain very tiny hard positively charged centers. To encounter the gold nuclei, they had to use similarly small hard particles, the radiation from

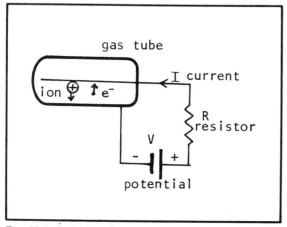

Fig. 11-1. Ionization detector.

A second particle detector, called the *cloud chamber*, was developed by Charles T. R. Wilson in 1912. It is so-named because a vapor is held in a chamber in such a state that it will condense as clouds under the slightest provocation. The cloud chamber is a good detector for charged particles such as alpha and beta because these leave a trail of ions around which the vapor condenses as tiny droplets. With the addition of photography and magnetic fields, the chamber became quite a useful means of showing particle tracks. For this detection method Wilson received the Nobel Prize in 1927. Later refinements of the cloud chamber (and discoveries in nuclear physics) earned Patrick Blackett the Nobel Prize in 1948.

Rutherford also found in 1919 another nuclear phenomenon, *transmutation*. Originally this term was used by alchemists to describe the conversion of one substance (chemical element) to another. Science fiction became science fact centuries later when Rutherford fired alpha particles at nitrogen, and protons (becoming hydrogen) came out, leaving new atoms of oxygen. The reaction is diagrammed in Fig. 11-2, where "?" denotes other neutral but unidentified mass known to be present in the nuclei. Charge is of course conserved in this and all nuclear reactions. This reaction among light elements to form other elements was the same sort of new transformation

radium. These were the alpha particles, found identical to helium nuclei in 1909 but observed to come from heavy radioactive atoms, not from helium gas. The alpha particles were occasionally found to return almost straight back after encountering gold atoms as if they had hit something small and hard (recall Fig. 3-12). But usually they passed through almost unaffected. The deflection, we already know, is by means of the electric repulsion between the alpha particle and the nucleus and does not indicate that a nuclear force has been detected. The nuclear model was formally proposed by Rutherford in 1911.

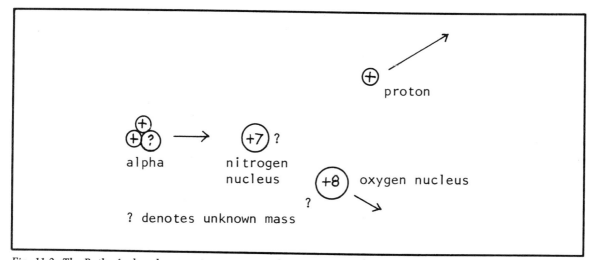

Fig. 11-2. The Rutherford nuclear reaction experiment showing transmutation.

in the chemical periodic table that had been found with the heavy radioactive elements. Soon Rutherford and others were able to make many elements from lighter elements by bombarding them with alpha particles. By 1925 an energy analysis of the alpha-nitrogen reaction was possible in the cloud chamber, made by P. Blackett.

By 1927 the proficient Rutherford and others had proven that electric forces were not the only contributor to the deflection of alpha particles by nuclei. The new force came to be known—naturally—as the "nuclear force." They established the maximum range over which it seems to act as less than 10^{-15} meters (using other units at that time). This can be done because the impact parameter of a particular collision can be deduced from the direction and energy of the outgoing particle.

Meanwhile attention was focused on the production of alpha particles. How could a helium nucleus escape suddenly and unpredictably from a heavy nucleus such as radium? Using the recently established notion that the wave function of a particle leaks past energy barriers, George Gamow and separately R. Gurney and E. Condon proposed in 1928 that quantum tunneling lets the alpha escape the nucleus. The nucleus was known to hold very tightly to its constituents in order to remain together despite the large disrupting force of electric repulsion among the protons. If the nucleus is seething with energy, occasionally a group of two protons could acquire enough to escape. However the alphas detected in the laboratory did not seem to have as much energy (large though it was) as they should have had. The beta particle, an electron, also makes its escape from the nucleus and further mysteries developed when it was fairly clear that electrons normally do not dwell in the nucleus. Their wave functions would not fit. Where did the electron come from?

Better control over nuclear reactions was achieved by John Cockcroft and Ernest Walton in 1932 with the first true particle accelerator. They used a van de Graaf accelerator to fire high speed protons at lithium (nuclear charge $+3$) to produce two alpha particles. They received the Nobel Prize in 1951 for their work. Nuclear physics continued to stimulate development of new particle accelerators. It was the only "high energy" physics in existence at the time. Lawrence and Livingston used the cyclotron to fire protons and "deuterons" (to be defined) at other matter and study the results.

The *deuteron* was found by Harold Urey, F. G. Brickwedde, and G. M. Murphy in 1931 to be a "heavy hydrogen" nucleus, twice as heavy as a proton but with one positive charge; what made up the remainder was not known but was suspected by Rutherford in 1920 to be a heavy neutral particle called a *neutron*. Heavy hydrogen (deuterium) has one electron and occurs naturally in water and other substances, diluted to about one part in 5000 of normal hydrogen. It is readily separated electrically and can be ionized and accelerated just like any other atom. Urey received the Nobel Prize (in chemistry) in 1934.

In 1932 the long-suspected neutron was found by James Chadwick who correctly interpreted experiments in which a new penetrating radiation was observed. He experimented with neutrons, knocking them out of beryllium nuclei with alpha particles, and measured their mass. It was nearly the same as the proton mass and finally explained the extra mass long known to be in nuclei. Detecting a neutron could only be done indirectly since its lack of charge means that it leaves no trail in the popular detector of the day—the cloud chamber. The invisible particle was literally tracked down by means of the tracks it caused known ions to make when neutrons hit them. Chadwick received the Nobel Prize in 1935.

Progress was rapid once the neutron was known. Heisenberg wasted no time in proposing that all nuclei consist of protons and neutrons, with neutrons making up all the extra needed mass that could not be protons. In 1933 Frederic Joliot and Irene Curie found the positron (the positive electron) to be another form of beta particle emitted by radioactive nuclei. (For other work with radioactivity they received the Nobel Prize in chemistry in 1935.) The positron, besides vindicating Dirac's relativistic quantum mechanics, will lead us to another chapter in particle physics. But its existence is important for understanding nuclear physics, because it is evidence that the neutron and proton are essentially the same kind of particle. They can change into each other by emitting a positive or negative beta particle.

The first influences of astrophysics on particle physics began when physicists and astronomers pondered the source of energy that powered the stars. It was known that chemical energy—e.g., burning a star made of coal—was neither hot enough nor would last long enough to satisfy the suspected ages of stars—billions of years. A new energy source had to be found in matter somehow, and it might be identified in processes accessible to scientists bound to our planet. The possibility that the energy came simply from a gradual gravitational contraction of the star was shown to be false with new stellar evidence. Arthur Ed-

dington calculated (correctly) that a star like the sun had to have a temperature of 20 million degrees (K) at its center to explain why it did not collapse but shone stably for so long. He considered a balance of pressures due to gravitation, radiation (light), and electrons (Pauli exclusion). At the expected temperature in the sun, the particles (mostly hydrogen atoms) were not only ionized but had speeds sufficient to cause nuclear reactions as seen in the laboratory on Earth. The Nobel-winning explanation will be given later.

11.2 NUCLEI

The nuclei of elements often come in several forms (isotopes) for the same element. Fig. 11-3 illustrates the ways in which small numbers of protons and neutrons can be combined to form nuclei. The symbol Z conventionally denotes the count

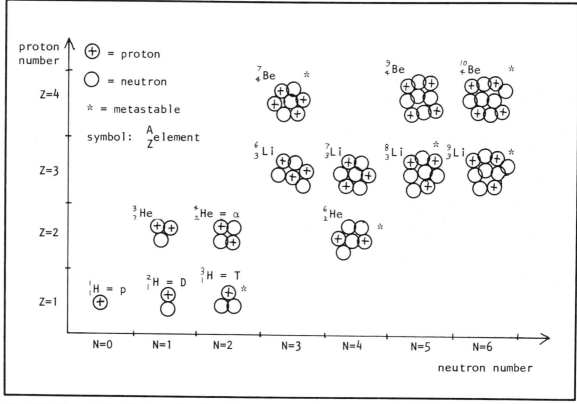

Fig. 11-3. The periodic table for small nuclei. (The actual shapes of nuclei are more spherical. Nuclei with lifetimes less than 10^{-15} second are unstable and not shown.)

of proton charges in a nucleus, and N gives the number of neutrons. The total $A = Z + N$ (the "atomic weight") is approximately the mass of a nucleus, in units of approximately a proton mass. At first, nuclei are built up on a one-for-one basis between protons and neutrons. A is approximately twice Z for many nuclei. As the nuclei become heavier there is room for a few extra neutrons, which is another way of saying that the increasing repulsion among the protons must be diluted by more neutrons if the nucleus is to stay together. While one could postulate an endless number of neutrons added to a given nucleus with Z protons, only those are shown which have been found in nature or have been created artificially and live for an appreciable time. An asterisk (*) shows which nuclei are metastable in Fig. 11-3. For example, a third radioactive isotope of hydrogen with two neutrons (tritium) exists but decays too rapidly to be found in nature. A combination of five neutrons and protons cannot be made to hold together longer than 10^{-21} second. Reasons will be given which limit the possible combinations. The schematic way of showing nuclei avoids any real indication of how a nucleus appears; the best guess would be that nuclei are spherical, possibly lumpy.

Nuclei form their own periodic table and have magic numbers. Figure 11-3 represents the beginning of the periodic table for small nuclei. The isotopes are arranged in two dimensions according to proton count Z and neutron count N. All the isotopes of a given element are in a row for the atomic number for that element, arranged in order of increasing mass. An isotope of a given element is written, for example, $^{238}_{92}U$, where U is the element symbol (Uranium). The preceding superscript gives the total count A of nuclear particles, and the preceding subscript gives the element number Z.

In nuclear physics isotopes rather than elements are referred to, since it does not usually matter what the element is, only the composition of its nucleus. There are 263 stable isotopes and a few naturally-occurring long-lived metastable ones. About 2200 metastable isotopes (living longer than a millisecond) have been made in laboratory experiments, with accelerators, or in research nuclear reactors. About 7000 more isotopes could be made that stay together long enough for high speed detection. Much work has gone into organizing and making sense of what must be very complex nuclear structure. "Magic numbers" have been found. Nuclei with multiples of four particles (multiples of alphas) are more stable. Nuclei with 8, 20, 28, 50, 82, or 126 total count of protons and neutrons are unusually stable. There would seem to be states or energy levels in the nucleus, in analogy to the atom and the states of its electrons.

The analogy to atoms extends further: radiation (gamma or a particle) is emitted when a nuclear particle changes state. There are a set of very specific energy transitions possible for a given nucleus. Nuclei can be identified by the various energies emitted when they are excited. This constitutes an extension of the concept of "spectroscopy." When matter can be arranged in a periodic table because it has quantum states, then there is an associated spectroscopy of emission energies. For atoms, only light is emitted and spectral energies are in the range of a fraction of an eV up to (in unusual circumstances) tens of keV. For nuclei, emission energies start in the MeV range, a million times higher than ordinary light. When electromagnetic, these are gamma rays. We will encounter another and still more energetic spectroscopy later.

A nucleus is stable, metastable, or unstable; in the latter two cases it is radioactive, emitting alpha, beta, or gamma radiation (or certain other particles) at some time after it is formed and changing into another nucleus. There is a characteristic time after which the decay occurs, called the *half-life* (τ). This is the time by which half of the nuclei of one species in a given sample have undergone decay. In Fig. 11-4A decay over four half-lives is illustrated. After each period of one half-life, half of the preceding quantity remains, and half of the preceding count of decays can occur. Less and less radiation is emitted, in proportion. If the reader has worked with a

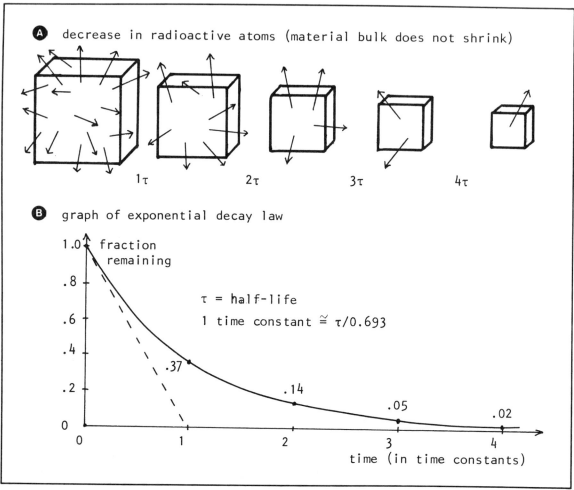

A decrease in radioactive atoms (material bulk does not shrink)

1τ 2τ 3τ 4τ

B graph of exponential decay law

1.0 fraction
remaining

.8

.6

.4

.2

0

τ = half-life
1 time constant $\cong \tau/0.693$

.37

.14

.05

.02

0   1   2   3   4

time (in time constants)

Fig. 11-4. Radioactive decay.

counter and a radioactive material in a school lab, it may be wondered why the material does not quickly run out of radioactive particles and the "activity" noticeably decrease. Half-lives range from femtoseconds (10^{-15}s) to billions of years. Most common sources of radioactivity, such as the naturally-occurring ones, must necessarily have lasted billions of years to be still available.

The isotope $^{238}_{92}$U has a half-life of $4.5(10)^9$ years. Much of it, formed five to ten billion years ago in nearby supernova, is still present in the Earth. It must necessarily have weak radioactivity or all would have decayed long ago. A piece of uranium ore containing about one kilogram of

uranium (most of the ore is oxygen and other elements) will emit about 10^7 particles per second. We might detect only those few which reach the surface of the rock. Since it contains about 10^{24} atoms and nuclei, it will be mostly converted to other elements in about 10^{17} seconds, or 10^{10} years.

There is a simple law for radioactive decay. If n_0 is the number of original nuclei at any given starting time, the number n remaining after a time t is given by:

$$n \cong n_0 \, e^{-0.693 \, t/\tau}$$

where e is the mathematical constant (value

2.71828 . . .). This expression is approximate since 0.693 is the endless decimal *ln* 2. The law of decay is called an exponential law and has several useful properties. A graph of the number n versus time is shown in Fig. 11-4B. The number falls rapidly at first and later decreases more and more slowly. It does not matter when one starts recording this graph. The sample does not have to be "freshly made"; the shape of the curve is the same. If the sample were to decay at the same rate it initially did (the dashed line) it would all be used up in the time $\tau/0.693$ (also called one "time constant"). As decay goes one, there are less nuclei to decay and the rate slows. After one time constant the fraction remaining is approximately 0.37 of n_0. After two time constants there is about 0.14 n_0. After any elapse of one time constant, 0.37 of the preceding amount is left. Becquerel has been honored with the name of a disintegration unit: 1 becquerel (Bq) is one disintegration per second, in SI units. In older units there is the curie equal to $3.7(10)^{10}$ disintegrations per second.

Recently the selection of nuclear decays has been extended. As more unstable nuclei are assembled in collisions in accelerators, they tend to decay by emitting bare protons or neutrons. Sometimes these emissions are delayed while a beta decay first rearranges the nucleus. Occasionally two protons or neutrons are emitted, whether simultaneously is not yet known. The first proton radioactivity was found by S. Hofmann in 1982 with the lutetium isotope $^{151}_{71}$Lu, which also exhibits beta-delayed proton emission. Neutron emission from nuclei in their ground states has not yet been observed. Very recently it was found that one decay in a billion of the natural isotope $^{223}_{88}$Ra results in a $^{14}_{6}$C nucleus being hurled out. This radium isotope had been long known as an alpha emitter.

11.3 NUCLEAR PARTICLES AND FORCE

To discuss nuclear particles we must settle what units the mass will be measured in. We could use SI units, in which case masses will be about 10^{-27} kg, but nuclear physicists have been able to obtain greater accuracy by using masses referred to some standard nucleus. This is possible by virtue of the mass spectrometer, the instrument which balances electric and magnetic forces to sort moving charged particles according to their mass and charge. Figure 4-13 showed a simplified schematic of a mass spectrometer. The nuclear isotopes of interest are produced as ions, preferably with all electrons removed. The instrument is very sensitive and can distinguish any isotopes easily. Among its many applications is a new one to 14C dating, where only tiny samples of archeological materials are needed since the 14C nuclei are counted directly. (Large scaled separation of isotopes has been done mechanically, using gaseous diffusion and, more recently, centrifuges.)

In 1960 the mass of $^{12}_{6}$C, the common isotope of carbon, was adopted as the basis for the *unified atomic mass unit* (u). This nucleus is defined to have 12 u. As we shall see, one u is slightly more than the mass of a proton or neutron because a nucleus possesses additional mass-energy of binding. Twelve grams of $^{12}_{6}$C does contain one Avogadro number (one mole) of these carbon atoms, thus making the connection between "micro" mass units and "macro" mass units. Also, 1 u can be shown to equal $1.660531(10)^{-27}$ kilograms.

One proton has mass 1.007277 u, equivalent to $1.672614(10)^{-27}$ kg. In energy units the proton is about 938.2 MeV. To construct a hydrogen atom, one electron of mass $0.000911(10)^{-27}$ kg would be added (1/1836 of a proton). The proton also has a spin of 1/2 unit, giving a magnetic moment. The neutron has mass 1.008665 u (939.5 MeV, slightly heavier than the proton). It also has spin 1/2 and, surprisingly, a magnetic moment. The size of the proton and neutron cannot be given more precisely than to say about 10^{-15} meter (1 fm).

A property which should be considered for every particle is whether it decays, and if so, how soon and into what. The proton has so far proven to be stable but experiments are underway to find if it has an extremely long half-life of decay. Details of this are in a later chapter, where the proton figures in the latest revolution in physics. The neutron decays into a proton, electron, and "something else" (to be shown later to be a neutrino). Its half-life is known, not very accurately, to be about 918 seconds. Since an electron rest mass added to a proton rest mass does not give all of the neutron rest mass, some of the missing mass is released as energy and another particle in the decay. The details are given later. Despite the decay of the neutron, slow though it is, most nuclei are stable. Those that decay usually do so for reasons unrelated to neutron instability.

It has been found that the neutron and proton are indistinguishable as far as the nuclear force between them is concerned. One clue is that the neutron can change into a proton by emitting an electron and that a proton, when struck with an electron, forms a neutron. Thus the notion of "state" has been generalized and neutron and proton are considered as two different states of the same particle, a *nucleon*. The nuclear force itself is blind to charge. Nevertheless, the nucleon is distinguished into its two states by means of a quantum number called *isotopic spin* (*I*). The proton has value $I = +1/2$ and the neutron has $I = -1/2$. This "spin" is not a real one but an imaginary one in another dimension, which physicists can imagine. This is only the first instance of a useful quantum number based on an imaginary or unmeasureable property. Isotopic spin and other such quantum numbers have quite real consequences in helping explain how nucleons order themselves in the nucleus.

The simplest nuclear system consists of two nucleons, either nn, pp, or np. To be bound (by some force) in the lowest energy state, it should have zero angular momentum, analogous to the hydrogen atom. Both neutron and proton have spin 1/2 and are fermions. Therefore Pauli exclusion keeps them from occupying the same ground state, unless some other new quantum number provides additional distinguishability of states. Such has not been established. Experiments have shown that nn and pp do not exist. It might seem that electric repulsion would keep pp from

binding, but the nuclear force is capable of holding two protons together.

What does exist is the deuteron (np). It does not have a ground state with zero angular momentum. Indeed it has been found to have only one bound state, possessing one unit of angular momentum ($l = 1$) and an energy level of -2.21 MeV. At some risk the reader might temporarily imagine the deuteron as two particles orbiting each other very closely. A better but more abstact picture for objects at such a small scale is shown in the state diagram of Fig. 11-5. The spins are parallel in this state. It also has a magnetic moment which is well explained from the moments of its constituents and their orbital motion. The distribution of charge in the deuteron has been found to be flattened, not uniformly spherical. It is said to have an "electric quadropole moment."

Quantum mechanical wave functions are readily calculated for two-particle systems, and the deuteron is found to be represented by one that is 96% spherical and 4% four-lobed, with even parity. Since this ground state is not purely

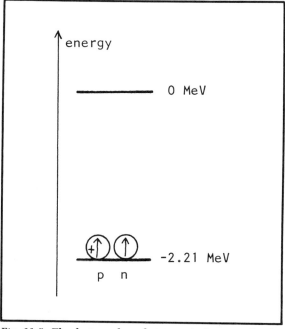

Fig. 11-5. *The deuteron bound state: a single energy level.*

spherical, the nuclear interaction cannot be a central force—that is, one which depends on distance from the center only. Therefore the nuclear force, whatever it is, takes into account spins or angular momentum. Thus does quantum mechanics combined with experimental measurements allow physicists to deduce properties of an unknown force. However, the analysis of three nucleons is as different from two as night from day and has no simple solutions. The systems pnn (tritium, a third hydrogen isotope decaying with half-life 12 years) and ppn (a helium isotope) do occur in nature.

Nuclear force theory began in 1935 when Hideki Yukawa showed that a new particle called a meson, with a mass intermediate (hence its name) between an electron and proton could serve as a "carrier" of force between nucleons. This was an analogy to the way photons serve as carriers of the electromagnetic force between charges. Since the electromagnetic interaction has an infinite range, uncertainty permits only massless carriers (photons). The nuclear force is known to have a very small and fairly definite range. The uncertainty principle applied to energy and time then allows the mass of the appropriate carrier to be estimated from the range and fundamental constants.

If it is assumed that the carrier may move as fast as lightspeed c, covering the range r in time $t = r/c$ and representing mass-energy $E = mc^2$ borrowed during that time, the uncertainty principle gives:

$$m \cong h/rc$$

The mass m of the carrier works out to be about 250 times heavier than the electron for the observed range of about $1.5(10)^{-15}$ meter. Yukawa received the Nobel Prize in 1949 for his theory of nuclear interaction, which included a full wave mechanics approach. It was some time before the predicted meson was correctly observed by C. F. Powell, C. M. G. Lattes, and C. P. S. Occhialini in 1947 and called the pi-meson or pion.

It has a mass 273 times the electron mass (mass-energy 140 MeV). Thus the Yukawa theory was verified, at least in part. However, further progress was slow and no simple accurate mathematical model of the nucleus has been forthcoming. More recently a different approach has arisen, given in a later chapter.

Meanwhile investigation of the nuclear force continued by means of scattering experiments, the collisions of two protons or proton and neutron. (Until recently neutrons could not be held still for n-n collisions.) This work has confirmed that the nuclear force (or potential) must depend on spins, angular momentum, exchange, and possibly particle speeds. No simple formula for it has resulted. The spin-dependence of the nuclear force has been studied recently by scattering electrons or protons from a solid target in which most of the nuclear spins are aligned the same way (no easy trick). A spin up electron is deflected to the left by a nucleus. Small changes in the deflection are determined by how the nuclear force interacts with the particle spin. The fact that the deuteron has both nucleons spin up in the ground state is another indication of how strong the spin part of the nuclear force might be. It has overcome the tendency of fermions to pair up-down. Another aspect of the nuclear force is that it is extremely repulsive at very short distances less than about $0.5(10)^{-15}$ meter (not much shorter than the range for which it is attractive). This is another way of expressing the fact that nucleons are virtually incompressible.

11.4 NUCLEAR STATES AND BINDING ENERGY

Aside from certain fundamental questions, nuclear research has been very fruitful. Careful measurement of the masses of isotopes has shown that all are slightly more than some multiple of the mass of a proton (or neutron). Moreover the masses of light and heavy ones are both heavier than expected, and the masses of intermediate isotopes are lighter than they should be. These findings were made by Francis Aston in 1927. (He received the Nobel Prize in chemistry in 1922 for finding isotopes by means of the mass spectrometer.) He called the deviation from a simple calculation of mass the "packing fraction." A low packing fraction indicates stability of the isotope. Elements around iron were found especially stable in the sense of tightly bound. This is also true of a few light mavericks ($^4_2$He, $^{12}_6$C, and $^{16}_8$O) but not for $A = 8$).

If the mass-energy of each individual nucleon is simply added, the result is greater than the measured mass-energy of the nucleus that is formed. The difference is called the *binding energy*. It is the energy equivalent of the packing fraction. Somehow in interacting the nucleons find a lower energy state than they would have as separate particles. Since there is no reason to assign the loss of energy to any particular nucleons, the negative binding energy is simply divided among all nucleons and expressed as binding energy per nucleon. It has been carefully measured for all possible nuclei that live long enough. As expected, it is largest (about 8.7 MeV) for elements near iron ($A = 55$), and is about the same regardless of the stability of particular isotopes. The binding energy per nucleon falls gradually as A increases, reaching about 7.5 MeV for uranium ($A = 238$). At the low end, binding energy per nucleon falls rapidly for elements lighter than carbon (7.7 MeV per nucleon), going as low as about 1 MeV for deuterium. The alpha particle $^4_2$He has an unusually high value for its position, 7.1 MeV per nucleon, in accord with its exceptional stability. Two neutrons and two protons must find an especially comfortable arrangement. It is no wonder that alphas are the only multiple-nucleon object frequently emitted from, or knocked out of, heavier nuclei.

One of the best ways to measure the size of a nucleus is to fire neutrons at it. The only interaction is nuclear, and if the nucleus is truly hard-edged, the neutron either hits or it does not. The resulting sizes are usually called cross-sections, measured in terms of area. (The units were once whimsically named "barns." The experiments were like trying to hit a barn door from 0.1 light year away.) We will translate the results to diameters. Uranium nuclei have diameters a little larger than 10^{-14} meter, and lighter nuclei proportionately less. The size seems to vary with A in such a way as to imply that the nucleus packs together as if nucleons were marbles. The nucleus has a constant density of nuclear matter.

Nuclear sizes can also be found by scattering relativistic electrons from nuclei. As we shall learn, electrons do not feel the nuclear force and thus scatter only from the electrical aspects of the nucleus (mainly the proton charge). More exotically, muons (200 times heavier than electrons) have been placed in atoms, where they dip so low as to enter the nucleus and provide further information.

Nuclei with equal numbers of protons and neutrons, or with even numbers of either, are very stable. This indicates that the energy states do not distinguish appreciably between protons and neutrons, and further that protons and neutrons tend to pair up with opposite spins. The slight energy differences are due mainly to the electromagnetic difference between proton and neutron. The protons are distributed approximately symmetrically in the nucleus. The effect of these and other factors on binding energy was formulated by C. F. Weizsacker in 1935.

Instability is a symptom of inadequately-filled energy states and gives further clues. The result is a change of state and an emission, as illustrated in Fig. 11-6. Protons and neutrons have not been observed to be emitted by natural nuclei but are emitted from some artificially-prepared nuclei. There should be no unpaired protons in the nucleus; hence no proton would help things by being emitted. It is also not energetically favorable to emit a neutron, as that leaves protons closer together and makes the nucleus less stable.

The most common decay is beta, resulting in a change in Z of one unit and no change in A. Study of beta decays gives a set of rules regarding how many stable isotopes there are with odd and even A and Z. Occasionally, as a result of natural emission or of being struck by a particle, a nucleus will emit a gamma ray, signifying some internal

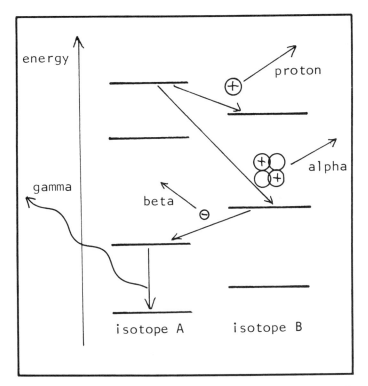

energy

gamma

beta

proton

alpha

isotope A   isotope B

Fig. 11-6. Illustrating possible changes of state in nuclei, resulting in emission of energy. (The nuclear state diagram itself necessarily changes after any transition except gamma emission.)

rearrangement. Some nucleon has dropped to a lower energy level. Another possibility is that an atomic electron is kicked out ("Auger transition" or "internal conversion"). This direct process is a result of the fact that the electrons around the nucleus have some interaction with it. Kai Siegbahn received the Nobel Prize in 1981 for his studies of the spectroscopy of beta emissions and internal conversions.

Since each particle or gamma ray emitted by a nucleus has a very well defined energy (a correction to this will be given later), each represents a change in nuclear energy state according to a definite pattern. Each emission has a "line width"—that is, an energy spread—related to the lifetime of the state that has decayed, in accord with uncertainty. The shorter the lifetime of the state, the more uncertain the size of the energy emitted.

No clear regularity has been found in nuclear energy levels except that involving spin and parity. Being fermions, the nucleons must each have their own quantum states, with proton and neutron

paired with opposite spins where possible. The energy levels tend to have spacings measured in MeV, a million times higher than typical atomic spacings. Fig. 11-7 illustrates the energy levels of a carbon-12 ($^{12}_{6}$C) nucleus. The ground state is 12 times 7.7 MeV, or about 92 MeV below zero. It would be quite a challenge to excite nucleons to selected upper states. Parity will be discussed in the next chapter; it expresses the left or right handedness of a quantum state, a kind of symmetry. Some of the states for carbon-12 are labeled with the total angular momentum J and the parity (+ or −) according to convention in Fig. 11-7.

That nuclei have angular momentum (which may be viewed as spin when the nucleus is considered as a particle) is known from the hyperfine splitting of atomic energy levels. It can also be inferred from beta and gamma emissions. The spin and magnetic moment can be measured directly with *nuclear magnetic resonance*, where the nuclei are placed in a strong constant magnetic field and their spins are made to resonate (actually, to precess) with a varying magnetic field at high frequency. The method was developed by Isidor Rabi in 1938, for which he received the Nobel Prize in 1944. Felix Bloch and Edward Purcell received the Nobel Prize in 1952 for their measurements of nuclear magnetic fields. The magnetic moment of nucleons may be weak, but very close to them the field is extremely strong.

11.5 NUCLEAR REACTIONS

A nuclear reaction is defined as a process in which one or more particles or nuclei are initially present and another set of particles or nuclei is present in the final state. What happens between

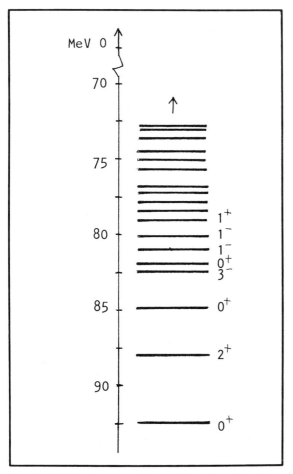

Fig. 11-7. The lower energy states for the carbon-12 nucleus.

is hidden from us; we only see input and output. A decay, where a nucleus changes into other products, is classified as a reaction even though nothing struck the original nucleus. In every reaction atomic electrons must be involved. Fast alphas usually carry no electrons; they were stripped off in producing the ion. But after the reaction the new element gobbles up any nearby electrons needed to balance its charge to zero. Electrons get knocked around during the reaction with negligible effect, since they are much lighter. Things always straighten out afterward.

We have met the alpha, beta, and gamma decays in several contexts. The alpha particle "tunnels" out of the nucleus, the time it takes to do so depending on the depth of the energy well it is in. The alpha's wave must vibrate an enormous number of times in the nucleus before escaping. It faces a barrier of 25 MeV in $^{238}_{92}$U so must wait billions of years for the chance to escape with only about 4 MeV to its name. The beta decay is also slow by nuclear standards, taking at least 10^{-3} seconds and as long as 10^5 years. The gamma ray (photon) is light emitted as a direct result of a charged particle changing state. Except for the photon, which must carry away one unit of angular momentum, the particles emitted during decay can carry away almost any amount, so the nucleus can undergo a drastic change in angular momentum. Alpha is a boson with spin 0, and beta has spin 1/2. Additional angular momentum is removed by having the particle leave at a certain radius from the center of the nucleus (in classical language).

There are various ways to diagram these decays with varying degrees of logical clarity. The following method is chosen:

alpha decay: $(Z,A) \rightarrow \alpha + (Z-2, A-4) + E$

example: $^{238}_{92}U \rightarrow \alpha + ^{234}_{90}Th + 4.18$ MeV

beta decay: $(Z,A) + \beta (Z+1, A) + E$

example: $^{234}_{90}Th \rightarrow \beta + ^{234}_{91}Pa* + 0.19$ MeV

gamma decay: $(Z,A)* \rightarrow \gamma + (Z,A) + E$
example: $^{234}_{90}\text{Pa}* + ^{234}_{91}\text{Pa}$

In abstract terms, a nucleus is designated by (Z,A), showing its proton number and total number. In the examples, the element symbol carries Z and A. Several other notations for attaching Z, A, and N to element symbols have been used and are now superseded. The use of N has been suppressed here for clarity. The alpha (α) can also be written ^4_2He. The beta β^- is understood to be negative unless shown β^+. As often happens after a beta decay, the resulting nucleus is left in an excited state denoted by *. A nucleon is not in its lowest energy level and will jump after a time (the lifetime of the state), emitting a gamma ray. Each decay occurs because a lower energy state will occur. Therefore excess energy E in MeV remains as kinetic energy of the products of the reaction. Often it is shown in the reaction equation. A nucleus can decay by two different routes. For example, 99.85 % of the time $^{234}_{91}\text{Pa}*$ decays by beta to form $^{234}_{92}\text{U}$ and only 0.15 % of its decays result in emission of gamma as it drops to an unexcited form of $^{234}_{91}\text{Pa}$.

A reaction such as Rutherford's collision of an alpha with a nitrogen nucleus is written thus:

$$^4_2\text{He} + ^{14}_7\text{N} \rightarrow ^1_1\text{H} + ^{17}_6\text{O}$$

A rare but stable isotope of oxygen is formed with one extra neutron. Without knowing the energy of the alpha, one cannot be sure how much kinetic energy is carried by the products. The reaction does not necessarily produce products with more binding energy (lower energy state), although that is usually the case when alphas are added to light elements.

There are four and only four radioactive series where an initial unstable heavy element changes into a sequence of unstable lighter ones until a stable one is reached. Sometimes the changes can occur by various routes. $^{238}_{91}\text{U}$ goes through the most complex series, involving 10 different routes and 17 intermediate elements. The

preceding set of examples were taken from the beginning of the decay series. By one route or another 11 alphas and 9 betas can be emitted. The end result for this and two other natural series (starting with $^{235}_{92}\text{U}$ and $^{232}_{90}\text{Th}$) is an isotope of lead. None of the series involves a β + emission. A β^- emission results in a (temporary) increase in Z.

The heaviest element found in nature is uranium ($^{238}_{92}\text{U}$ in particular). Since 1940 elements heavier than uranium have been synthesized. They are called "transuranium" elements because their chemical properties match the series that includes uranium in the periodic table. At first they were made in nuclear reactors, where the abundance of free and fast neutrons penetrate uranium nuclei to build up the nuclear mass. Extra neutrons convert to protons by beta decay, raising the proton number and making higher elements. The cyclotron also can make them by firing fast alphas at uranium and at the heavier elements so produced. To build mass more efficiently neon nuclei ($A = 22$) are now used. The heavier the element, the shorter its lifetime. Progress has stalled recently at about number 108, with the quantity made being only one atom and the lifetime about a second. It is suspected more stable nuclei exist at about $Z = 114$, a new magic number in conjunction with $N = 184$, another magic number. The Nobel Prize (in chemistry) was received by Edwin McMillan and Glenn Seaborg in 1951 for their work on transuranium elements.

Fission is the tendency of heavy unstable nuclei to split into two large parts. It was discovered by Lise Meitner and Otto Hahn with barium in 1938. The energy of each part is less than the whole, and binding energy is released as kinetic energy. Occasionally fission occurs naturally, but more often when another particle enters the nucleus, such as a neutron. Enrico Fermi received the Nobel Prize in 1938 for nuclear work involving isotopes created by neutron bombardment. $^{235}_{92}\text{U}$, for example, will break into almost any two fragments whose atomic weights sum to 235 plus the 1 neutron that started it. Most commonly it will break into about $A = 140$ and $A = 95$ and a few neutrons (usually one). A reaction is shown in Fig. 11-8A. The products could be stable isotopes of cerium and molybdenum or metastable isotopes of other elements which then emit alpha or beta particles until stable isotopes result. They fly apart at high speed because they carry the excess of about 1 MeV per nucleon or a total of about 200 MeV. This is the source of nuclear power by fission. The fission fragments are still much slower than light, but they are very "hot" by ordinary chemical standards. Here we refer mainly to temperature, but they are often radioactively "hot" as well.

The fission reactor is both a research tool and a power source. It is constructed by piling together large quantities of radioactive uranium, not the common isotope but an enriched mixture with a rare and more active isotope concentrated. The extra neutrons emitted as the alphas bounce around among the uranium nuclei causes some fissioning and more neutrons to be emitted, which then cause more. The result is a *chain reaction*, a buildup of neutrons and fissioning. It will run away if not controlled by capture of excess neutrons. Fermi's original reactor in 1942 was called an "atomic pile" but we now know the essential process should be called nuclear. When

the reaction is started in such a way that it can run away, the resulting catastrophe takes place in milliseconds and is called a nuclear fission bomb (not "atomic" bomb). The so-called "neutron bomb" is a slight variation on the fission bomb, emitting a few more neutrons and a little less energy, but all nuclear weapons give off huge amounts of neutrons, gamma rays, and many highly radioactive fission fragments of intermediate Z.

Fusion, the combining of light nuclei to form one of intermediate mass, does not occur spontaneously. The nuclei must approach at high speed to overcome the electric repulsion of the protons, at least 1 keV of kinetic energy per nucleon, and higher when there are more protons in the nucleus. This corresponds to a temperature of millions of degrees Kelvin. Nevertheless, the energy released when the fusion occurs more than makes up for the initial energy needed. In fact, the input energy is simply needed to take one particle over the potential "hill," after which it falls into the nucleus, releasing both binding energy and the original KE.

The goal of fusion is usually to combine the abundant protons of hydrogen into helium, a much lower energy state, releasing many MeV of energy per particle. It has to be done one proton

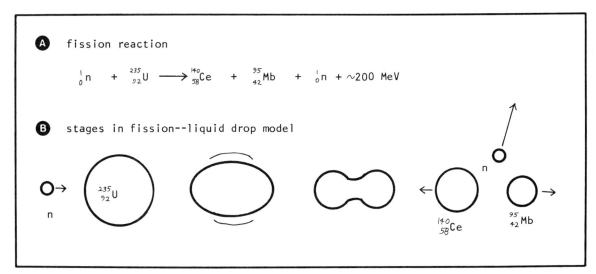

Fig. 11-8. *One of the most likely fission reactions and the liquid drop model of it.*

at a time as the chances of three separate protons simultaneously colliding together is too low. Some possible fusion reactions are as follows, where the subscript Z is omitted:

$$\text{proton-proton:} \quad p + p \rightarrow {}^2H + e^+ + ? + 0.42 \text{ MeV}$$

$$\text{proton-deuterium:} \quad p + {}^2H \rightarrow {}^3He + \gamma \rightarrow + 5.49 \text{ MeV}$$

$$\text{helium-helium:} \quad {}^3He + {}^3He \rightarrow {}^4He + p + p + 12.83 \text{ MeV}$$

$$\text{deuterium-deuterium:} \quad {}^2H + {}^2H \rightarrow {}^3He + n + 3.27 \text{ MeV}$$

$$\text{deuterium-deuterium:} \quad {}^2H + {}^2H \rightarrow {}^3H + p + 4.03 \text{ MeV}$$

$$\text{deuterium-tritium:} \quad {}^2H + {}^3H \rightarrow {}^4He + n + 17.59 \text{ MeV}$$

Here the deuteron is symbolized ^2H. The mysterious neutrino emitted during beta decay is again shown as "?". Neither it nor beta carries useful energy, so a small part of the energy shown does not appear as KE of the other particles. The energy must be captured as heat in order to be useful. The first three reactions occur in series in the sun, rather slowly and limited by the proton-proton reaction. The reaction needs a temperature of about 10^7 K to occur at the density of the sun's core. The last three reactions are proposed for controlled fusion reactions on Earth. It can be seen that they use exotic materials—the somewhat rare deuterium and the very rare and radioactive tritium. However, much less density is needed to run these if well over 10^7 K can be achieved.

Another possible stellar fusion reaction uses carbon nuclei as a catalyst. Hans Bethe worked out the process in 1938, receiving the Nobel Prize for it in 1967. The reaction is called the "carbon cycle" and involves 6 steps. Protons react successively with carbon and nitrogen isotopes until a nucleus is built that splits into carbon and helium. Five gamma rays, 2 electrons, and 2 neutrinos are also released. The thermal energy released is about 26 MeV during one cycle, similar to the proton-proton cycle. Since about the right amount of carbon is in the solar spectrum, it was thought that the carbon cycle produced the sun's energy, but debate occurred for many years until the simpler proton-proton cycle was decided upon. Stars hotter than the sun which have used up their hydrogen combine three helium nuclei in one step to form carbon.

Attempts are being made to achieve controlled fusion in reactors that use magnetic fields to confine a hot gas, usually deuterium. The most successful experiment uses a large device called a "tokamak," after its Russian origin. In 1983 the Princeton tokamak achieved deuterium-tritium fusion on an experimental basis. Uncontrolled fusion has been achieved with the "hydrogen bomb," which uses a fission bomb to heat hydrogen or deuterium to the temperature at which it will fuse. Because there is no limit to the amount of fusion fuel, the energy yield can be much greater, and we have the enormously destructive nuclear weapons that threaten human survival. The enormous amount of energy locked in the nucleus can be partly and rapidly released, resulting in energy that is about a million times greater than the equivalent weight of TNT, a typical chemical fuel. Many of the physicists mentioned as working on nuclear physics or quantum mechanics participated in the development of these weapons during the panic of the war against nazi Germany and the following distrust of the USSR. Many of them are now working toward the elimination of nuclear weapons.

11.6 NUCLEAR MODELS

Various models of the nucleus as a whole have been proposed. The *liquid drop* or *collective* model, originally due to Neils Bohr, is one of the most popular, as it simplifies the nucleus while explaining fission. In this model each nucleon interacts with its nearest neighbors only, as do water molecules in water. There is a kind of surface tension, since the nucleons on the surface have less neighbors. The protons are always pushing outwards, unlike water molecules. The whole has no solid structure but wobbles like gelatin. It can vibrate in various modes, as confirmed by experiment. If an extra neutron enters a large nucleus (Fig. 11-8B), it adds extra energy but does not find a stable state. The nucleus as a whole starts vibrating wildly and suddenly finds existence in two parts more feasible. It flies apart at high speed. This fission was a foregone possibility because the two parts have a lower total energy (higher binding energy per nucleon), and the neutron simply catalyzes the change.

The *shell model* is, of course, analogous to the atomic model and uses a quantum energy level diagram. It supposes that each level can hold a certain number of nucleons, after which it is complete. Since the nucleus remains tightly packed, the nucleon spacing should remain close and more uniform. Their wave functions represent orbital angular momentum states and the energy level spacing should be determined partly by the spin-dependence of the nuclear force. The shell model was developed by Maria Mayer and J. H. D. Jensen in the late 1940's, for which they received the Nobel Prize in 1963, along with Eugene Wigner. The atomic shell model was successful because the interactions of the electrons was much weaker than the attraction of all of them by the nucleus. The nuclear shell model cannot be as successful because the nucleons interact strongly with each other and there is no central organizing force.

When the number of protons or neutrons is equal to 2, 8, 20, 50, 82, or 126 (a different set of magic numbers than the atom), the corresponding shells seem to be complete and the nucleus is especially stable. The lead isotope $^{208}_{82}$Pb is especially stable. The existence of shells has been deduced from the natural data on stability of nuclei. Measurements of nuclear shape (through the electric quadropole moment) have shown that closed-shell nuclei are the most spherical. Each shell contains subshells of various angular momenta, patterned after those for atoms. To deduce possible energy levels, simple potentials such as a "square well" or harmonic potential have been tried, not the Yukawa potential. Fairly good predictions of existing nuclei, their spins and

parity, and their magnetic moments have been obtained from the model.

Aage Bohr, son of Neils Bohr, combined the collective and shell models into a **unified model** in 1952. For this he received the Nobel Prize in 1975, along with Ben Mottelson and James Rainwater who also worked on nuclear structure.

11.7 NUCLEAR FRONTIERS

Work continues in many directions to understand the nucleus and its constituents. Attempts are being made to construct stable superheavy nuclei at about $Z = 114$. Collisions between very high speed nuclei heat and compress the nuclei so much that the "liquid drop" is converted to a nuclear "gas." The energy is about 2 GeV per nucleon, and the equivalent temperature is about 10^{12} K. It is being attempted to find the equation of state of nuclear matter—that is, the relation of its temperature, pressure, and density. The results of high temperature collisions indicate that the liquid drop model is still the best predictor of overall nuclear behavior.

Mossbauer spectroscopy involves studying the gamma ray emission from nuclei which are part of atoms in a crystal. The nucleus is held in place and can no longer recoil when it emits the photon. The spread in wavelength is much reduced. Rudolf Mossbauer received the Nobel Prize in 1961 for discovering this method. The Mossbauer effect has been a useful tool in itself for finding how ordinary atomic effects (the electron changes in molecules) affect the nucleus. The effect is very small, but the atomic electrons do affect nuclear energy levels. The effect is also useful in other sorts of research.

There are about 7000 different nuclei which theoretically could exist but have not been made or detected in nature. Most would not live very long, but much can be learned if any can be made and studied, however briefly. The nuclear periodic table has recently been extended in regard to light nuclei by using particles such as protons, alphas, and ^{3}He nuclei against target nuclei such as helium, carbon, and neon. For example, boron isotopes ^{14}B, ^{15}B, and ^{17}B have been made. ^{9}B lives too briefly to detect but ^{8}B has been found.

As one might expect, nuclei with an excess of neutrons are much more stable than those with an excess of protons. The less stable ones decay by a previously unknown process, prompt emission of a proton or neutron. This occurs much much faster (about 10^{-16} second) than beta decay. Protons must escape a nucleus by tunneling through the energy barrier, and the process has been found to follow the same laws as alpha decay.

Chapter 12

Leptons and Quantum Electrodynamics

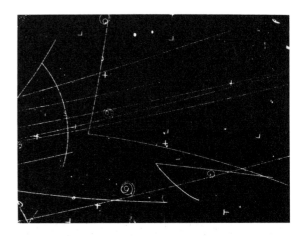

E LEMENTARY PARTICLES AND MODERN THE-
ories of their interactions are the principal
topics of this chapter and the rest of the book. Here
we find a place for a familiar particle, the electron, and meet some new particles, all in the *lepton* family (from the Greek for "lightweight").
This group does not take part in the nuclear
(strong) interaction but instead interacts by means
of the three other fundamental forces. Modern
theories have been most successful for these particles and their interactions because they are
relatively simple particles. The "Quantum Electrodynamics" in the chapter title refers to the
theory of the behavior of the electron. It is a quantum, it is electrically charged, and it moves according to certain laws. One might not think the
"simple" electron could lead so many physicists
on such a long chase for an accurate theory, but
nature has many surprises in store. Some of the
topics to be covered review and integrate previous
subject matter. Quantum electrodynamics is another step farther from classical mechanics, since
it embodies relativity as well as quantum mechanics.

12.1   THE ELECTRON AND POSITRON

The electron was the first elementary particle to be definitively discovered (by Thomson,
1897). Its mass and charge are well-known,
$9.109558(10)^{-31}$ kg and $1.602191(10)^{-19}$ coulomb, respectively. It has the smallest mass of any
particle whose mass has been established to be
non-zero. Its charge is a standard unit of charge
in the quantum world, and we shall usually refer
to charges in terms of the quantity of the electron
charge or the "unit charge," understood to be
positive although the electron's is negative. Its size
(as a diameter) is much more debatable and
depends very much on what one is doing with the
electron. When electrons are scattered from each
other or from other particles, they behave as point
particles with no size. Physicists deduce this from
certain simplifications that appear in the results.

There is no sense in worrying about parts of the charge in the electron repelling other parts if it is all at one point. The mathematics is easiest for sizeless particles also. At present there is no evidence that the electron contains any structure. It is truly elementary in the sense that it has never been broken down. The electron is a sturdy particle, retaining its identity in 20 GeV collisions (ultra-relativistic). Nevertheless, when the electron is bound in an atom, or released in a nucleus, it seems to be bigger than the nucleus. Its influence when an atom radiates light extends far outside the atom.

The electron has spin of amount $\frac{1}{2}$ h/2π, making it a fermion with certain restrictions on it when mixed with other electrons. Because of this inherent angular momentum—which cannot be visualized as a spinning ball—it acts as a tiny magnet, having a magnetic moment. Measurement of this property has been of great interest in advancing the theory of the electron. It can be both measured and predicted to about 11 decimal places at present, and the numbers agree! This is the greatest triumph of physics or of any other theoretical science, in terms of accuracy.

One can visualize the magnetic moment as a tiny current flowing in a loop around a tiny area. The measured value is $9.2837(10)^{-24}$ ampere square meter. The relation of magnetic moment μ to spin S for rotating charge is given classically by:

$$\mu = \frac{-geS}{4\pi mc}$$

where the g-factor is a constant that tells how much moment results from the given spin, mass, and charge. For an electron the minus sign shows that μ points in the direction opposite to the spin because the charge is negative. The distribution of charge is expressed by g. It has the value 2 for a quantized electron as compared to a classical rotating sphere of charge. Measurement gives $g = 2.0023193044$. Theory (QED, given shortly) predicts very nearly this value, falling short by a few parts in 10 billion. In other words, modern theory does not give exactly 2 but a small precise correction to 2. As a fermion, the electron seems to rotate twice when the world turns once, another explanation of its g. Polykarp Kusch received the Nobel Prize in 1955 for measurement of the electron magnetic dipole moment.

When the electromagnetic field around the electron is considered, difficulties soon arise. Particularly when the electron moves and radiates, there must be a reaction on the electron by the departing waves or photons. This affects how the radiation is emitted. Another way to say it is that the electron moves around in its own moving field. The effect of this self-field or self-energy of the electron cannot be ignored, yet much accurate physics has been described without such a consideration. The self electric field at the "surface" of an electron is infinite if the size of the electron is zero. This is not a useful result.

A simple classical model of the electron can be made in which the potential energy involved in holding the electron's charge together as a sphere of a certain small radius is equated to the rest-mass-energy (m_0c^2). The result for its radius is about 10^{-15} meter, about the same as a proton. Not bad, so far. We might have expected it to be smaller because of its smaller mass, but instead we should think of the electron having normal charge smeared over this amount of volume but with very little mass to accompany it. A difficulty arises in carrying this thinking over to the quantum world where h plays a major role. The quantity hc has the same units as $e^2/2\epsilon_0$ for electric energy. The ratio of these is a famous number called the fine structure constant, since it enters into the theory of spectral line splitting. It has the value of about 137, and we will refer to it as such. Thus the quantized aspects of the electron are expected to be about 137 times more important than its classical properties, and we must seek an entirely different quantum theory for the electron, despite earlier success. Dirac's relativistic quantum mechanics included a new theory of the electron taking this into account, but certain experiments showed there were small errors.

Dirac predicted the first new particle, the positron, as well as the existence of *antimatter*. The positron has all the properties of the electron except that it has positive charge and a reverse magnetic moment. It could be viewed as a mirror image of the electron except that we must think in terms of a charge-reflecting mirror (Fig. 12-1), not a spatial one. The operation of reversing the charge is called *charge conjugation*. In general, an antiparticle, which makes up antimatter, is identical to a normal particle except it has the opposite charge. (There is a bit more to being "anti-" than this, as will be seen later.) The positron was first found in cosmic rays by Carl Anderson and P. Blackett in 1932. The positrons appeared as the result of collisions by high energy particles coming from cosmic rays striking the upper atmosphere. Anderson received the Nobel Prize in 1936 for the positron. Although some beta decays result in positrons, they were not observed until 1933, when Frederic Joliot and Irene Curie found positrons coming from nuclei induced to be radioactive by bombardment with alphas. For this work they received the Nobel Prize in chemistry in 1935.

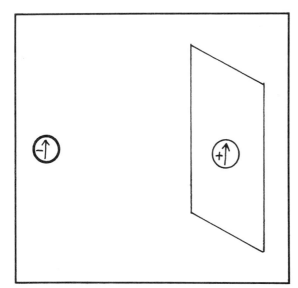

Fig. 12-1. The charge conjugation "mirror" in which a particle is reflected as an antiparticle, reversing charge and magnetic moment. (Spin is not reversed.)

Positrons had broader effects on the development of particle theory. They were originally envisioned as electrons with negative energy, then as "holes" in space left when electrons are formed. Thus empty space, also called the *vacuum state*, could have a strange new property. The uncertainty principle applied to energy allows energy to be borrowed from the vacuum (nowhere?) in violation of conservation as long as the energy debt occurs for less than the time we would need to detect it. To borrow an electron (0.5 MeV) would require that it be paid back in less than 10^{-21} second. This sounds like a ridiculously short loan, but it is a hundred times longer than it takes light to pass by an electron.

12.2 PAIRS

A photon (a gamma ray) with more than about 1 MeV of energy happens to possess enough energy to be equivalent to two electron rest masses. Nature has the choice of having the two electrons instead, since energy would allow it. In this case nature exercises the option, and the gamma ray spontaneously changes into an electron-positron pair. This is called *pair production*, an electrodynamic reaction shown in Fig. 12-2A. The process does require the presence of another electromagnetic field, usually that of a nearby particle or atom, and the atom receives a small kick. But most of the energy and momentum of the photon appears as kinetic energy and momentum of the electron and positron. Obviously there is no problem with charge conservation since the net charge created is zero. One must consider spin as well. The photon carries one unit, and the two particles together carry away one unit. Since everything balances, nature finds this reaction possible. The produced pair appears at first to move together in the lab frame, as shown in the sketch from a typical photograph of the event given in Fig. 12-2B. However, each member possesses a large momentum in their CM frame (Fig. 12-2A), and they gradually move apart in opposite directions. For purposes of identification and measurement, a strong magnetic field is used, so that the two spiral away in opposite directions,

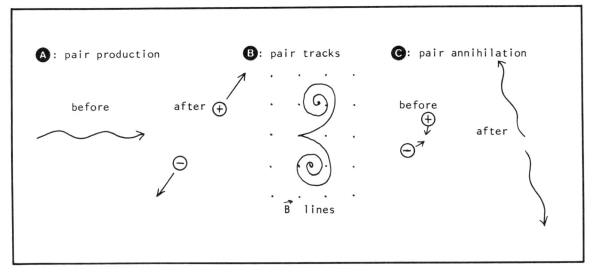

Fig. 12-2. Electron-positron pair production and annihilation.

losing energy to the medium in which the event was recorded (Fig. 12-2B).

Any reaction that goes one way may go the other way. What can happen if an electron and positron encounter each other, either by colliding or by collapse of a positronium atom which they had temporarily formed? As with any particle and antiparticle, they *annihilate* each other, leaving two photons. Because momentum must be conserved, the two photons may depart in opposite directions, depending on the initial momentum (Fig. 12-2C). The total energy of the two gamma rays must equal two electron rest mass-energies plus whatever original kinetic energies there were. Like pair production this reaction is essentially an electromagnetic interaction.

It has recently been realized that the electric field near a heavy nucleus is almost large enough to materialize electron-positron pairs out of the vacuum. This should begin at $Z = 173$. Such a nucleus is quite difficult to make, but two existing heavy ones need nearly approach one another closely in a high energy. The electrons would fall into an extremely deep orbit around the nuclei and the positrons would spew outward. The vacuum will not be a source of free energy, however, as the electrons will shield the nucleus and reduce its effective charge.

12.3 WEAK INTERACTIONS AND THE NEUTRINO

Unavoidably in considering the behavior of nuclei we have encountered the weak interaction at work. Many nuclei "choose" to eject their excess energy and rearrange their neutrons and protons by emitting a beta particle (usually an electron) and becoming a different nucleus. The beta particle often has a variety of energies, yet supposedly it came from a well-defined energy transition. Also, the electron carries away a half unit of spin when the nucleon is known to have no change in spin. This was puzzled over for some time, and Pauli first suggested an unknown neutral particle must be involved. Fermi worked out a theory for beta decay and named in 1933 the invisible, undetectable particle that was stealing away the missing energy, momentum, and spin. The vogue of the day was to propose new particles, so the *neutrino* ("little neutral one") was readily accepted. It also helped explain the decay of the neutron, where energy and spin were unaccounted for, and K-capture, where the nucleus captures an electron from its atomic K shell. In the latter case, the nucleus was observed to change, yet nothing (!) was observed to come out, not even a gamma ray.

The neutrino (symbol ν, Greek nu) has no charge, and negligible mass, but it can carry substantial energy. It must be moving near lightspeed in every case. It seems to have no cross section since nothing was able to stop it or detect it. Finally in 1956, using the immense flux of neutrinos from a nuclear reactor, neutrinos were detected. Occasionally one does strike a nucleus and causes a detectable change in nucleon count, but about 10^{40} collisions are needed before there is a good chance of it. The neutrino can pass easily through the Earth or sun, so that nearly all neutrinos produced during fusion in stars depart, carrying away substantial unseen energy. Three light years of liquid hydrogen (a commonly used substance in particle detectors) would be needed to have half a chance of catching one. The neutrino also seems to have negligible magnetic moment, if any. (It would be quite disturbing to find charge at play in so minute and ephemeral a particle.) With no charge it does not interact electromagnetically, but it is attracted by gravity as all particles are. It was found not to be affected by the nuclear force.

Evidence mounted that a new force or interaction was working in nature. It appeared in the selective way that electrons interact with nucleons. The new interaction is slow; the neutron needs about 16 minutes to decay to a proton, electron, and neutrino. The range is shorter than the nuclear force, eventually established at about 10^{-17} meter. This helps explain why the interaction is weak; rarely do particles come so close together. In fact, as we shall see, a great deal of energy must be involved. Other leptons were discovered which react or decay weakly. All reactions seemed to involve neutrinos. (Later a new kind of particle was found that interacted weakly without neutrinos.) Fermi proposed the weak interaction in 1933. The slow time scale of the interaction is another clue to the weakness of the interaction; it was estimated to be 10^5 times weaker than the nuclear "strong" force. Eventually the question arose whether the force had a carrier. The neutrino is a poor candidate because it is so light and is a fermion. The boson that

would carry the weak interaction had to be rather heavy, certainly more than a proton. The story of how long it took to find the carrier is reserved for a later chapter.

Like every particle, the neutrino has to have an antiparticle. Being neutral, can the two be distinguished? Unless the neutrino turns out to have measurable magnetic moment, neutrino and antineutrino cannot be distinguished. For orderliness the antineutrino is defined to be that which is emitted by a decaying neutron. It is a right-handed particle, with spin parallel to the direction of travel. The "normal" neutrino, also everywhere in great abundance, can interact with a neutron (usually in a nucleus) to form a proton and electron through stimulated neutron decay thus:

$$\nu + n \rightarrow p + e^-$$

Lest the impression be left that the weak interaction can have little to do with ordinary events, some consequences of beta decay should be mentioned. The weak interaction provides a route for nuclei to change state that otherwise would not happen. Beta decays are prominent in the three natural radioactive series. While beta decay itself does not contribute much energy, this weak interaction allows the formation of nuclear states which can then decay by alpha, emitting substantial energy. The warmth of the interior of the Earth's crust and mantle is due primarily to the alpha decay of uranium series elements. We would not have volcanoes and continental drift if it were not for the heat released by radioactivity. Beta decay is also the way $^{14}_{6}$ decays (half-life 5568 years), providing us with a way of dating organic remains and contributing enormously to our understanding of human history. (How $^{14}_{6}$ is formed is another story.) Many other examples of the effects of the weak interaction through beta decay could be cited; its role in stellar fusion is given later.

12.4 MUONS AND TAUONS

A heavier member of the lepton family,

sometimes called a "heavy electron" because it behaves the same as an electron except for its mass, was found in 1937 by Anderson and S. Neddermeyer in the attempt to find Yukawa's carrier for the nuclear force. M. Conversi, E. Pancini, and O. Piccioni worked with the particles resulting from cosmic ray collisions with the Earth's atmosphere. They discovered that this particular kind did not interact with protons and neutrons and therefore would not make a good nuclear carrier. It was called a mu meson (Greek letter μ) but later shortened to muon because it did not act like a meson.

The muon has 206.84 times the electron mass, or 105.7 MeV, and has spin 1/2 and a magnetic moment. It comes in both positive and negative forms, with one unit of charge. It is unstable and decays after an average of 2 microseconds to an electron and two neutrinos. The decay proceeds as follows:

$$\mu^+ \rightarrow e^+ + \nu_e + \bar{\nu}_\mu$$
$$\mu^- \rightarrow e^- + \bar{\nu}_e + \nu_\mu$$

The antineutrino is shown with a bar over its symbol $\bar{\nu}$. The neutrinos here are given subscripts because it has been discovered that there are several kinds of neutrinos. One kind is associated with muons, and one kind with electrons. We see one of each in muon decay.

One of the muons is an antiparticle. The one that decays to the normal electron (e^-) should be the particle, so that μ^+ is the antimuon. Muons do not interact strongly with nuclei, and they are too heavy to be scattered much by nuclei through electric forces. Therefore they lose energy mainly by ionizing atoms, a slow process. This is why they are the principal surviving component of cosmic rays that reach the ground. They can travel through hundreds of meters of concrete or steel and therefore are second in penetrating ability to the virtually unstoppable neutrinos.

Despite the elusiveness of the neutrinos, it was possible as early as 1962 to show there were two kinds by observing the results of letting neutrinos of both kinds strike neutrons and protons. The

electron type ν_e {nu sub e} invariably leads to electron production, as in:

$$\nu_e + n \rightarrow p + e^-$$

and the muon type ν_μ leads to muon production, as in:

$$\nu_\mu + n \rightarrow p + \mu^-$$

A negative muon that is captured by an atom drops rapidly through all possible energy levels and enters the nucleus. Then it may interact weakly with a proton to form a neutron and neutrino:

$$p + \mu^- \rightarrow n + \nu_\mu$$

The Yukawa nuclear force carrier, a charged pion (symbol π^+ or π^-, greek pi) decays to a muon after an average of 0.02 microseconds, yielding a neutrino also. We will see other aspects of the pion later. It is not a lepton.

The story of leptons is brought up-to-date by the finding (1976) of a third kind, a heavy-heavy electron called the *tauon* (symbol τ, Greek letter tau). It was found in the collisions of electrons and positrons by means of a new method in accelerators. Beams of electrons and positrons rotate in opposite directions in the same *storage ring* so that when they are made to collide, all the energy goes into the collision. The tauon comes with both positive and negative charges, has a mass of about 1850 MeV, about 3600 times that of the electron, and lives less than 10^{-13} second. It has its own neutrino, symbol ν_τ.

It is possible that still heavier leptons will be found as more powerful accelerators for electrons and positrons become available. It is difficult to achieve as high energies with leptons as with protons, but proton experiments would produce non-leptons much more copiously than leptons. A summary of the known leptons is provided in Table 12-1. In reactions that involve any leptons, there is a complicated conservation law at work. What is conserved is *lepton number*. A quantum number is assigned to the count of each kind of

Table 12-1. The Leptons.

Name	Symbol	Rest Mass (MeV)	Charge	Spin	Life (seconds)
electron	e^-	0.511003	-1	1/2	stable
neutrino	ν_e	0?	0	1/2	stable?
muon	μ^-	105.6595	-1	1/2	$2.197(10)^{-6}$
neutrino	ν_μ	0?	0	1/2	stable?
tauon	$\tau^{\pm}$	1850?	-1	1/2	$3(10)^{-13}$
neutrino	ν_τ	0?	0	1/2	stable?

Notes: Each particle has an antiparticle (shown with a bar over the symbol) with opposite charge. All leptons are fermions.

lepton; for example $L_e = 1$ for the electron and its neutrino, and $L_e = -1$ for the positron and associated antineutrino. A separate accounting is done for the muon types, and so forth. No violations have been found, so the universal rule is that the total L_e and total L_μ cannot change in any interaction. For muon decay, for example, two neutrinos must be formed, one of each kind so that the lepton count for muons and separately for electrons can be conserved.

12.5 PARITY

In 1927 Eugene Wigner proposed a new law called the *conservation of parity*. What this means is that particle behavior when viewed in a mirror (a real if idealized one this time) should follow the same laws as without the mirror. Other terms are "handedness" or *chirality*. By comparing its spin with its direction of travel, a particle can be labeled as right handed or left handed (or " + " and " − ", respectively). A right handed particle has its spin pointing in the direction of travel. Spin appears reversed in the parity mirror (Fig. 12-3). Regardless of parity, the laws for particle behavior should be unaffected and there should be no physical distinction. Parity was found to be conserved experimentally and theoretically with the older classical forces, gravity and electromagnetism, but the new weak force has proven insecure in this regard. When it is conserved, the invariance principle applies to parity. The invariance is simply whether one takes a right handed or left handed (mirror image) view of

nature. Natural laws should be invariant under this transformation.

Based on suspicions about nuclear physics data Tsung-Dao Lee and Chen Yang suggested in 1956 ways of testing the conservation of parity. Soon C. S. Wu *et al.* developed a way to line up the spins of cobalt nuclei. The $^{60}_{27}$Co nucleus beta decays, and it was found that more electrons came out one side than the other. Almost simultaneously Richard Garwin, Leon Lederman, and Marcel Weinrich found that the muons emitted by decaying pions had their spins preferentially pointing along the line of travel. For the first time a natural way had been found to distinguish left and right in nature, and only the weak interaction

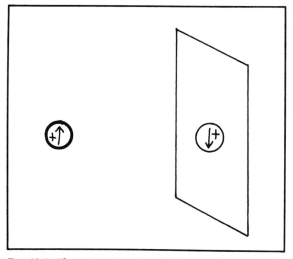

Fig. 12-3. The parity mirror in which a particle is seen with the opposite handedness (chirality).

could make the distinction. The nuclear and other forces have never been found to violate parity conservation. The consequences will be shown later to be larger than simply a violation of a conservation law. Lee and Yang received the Nobel Prize in 1957 for disproving the conservation of parity.

That the conservation of parity is a useful law can be considered in its application to a nuclear reaction, where it has not yet been found violated and is not likely to be. Because a spin or any unit of angular momentum is reversed by the parity mirror, a spin or $l = 1$ state has negative (" − ") parity. The parity of any angular momentum state l can be calculated by $(-)^l$, which gives " + " for spherical and even-numbered l states and " − " for odd l states. Consider a nucleus in an $l = 1$ state, struck by a slow neutron, necessarily in an $l = 0$ state. The nucleus of odd parity is combined with a particle of even parity. Necessarily odd and even give odd, so the resulting nucleus must have odd parity. Thus does the law determine what results are allowed and what are forbidden.

12.6 QUANTUM ELECTRODYNAMIC AND FEYNMAN DIAGRAMS

As discussed much earlier, the Dirac quantum mechanics predicts that hydrogen will have both electrons in exactly the same ground state, making it doubly degenerate. In 1947 Willis Lamb and R. C. Retherford [not Rutherford] measured a small shift (about a millionth of an eV) so that there are two closely spaced energy levels. This could have been expected as nature always seems to find a way of breaking degenerate energy levels. Dirac must have overlooked something. For this discovery involving a discrepancy of less than one part in a million, Lamb received the Nobel Prize in 1955 as well as having the result named the *Lamb shift*.

Some of the principle physicists who began to uncover and solve this mystery were Sin-itero Tomonaga (whose work in 1943 went unknown for a while) and, independently, Julian Schwinger. The basic realization was that any electron, such as the one in hydrogen, modifies the vacuum around it. Borrowing energy briefly, positrons, electrons, photons and perhaps other particles start popping in and out of existence around the real electron (Fig. 12-4). The real electron trades photons back and forth with them; in field language, their fields interact. The temporary particles are called *virtual* particles as we cannot measure their existence directly. They exist too briefly. All these virtual particles buzzing around the real electron add to its effective mass. In fact, there seems to be no limit so the mass becomes infinite. Of course electrons have been handled a lot and are known to have a normal, finite mass. Decades of laboratory measurements had already been made on electrons supposedly surrounded by massive "junk."

The process of mathematical *renormalization* was invented so that the mass of the electron that went into theories was a new corrected or renormalized value, based on what is actually measured. The correction happens to be infinite, which certainly bothered physicists immensely. The mass of the electron without the cloud of virtual particles is set to be that mass which, when infinity is added, gives us the observed mass. All this is done with advanced mathematics, not with ordinary numbers, so there are not the difficulties that the non-physicist reader might expect. In the same way the virtual charges surrounding the electron are corrected for so that the observed electron charge is correct, even with the cloud around it. The result is that predictions for one-electron atoms and quantities like the magnetic moment of the electron can be made to an astonishing degree of accuracy, up to 11 decimal places and be in agreement with experiment. The theory is called *quantum electrodynamics*, or QED for short. It is still *the* most triumphant theory of all physics, covering all behavior of electrons as quantum particles.

Some order and new insight was brought to QED by Richard Feynman. Tomonaga, Schwinger, and Feynman shared the Nobel Prize in 1965 for their accomplishments in QED. Feynman realized that a positron resembles an electron going backwards in time. This is a general sym-

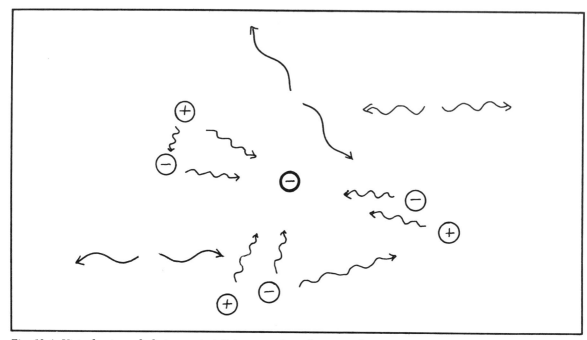

Fig. 12-4. Virtual pairs and photons materializing around an electron and interacting with it and each other.

metry involving antiparticles; they move like normal particles with time reversed. Feynman drew simple graphs of particles on time-space axes (Fig. 12-5), letting an X axis stand for all three spatial dimensions so that the time behavior could be examined in a simple way. In Fig. 12-5A, we see at first an electron progressing in the X direction as time goes by. Suddenly the particle is moving backwards in time, having become a positron, but still progressing along X (probably at a different speed). The arrow is always shown for the electron's direction. Later it is again moving forward in time and has reverted to being an electron.

This may sound as crazy as the virtual particle cloud, but let us recall the extreme accuracy that results from QED and also note that this picture of backwards particles is another way of expressing QED. There is also the disturbing possibility that the world line of the particle shown is just a tiny part of its general behavior. Being able to carry on a lot of motion while little time passes, it is possible that this single particle manages to get everywhere in the universe, being the path for all electrons and all positrons

everywhere, nearly all at once. This wild idea would certainly explain the electron's indistinguishability and a few other features.

We should know from the laws of physics that electrons do not suddenly, by themselves, turn into positrons. For example, two gamma rays (shown as wiggly lines) must be emitted as the electron and positron collide in the CM frame in Fig. 12-5B. A single particle or photon departing would have momentum in its CM frame, in conflict with the initial zero momentum. A more correct diagram is Fig. 12-5C, where the two photons are emitted separately. There is another way this diagram can be drawn, with the positron emitting the first photon.

If we examine the simplest interaction, called a *vertex* and shown in Fig. 12-5D, two particles come together and a third results. This interaction violates conservation of momentum or energy or both. Conservation of momentum states that the components of the velocities times the masses of the two incoming particles must equal the components of the outgoing particle. Conservation of kinetic energy would have the sum of initial masses and speeds squared equal to the final mass times speed squared. In a general case these two requirements are in conflict. Nature requires some fourth particle to depart and enable both conditions to be simultaneously satisfied.

At a vertex, however, there is the possibility of borrowing energy for a short time and borrowing momentum over a short distance, as long as the uncertainty principle is satisfied. This indeed happens, and anything that can be borrowed is borrowed, on a statistical basis. For example, the electron innocently moving along in Fig. 12-5E suddenly emits a virtual photon. The energy of the photon is borrowed for a correspondingly short time. The curved path shown for the photon (or any other particle) is simply a convenience in drawing the diagram, and the changes in direction of travel for particles is only shown in an approximate way. The electron changes direction (accelerates) in recoil and soon meets the virtual photon again and absorbs it at a second vertex. In this way the electron has interacted with itself, corresponding to its self-field energy. This process can continue over and over. What is worse, a virtual photon might decide to become a pair temporarily, which then revert to being the photon. Or other vertices might occur, all related to the single electron. All this activity following the electron around constitutes its field, or the *radiative corrections* to the "bare" electron.

As another example, let us examine the interaction of two electrons, as shown in Fig. 12-6A. This is called a *first order* interaction, with one photon exchanged. There are many possible ways (nine in all) that two photons can be exchanged, giving *second order* corrections to the process. Some are shown in Figs. 12-6B through 12-6F. The probability of two photons is lower, but there are many more ways for this to occur than for one photon. Each possibility does occur and contributes to the energy exchanged between the two electrons. Fig. 12-6C shows a pair temporarily produced between the two photons. The positron must be traveling backwards in time. The symmetry of a Feynman diagram not only reflects the possible symmetry of the physical interaction but also aids in calculating the effect of all possible photon exchanges. There is no end to the process. One can add a third photon in many ways to each of the second order diagrams. The more processes examined, with progressively lower probability, the more accurate the calculation of the interaction. The diagrams help keep track of just how many calculations must be made, too. The electron cannot be tracked during the interaction. Electron 1 may not come out as electron 3 in Fig. 12-6A but rather as electron 4. A bonus from Feynman's QED is that disturbing infinities no longer appear in calculations.

Feynman diagrams with their radiative cor-

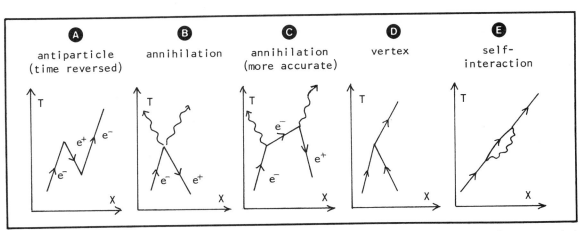

Fig. 12-5. Basic Feynman diagrams illustrating electron, positron, and photon interactions. (Antiparticles move against the flow of time.)

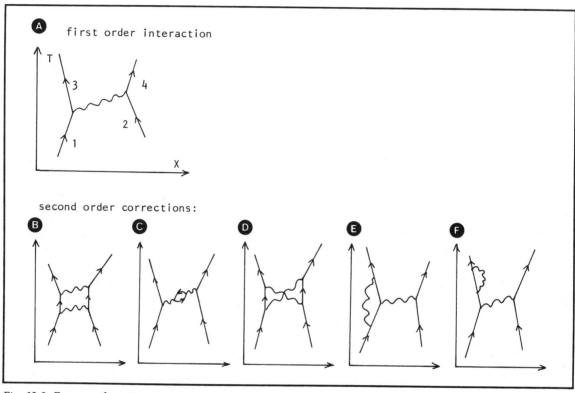

Fig. 12-6. Feynman diagrams for an electron-electron interaction.

rections to any order can be drawn for any particles that interact. Compton scattering, brehmsstrahlung, ionization of an atom, and more can be diagramed, then calculated to great accuracy. The creation of pairs as an electron interacts with an atom serves partially to shield the electron from the full effect of the electric field of the atom. The effect is called *vacuum polarization*. As we shall see later, other particles, in-

cluding the carriers of other fields, can take the role of the photon in Feynman diagrams and calculations. Positronium recently has aided the testing of QED. The laser was used to measure the energy difference between positronium energy levels to about one part in 10^8. Comparison can be made with high order calculations in QED for this simplest of atoms.

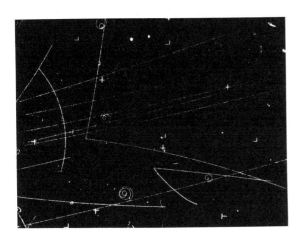

Chapter 13

Modern Accelerators, Hadrons, and the Strong Force

W E HAVE DISCUSSED THE SMALLER AND SIM-
pler of the families of particles, the leptons.
The mesons, proton, neutron, and many other
heavier particles belong to a family called
hadrons. This name, from the Greek for "strong,"
is appropriate because these are the only particles
which interact by means of the strong nuclear
force. They also take part in all the other forces.
We have met the carrier of the strong force when
discussing Yukawa's prediction of its size and the
search for it. Before discussing the hadrons and
their interactions in the world of high energy
physics, we must catch up with the history of particle accelerators, describing a few of the large recent ones developed initially for the exploration
of strong interactions. They have since been found
useful in studying weak interactions. Advances in
particle detection will also be described. Still
larger proposed accelerators and detectors will be
discussed later.

13.1 MODERN ACCELERATORS

Accelerators make clever use of electric and

magnetic fields to control and accelerate charged
particles. The physics of this was covered earlier.
Accelerator technology matured with the invention of the *alternating gradient synchrotron*,
allowing the magnetic field to be spread out along
a huge ring at reasonable cost. Magnetic fields,
which cost much power to create until recently,
are only provided where needed to bend the paths
of accelerated particles, as was shown in Fig. 7-11.
Special focusing magnets keep the accelerated
beam of particles small, so that the magnets can
be as small as possible.

An early large synchrotron was completed in
1954 at the Lawrence Radiation Laboratory at
Berkeley, California (see Plate 3). It was called the
Bevatron because it was the first proton accelerator to enter what was then known as the
"bev" (billion electron volt) range. Its energy is
now stated as about 6 GeV, and only about 2.5
GeV are available in the CM frame if proton sized
particles are involved. This is just high enough to
provide the mass-energy to create one or more proton sized particles in a collision. The beam in the

Bevatron follows a path about 16 meters in radius, but only where magnetic field is provided. This machine uses four bending magnets, giving a rounded square shape for the beam path. The maximum magnetic field strength is about 1.6 webers per square meter, nearly the limit that can be produced in iron cores.

During their early revolutions, the protons undergo substantial increase in speed, and the magnetic field is increased appropriately to bend them through the proper radius. This process is called *ramping*, and must be done for each bunch of particles. After the accelerated bunch is removed, the magnetic field is decreased for the next bunch. As with all accelerators, the particles are pre-accelerated by smaller accelerators; in this case a Cockcroft-Walton accelerator is followed by a linear one to bring the protons up to about 10 MeV. About 0.0015 MeV is added to them per revolution in the Bevatron, using an electric field in a straight section of the path. In other words, the protons are "dropped" through a potential of about 1500 volts during each acceleration. Typical accelerator beam currents are about 0.01 ampere, similar to the current used by a pilot light but lasting for only a brief pulse. The final effective potential is equivalent to billions of volts.

To design larger accelerators one can scale up the radius or the magnetic field strength. At Berkeley about 0.4 GeV was attained for each meter of radius. It is difficult to make much stronger magnets, so the radius is increased. This also provides more room to obtain more acceleration per revolution. The speeds are already ultra-relativistic. A 6 GeV proton has 6 times its rest mass-energy and therefore has $\gamma = 6$, about 98 % of lightspeed. It will not go much faster, but there is no theoretical limit to the energy it can be given.

A step up to 33 GeV in energy was achieved in the early 1960s at the Brookhaven National Laboratory on Long Island, with its alternating gradient proton synchrotron. The use of shaped magnetic fields to focus the beam allows the higher energies. The radius of the beam path is about 120 meters, and it uses 240 bending magnets. About 0.3 GeV is attained per meter of

radius. Only about 6 GeV are available in the CM frame, for proton sized particles. An important new hadron (the omega minus) was found with this machine, tying together the first major hadron theory (covered near the end of this chapter).

The most ambitious operating accelerator in the U. S. is at what is now called Fermilab (Fermi National Accelerator Laboratory), near Batavia, Illinois (see Color Plate 9). It was initially designed for 400 GeV, using conventional iron core magnets. It has been upgraded to 1000 GeV (1 TeV), using superconducting magnets, and the main accelerator is now called the Tevatron. The radius was chosen to be 1000 meters, so the original scaling for iron magnets was about as expected, 0.4 GeV per meter of radius. There are 774 bending magnets about 6 meters long and 216 focusing magnets in the main ring, which is underground.

Protons must go through three stages of acceleration before reaching the main ring. These stages parallel the history of accelerators and are shown in Fig. 13-1. Hydrogen atoms with an electron added, not removed, are first boosted to 0.75 MeV in a Cockcroft-Walton generator. Then they are linearly accelerated to 200 MeV in a space of 145 meters. Next they are boosted to 8 GeV in a Booster synchrotron, the equivalent of the Bevatron, where their electrons are stripped off. The bunches of protons from the Booster are collected into the main upper ring (see Color Plate 10), where conventional magnets are used to bring them to 150 GeV. A bunch travels around the ring (over 6 kilometers circumference) 50,000 times per second and receives an additional 3 MeV during each revolution as it passes through an electric acceleration section. Each bunch is sent to the lower main ring, recently constructed just below the original ring. The extra magnetic field strength available from superconducting magnets allows up to 1 TeV of energy. The beam consists of bunches of about 10^{13} protons confined to an oval region about 5 cm high.

Given a beam of protons very near the speed of light, representing enough power to operate a small town (but a small fraction of the 40

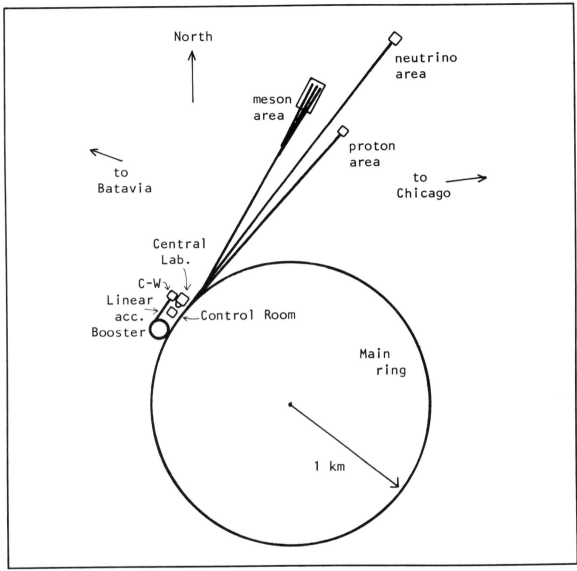

Fig. 13-1. The Fermilab accelerator and beam lines. (Adapted from the Fermilab map, courtesy of Fermi National Accelerator Laboratory.)

megawatts used to get the beam), what does one do with the protons? To be used for research, they must either collide with a target in the beam tube, or be extracted for use elsewhere. It is most convenient to use an auxiliary magnet to "kick" a portion of the beam out of the ring and down a tangential path. This is done at one of the long straight sections. The extracted beam must be in the same high vacuum. The main beam can also be made to wobble in its path, letting protons leak to a point where a magnet can bend them away from the main ring.

Secondary beams of particles can be formed by having the protons strike suitable targets, then collecting the desired products. Because of the huge momentum of the beam, most particles resulting from collision are directed nearly straight ahead and carry nearly their original energy. A complex switchyard of magnets and absorbers needs hundreds of meters to sort and direct the particles. At Fermilab beams of protons, neutrinos, and mesons are directed to three different experimental areas, located a kilometer away (see Color Plates 11, 12, and 13 and Fig. 13-1). The beam of neutrinos cannot be controlled by any means after formation, so the path is straight ahead. The neutrinos do not need an evacuated tube; instead a kilometer of dirt is used to filter out everything else from the beam.

After controversy over the site and design, construction of Fermilab began in 1968 and the beam was started in 1972. It was an unusually well-handled project completed ahead of deadline and under the proposed cost. This national facility is well worth the trip to the Chicago area, as the size of the tunnel containing the main ring is impressive. There is a large research and visitor center and tours are available. The main ring cannot usually be viewed as the area is highly radioactive. Particles strike the beam tube inside, causing other particles to be emitted that can penetrate to all parts of the experimental areas.

Fermilab's chief contender in the U. S. S. R. is IHEP (Institute for High Energy Physics, in translation) at Serpukov, where a 76 GeV machine is in operation. Earlier (1957) a 10 GeV synchrotron was completed in Dubna. DESY (Deutsches Electronen Synchrotron) in Hamburg, West Germany is another growing research center. The largest ring operating is PETRA, about 0.4 km radius (see Plate 4). It is designed for electrons, and is preceded by the smaller original synchrotron DESY. A larger ring is under construction.

On the border of France and Switzerland is a major growing international particle research center called CERN (translated as European Center for Nuclear Research). It started in 1959 with a 28 GeV Proton Synchrotron (PS), to which a 450 GeV Super Proton Synchrotron (SPS) was added, making PS a booster (see Color Plates 14, 15, 16,

and 17). The PS has 100 magnets in a ring 100 meters in radius, and is preceded by the usual linear accelerator and booster synchrotron. It is versatile, befitting its integration into other rings, supplying protons, antiprotons, electrons, and positrons. The SPS started in 1976 and uses 1000 conventional magnets in a ring 1.1 km in radius. There are two secondary beam lines, providing neutrinos and muons.

Building larger diameter accelerators at rising cost is not the only answer. At the 400 GeV level, 95 % of the beam energy is wasted in the lab frame, and only about 20 GeV is available in the CM frame where the action is. A new method of using all the available energy makes use of *storage rings*. Machines resembling synchroton accelerators not only accelerate but also store and keep focused bunches of particles. The beam could be formed in another accelerator and switched to the storage ring. Two rings are needed, with particles going in opposite directions. With the rings joined like a figure eight or stacked on top of each other, a collision of the two beams can be effected by using switching magnets. If the particles are of equal mass and opposite charge (particles and antiparticles), they can be stored in the same ring, moving in opposite directions with a small separation between beams.

It has taken much research work, but the problem of low intensity (low current) with colliding beams has been gradually solved. Increasing the density of particles by collecting bunches over a long time is one of several measures taken. High currents in the ampere range are being achieved. Two bunches of particles, which rarely interact, do not compete well with a beam striking a solid target, in terms of number of collisions per second. Problems such as stability and focus have also had to be solved. The rate of particle interaction ("luminosity") has been increased to the point that a reaction with a cross-section of 10^{-38} square meters can be obtained every few minutes. That is getting close to the tiny cross-section of neutrino reactions. By comparison, two protons have a cross-section of 10^{-30} square meters and will react with each other millions of times more often.

Gerard O'Neill, among others, developed the idea of the storage ring in 1956. The first experimental test with colliding electrons and positrons was in 1963. In 1971 CERN started up its pair of 28 GeV proton rings, making available 56 GeV in the CM frame and consternating the expectations of Fermilab. CERN now has proton-antiproton collisions available at 540 GeV in its SPS. Physicists led by Simon Van der Meer there have developed an Antiproton Accumulator to concentrate and store the antiprotons (see Color Plate 18). Since antiprotons are produced erratically in collisions of protons with a target, methods are needed to produce an ordered beam of them, a process called "stochastic cooling." A clever and seemingly simple technique is to detect errant antiprotons at one position along the beam tube and send signals across the ring to control "kicking" devices ahead of the beam, the only way to beat the near light-speed of the beam. Work is also proceeding on the Large Electron-Positron Storage Ring (LEP), to provide 120 GeV in the CM frame. Because electrons radiate so strongly in curved paths, the LEP is very large, about 4.3 km in diameter. Its magnets must bend the beam gently, so they are used at low field and low power. Most of the power is used in acceleration. The Russian center IHEP at Serpukov is starting work on a proton-antiproton machine to achieve 6 TeV. It will be over 3 km in radius.

The two-mile linear accelerator at the Stanford Linear Accelerator Center (SLAC) has the ability to produce beams of electrons or positrons at 33 GeV (see Plate 5). Its capabilities were further expanded when the SPEAR electron-positron storage ring was finished in 1972, providing 8 GeV in the CM frame. Like all storage rings, it can hold bunches of particles moving in opposite directions in storage for many hours. Each bunch of hundreds of billions of particles is only a few centimeters long and travels around the ring a million times each second. Collisions between oppositely directed bunches of electrons and positrons are arranged at two places where the slightly offset travel paths intersect. In 1980 SLAC completed the PEP positron-electron collider with 36 GeV and 0.4 km radius. SLAC is looking toward 100 GeV with the SLC machine under construction with a half kilometer radius.

Work is under way at Fermilab to incorporate the means of producing and collecting antiprotons in the main ring. Then proton-antiproton collisions will be possible at a total CM energy of 2 TeV. As of October 1985 1.6 TeV collision energy was achieved in the first test run of the new system. It makes maximum use of the superconducting magnet ring, allowing large savings in electric power. Two of the many technological "spin-offs" from such a research facility are the enhanced development of the superconducting cable industry and the large-scale handling of liquid helium for cooling superconducting cable. This will enable the solution of electric power handling and other problems for society at large.

DESY has an electron-positron storage and colliding ring called DORIS with 10 GeV, and one called PETRA with 45 GeV total. Construction on a machine (HERA) which would store and collide 30 GeV electrons with 820 GeV protons has recently started. It will be 1 km in radius and have separate rings of the same size for these different particles. A new realm of interactions will be opened, with total CM energy of nearly 70 GeV in such a case. Continuing the proliferation of electron-positron colliding beam machines, Cornell University in Ithaca, NY has one called CESR running at 16 GeV.

Electron-positron (lepton) colliders are much more difficult to make than proton-antiproton (baryon) ones because of the severe synchrotron radiation emitted by the lighter particles and their lower cross-section of interaction. CERN's LEP, to use as much power as a city, may be near the practical limit. Ways are being sought to find other ways to accelerate electrons to higher energies, using lasers. Lepton colliders are mainly used to study interactions involving the weak force, but also to probe the structure of hadrons.

Since only protons and electrons are freely available in nature, special methods have been arranged to obtain beams of other more exotic particles. It is often necessary to have a beam of substantial intensity (many particles in a small volume) so that the desired interactions occur often enough to measure carefully in a reasonable time. There is an increasing conflict between the rarity of events studied and the lack of enough particles at sufficient energy. Special particles are only generated by collision of ordinary protons or electrons with selected targets or particles. Plenty of neutrons and neutrinos come from nuclear fission reactors, although not at high energy. Accelerators are the preferred way to work with these particles, and a proton storage ring (PSR) at Los Alamos National Laboratory has recently been dedicated for the production of moderate energy (eV range) neutron beams at high intensity. The neutrons are produced in abundance by *spallation* when the protons "chip" them from heavy metals.

After the particles are created, they have an array of directions and energies and must be either focused into a beam or selected for certain qualities. Absorbing barriers of various materials and thicknesses allow certain particles to be selected from the assortment that comes from high-energy collisions. Neutrons can be collimated through apertures and perhaps selected for speed with a chopper—a rotating wheel with slots in it. Almost every new experiment has its unique requirements and must usually be designed and constructed from basic components.

13.2 PARTICLE DETECTORS

As higher energy accelerators come "on line," the need for bigger and more efficient detectors grows. It would be desirable to have photographs of the entire interaction, and the *bubble chamber* was invented in 1952 by Donald Glaser. By filling a large container with liquid, often hydrogen, and controlling its pressure, an ideal target of protons is provided and the passage of any charged particle would appear as a trail of tiny bubbles (Plate 6). A strong and uniform magnetic field is applied so that all charged particles follow curved paths in relation to their charge and mass. The liquid was clear and well-lighted so that cameras could take stereoscopic pictures. Computer assisted picture scanners were invented to measure the tracks on the pictures (see Color Plate 19), classify them, and calculate the energy and momentum of each particle. "Invisible" neutral particles could then have their properties deduced. The bubble chamber can turn out 10 pictures per second; unfortunately it cannot be triggered by specific events and much unwanted data is recorded. Glaser earned the Nobel Prize in 1960 for the bubble chamber, and Luis Alvarez received it in 1968 for the scanning system.

A later development, by T. E. Cranshaw and J. F. de Beer in 1957 (among others) is the *spark chamber*. Closely-spaced plates or wires partly fill a region, and high voltage is put on some wires with respect to their neighbors (see Color Plate 20). The chamber is triggered; that is, when some other detector reports that particles with certain properties have passed through, the high voltage is switched on in less than a microsecond. The chamber contains an inert gas which is ionized by charged particles. Sparks jump wherever the gas is ionized, showing particle tracks, which can be photographed. The paths can also be recorded electronically by sensing which pairs of wires had sparks. This permits direct computer analysis of particle interactions. Again a strong magnetic field is applied to make charged particles follow curved paths determined by their charge and mass. The bubble chamber excels in accuracy, but the spark chamber is good for speed and selectivity.

Scintillation detectors are individually small and use a crystal of sodium iodide, a plastic, a liquid, or some other material. Particles or gamma rays passing through the material cause light to be emitted, which is detected by a photomultiplier and counted. Neutrons and gamma rays can be detected in addition to charged particles. The energy of the particle can be measured from the amount of light. These detectors have been assembled in huge numbers to form large scale detectors sensitive in all directions around an interaction region. Notable are the Crystal Ball, now at DESY, and the Plastic Ball at LBL (See Color Plates 21, 22, and 23, and Plate 7).

Another type of detector uses Cerenkov radiation, the light emitted by a charged particle as it exceeds the speed of light in a liquid or solid. The angle at which light is emitted is proportional to the speed of the particle, so this detector measures speed directly. An old but popular method is the photographic emulsion, first employed by Becquerel in 1896. The first electrical methods, the ionization chamber and proportional and Geiger-Muller counters, are also still in use. Semiconductors also make fast and accurate particle detectors, best for use with heavy particles. Plate 8 shows a Cerenkov detection system at SLAC.

Every high energy laboratory has been building and using huge multiple component detector systems to surround the interaction region, usually for colliding beams. As one of many examples, Color Plate 24 shows a system LBL is building for use at SLAC's new positron-electron collider. Also for PEP is a 200 ton set of parallel iron plates surrounding a drift (spark) chamber (Plate 9). The iron is needed to slow particles that result from a collision and to measure their energy, a process called "calorimetry." The Mark III detector at the SPEAR storage rings at SLAC is shown in Color Plate 25. A large magnet, producing a field along the beam, surrounds many scintillation counters which surround four layers of spark chambers. A shower detector can distinguish showers of particles due to electrons from those due to hadrons. Outside is an iron barrier so thick as to let only muons (and neutrinos) through where more spark chambers detect the muons. This system found the tauon. Under construction at Fermilab is the Collider Detector (CDF) weighing 4500 tons, designed and being built by almost 200 physicists for recording results from 2 TeV proton-antiproton collisions.

The selection of detection methods depends on what time resolution is needed, how much time can be permitted while the detector recovers, in what volume the particles must be detected, and the smallest distances that need to be distinguished. Other factors include the method of recording data and the cost. Modern electronics has enabled very sophisticated and complex detectors to be built. Some of their logic is inherent; the detector does not record unless there is *coincidence* between two detectors, indicating that a particle of interest has passed between two points or that two different particles were produced at the same time. Coincidence arrangements are good for obtaining statistically correlated data, which is typical of measurement in the microworld. Walther Bothe received the Nobel Prize in 1954 for developing the coincidence method.

The methods developed for studying nuclear and particle physics are being found increasingly useful for particle-based medical diagnosis of patients and their treatment by means of radiation, as well as for the production and study of new materials. Neutrons and high-energy electrons and positrons and other more exotic particles have special properties for delivering energy to specific interior regions of the body or other materials. Also tracer radioactive isotopes such as $^{11}_{6}$C, $^{13}_{7}$N, or $^{15}_{8}$O can be introduced chemically into organisms and detected by means of gamma radiation resulting from the decay of the positrons which they emit. This new method is faster and safer.

13.3 PIONS, THE NUCLEAR CARRIER

After several false leads, the search for the particle carrying the Yukawa nuclear force was completed by Cecil Powell in 1947. His method was to let cosmic rays fall on photographic film left on mountain peaks where the intensity is greater. By analyzing the tracks found when the film is developed, he and C. M. Lattes and G. P. Occhialini found the traces of an intermediate mass particle which decays to muons. Mainly for his general technique Powell received the Nobel Prize in 1950. Called "pi mesons" (symbol π) and shortened to *pions*, these particles do interact strongly with nuclei, unlike muons. They also have the mass predicted by Yukawa and come in positive, negative, and neutral forms.

Any nucleon can interact with any other by means of an appropriate pion, changing charge

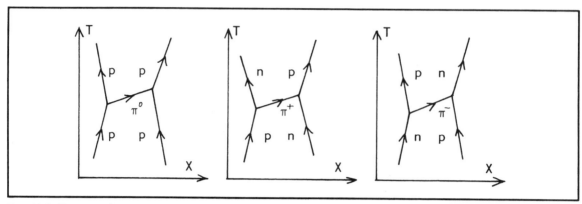

Fig. 13-2. *Feynman diagrams for two nucleons interacting via the pions.*

as necessary. Feynman diagrams of the possible first order interactions are shown in Fig. 13-2. Being the carrier of a field, the pion can appear as virtual particles in many other ways in these interactions, in analogy to photons in QED. However, the strong force is so much stronger that a similar mathematical theory has not been found yet. About the best that can be said non-mathematically is that an *exchange* of a virtual pion between two nucleons results in an attraction between the two nucleons by means of the attraction of each to the pion. To apply a classical picture of each nucleon recoiling from the emission and absorption of a pion would be misleading. Neither momentum nor energy is conserved during a virtual exchange.

The charged pions (π^+, π^-) have about 140 MeV rest mass-energy, and the neutral one (π^0) has less, about 135 MeV. The mass difference is probably due to the extra electric energy of the charged pions. All have spin 0, which is to say, no spin. This is convenient for explaining certain nucleon interactions. Pions have no magnetic moment, as one would expect. They do have *isospin*, a quantum number I which further distinguishes strongly interacting hadrons. Isospin has the value $+1$ for the π^+, 0 for the π^0, and -1 for the π^-. When viewed as projected in an imaginary dimension, these values correspond to a total $I = 1$. The charged pions decay relatively leisurely, taking $2.5(10)^{-8}$ second to turn into a muon and its neutrino. The neutral pion is in a hurry, decay-

ing in 10^{-15} second to two gamma rays. The decays have other rare modes. Occasionally a charged pion will decay to an electron (or positron) and its neutrino. The neutral may form one or more pairs.

The charged pions are particle and antiparticle; the neutral one is its own antiparticle. Although the pion can travel about 10 meters before decay, its mass nevertheless sets the range of the strong nuclear force at about 10^{-15} meter. As the carrier of a field, there is no limit to the number of pions that can appear in a reaction, in analogy to the role of the photon. This also follows from its status as a boson. The pions are the members with the lowest mass of a set of mesons which we shall meet next. A typical high energy experiment involving the collision of a pion with the protons in a hydrogen bubble chamber can result in seven different outcomes, as will be seen later. It should be kept in mind that the duration of a collision is less than 10^{-23} second, determined as the time it takes to cross the approximate size of the proton at the speed of light. Any event that takes longer involves some internal interaction that gives a metastable state. Decays by means of the strong interaction take only a little more than 10^{-23} second. Anything longer involves the weak interaction.

13.4 MORE MESONS

The mesons are bosons, all with zero spin. Beyond the pions, there is a heavier uncharged meson with spin 0 called the *eta* (Greek letter η). It has rest mass-energy 549 MeV and decays in about 10^{-18} second in a number of different ways to pions or leptons.

The rest of the meson family introduces several new features. Some have nonzero values for a new quantum number called *strangeness* (S). This property, named by Gell-Mann and Nishijima in 1952, describes the behavior of many newly made hadrons. They are created under the rules for strong interactions but they decay by weak interactions, although emitting a neutrino or other lepton. Strangeness is conserved for strong interactions, a fact discovered in 1953. The pion is defined to have $S = 0$. A strange particle cannot be created alone, but strangeness need not be conserved when a strange particle decays. S can change by only one unit—a new kind of selection rule. Sometimes another equivalent quantum number called *hypercharge* (symbol Y) is used. It is calculated from strangeness and follows the same conservation laws.

Besides having strangeness, one of the two *kaons (K* meson) exists in two different forms simultaneously, a clear quantum mechanical superposition of states. It also violates some symmetry laws. The "simpler" kaon is positive (symbol K^+) with a mass of 494 MeV. It decays in about 10^{-8} second to leptons or pions. It has strangeness $S = 1$, the only instance of positive S, which is strange in itself. The antikaon $\overline{K}^0$ has $S = -1$. The "tricky" kaon is K^0 with 498 MeV. It has $S = 1$ ($S = -1$ for the $\overline{K}^0$). The two are distinguished by their strong interactions, when S is conserved. The K^0 is a combination of two different states, K_S^0 and K_L^0 (formerly K_1^0 and K_2^0), with two different decay times. The S-state decays in about 10^{-10} second to two pions only, and the L-state decays in $5(10)^{-8}$ second to three pions or to a pion and leptons.

Because of its structure, the K^0 has some unusual changes as it travels. A beam of K^0 can be produced by colliding pions and protons. After about 10^{-10} second the beam is half as intense and consists of K_L^0. K_L^0 can be thought of as a combination of K^0 and $\overline{K}^0$. This is demonstrated

when the beam strikes more protons and a K_S^0 component is temporarily "regenerated." The neutral kaon behaves in a manner analogous to polarized light. The K^0 and $\overline{K}^0$ are the two components (matter waves) linearly polarized at right angles. Together they form the equivalent of a circularly polarized wave, left for K_S^0 and right for K_L^0.

The neutral kaon violates the symmetry of charge conjugation (C) and the symmetry of handedness (parity P). The operation of C on a reaction reverses the signs of charged particles, turning them as well as neutral particles into their antiparticles. It also reverses the magnetic moment. For example, it has been found that the decay of K^0 to $\overline{K}^0$ is not quite at the same rate as the charge-conjugated decay, $\overline{K}^0$ to K^0. Taken together, the two symmetries combine to form one called CP. When CP is violated, it is theoretically required that time reversal symmetry (T) also be violated. This has been experimentally verified as well. Val Fitch and James Cronin received the Nobel Prize in 1980 for their work on the violation of CP and T symmetry. They have thus proven that a universe of antimatter behaves (at least for kaons) slightly different from normal matter. These results are experimental only, and no satisfactory theoretical explanation is yet available. The mass of the kaon seems to differ from that of the antikaon by about 10^{-8} eV, to accompany the CP violation. The only other violation which has been verified at this time involves the eta meson.

Other mesons have been found which involve a further complication best left to the next chapter. Table 13-1 summarizes the properties of the mesons. The mesons have a spectroscopy in the sense that each individual meson represents an energy state of a generalized meson. Figure 13-3 shows the ones discussed placed on an energy scale and distinguished according to I and S. In analogy to angular momentum, the number of charge states is calculated from the value of I and gives the number of possible particles at each energy level. Some of the higher energy states (particles) will be discussed later. It will be seen that transi-

Table 13-1. *The Stable and Metastable Mesons.*

Name	Symbol	Rest mass (MeV)	Charge	Isospin	Strangeness	Life (seconds)
pion	π^+	139.5673	+1	1	0	$2.6030(10)^{-8}$
	π^-	139.5673	−1	1	0	$2.6030(10)^{-8}$
	π^0	134.9630	0	1	0	$8.3(10)^{-17}$
eta	η	548.8	0	0	0	$7(10)^{-19}$
kaon	K^+	493.667	+1	1/2	1	$1.2371(10)^{-8}$
	K^0	497.67	0	1/2	1	see note

Notes: Each has an antiparticle, shown with a bar over the symbol. Each is a member of a multiplet of different charge states, with the number of members calculated from isopin I by 2I + 1. All are spin 0 bosons and have parity "−". The antikaons have strangeness S = − 1. The neutral kaon is a combination of two states, K_S^0 decaying in $8.923\,(10)^{-11}$ second, and K_L^0 decaying in $5.183(10)^{-8}$ second.

tions (decays and excitations) among these states result in emission of particles.

13.5 NUCLEONS

We have discussed the proton and neutron as essential components of nuclei. They are susceptible to all four forces. Both proton and neutron can have weak interactions which take much longer than strong interactions. We are now interested in them as participants in the strong force, with the pion as the carrier of the interaction. The isospin quantum number I applies to nucleons and other hadrons. It has been mentioned that the proton is distinguished from the neutron as a nucleon by the proton having $I = 1/2$ and the neutron $I = -1/2$. Total isospin is conserved in strong interactions, including the count for any number of pions produced. The nucleon has zero strangeness.

The proton has an antiparticle, discovered by Emilio Segre and Owen Chamberlain in 1955 with the Bevatron after being expected for many years. They received the Nobel Prize in 1959. The antiproton annihilates the proton, producing a number of pions, which should be no surprise. Pions are expected, not just because pions belong with nucleons, but also because they are more strongly coupled with nucleons than photons would be through the electromagnetic interaction. The antineutron was found in the same experiments and it, too, annihilates with a neutron to produce pions. The basic interactions of neutron and proton were illustrated earlier in Fig. 13-2.

From the lab viewpoint, we see that in some collisions *charge exchange* will occur. If the pion is not detected, all that is observed is that a fast proton goes in and a fast neutron comes out. The flow of the charge from one to the other during an interaction is called a *current*. (Being of similar mass, the proton and neutron also exchange momenta in an interaction.)

Since the proton is charged (with what seems to be exactly the same charge as the electron, but positive), it acts as a tiny magnet. It has a magnetic dipole (magnetic moment) of 2.79 "nuclear magnetons." This classical unit is calculated from charge e and mass M as $\frac{eh}{4\pi M}$, similarly to the Bohr magneton for the electron. Thus the calculation of the magnetic moment of the proton from its spin give a result too large by a factor of almost 3. The proton magnetic moment is over 600 times weaker than that of an electron because of its large mass. Otto Stern received the Nobel Prize in 1943 for measuring the magnetic moment of the proton, among other achievements.

The neutron has a magnetic moment despite being neutral. It was first found in 1933. The value is − 1.9 nuclear magnetons, reversed from and weaker than the proton, but not much weaker. Its magnetic moment indicates that it has some charge in its structure, although the overall charge is zero. In classical terms the negative part of the charge seems to be on the "outside" and the positive part on the inside. The outer part would have the greatest effect on the moment. Despite the small value of their magnetic moments, both the neutron and proton are so small that the magnetic fields near them are a million times stronger than the highest artificial magnetic fields. The neutron's role as a fermion has been confirmed in the sense that it, like the electron, seems to turn twice (720°) during one rotation in the real world.

It is also wondered whether the neutron has an electric dipole moment; that is, whether the charge in it is distributed asymmetrically so that there is more positive on one "side" and more negative on the other (speaking loosely). Finding this would show that the neutron violates another supposed symmetry in nature, *time reversal* (T). Any observable part of an electric dipole should point in the same direction as the magnetic moment. But reversing time could reverse the relative direction of these two moments. Recently has it

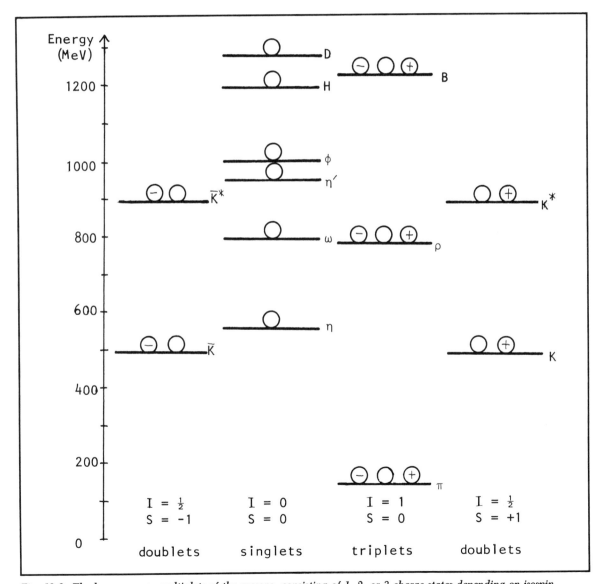

Fig. 13-3. *The lower energy multiplets of the mesons, consisting of 1, 2, or 3 charge states depending on isospin.*

been possible to slow neutrons down so much that they can be bottled up. Measurements on ultracold neutrons have established new lower limits for a possible electric dipole moment. Since "ultracold" neutrons can be observed for long periods, the decay time can also be better determined. The half-life of about 920 seconds is not known very accurately, and more accuracy is needed.

Reversing time is equivalent to reversing the direction of all motions. This would include spin, thus flipping over a spin direction. Elementary particles should have no intrinsic sense of time and not know whether they are going forward or backward in time. Time reversal has been found to hold for many reactions in an indirect way. We cannot truly reverse time flow, but we can transform reactions into the form that would occur backward and check to see if that reaction occurs in nature—forward, of course. For example, neutron decay under time reversal would involve an electron striking a proton to form a neutron thus:

$$e^- + p \rightarrow n + \nu e$$

One of the rules of time reversal is that a particle such as the neutrino on one side of the equation is equivalent to its antiparticle on the other side. Then we have the complete reversal of neutron decay. The weak interaction does not always violate time reversal, and the strong interaction does not necessarily preserve it. All other forces do preserve time symmetry.

There is evidence that the proton and neutron have structure, hinting at a deeper level of nature which is a major topic later. The proton in particular is not an elementary particle and may even be found to decay. Scattering experiments indicate that the proton may have a very tiny very hard core with the charge scattered more loosely around it. It may even have three hard spots. Robert Hofstadter was among the first to use the scattering of high energy (up to 1 GeV) electrons from protons and deuterons to explore the charge structure of protons and neutrons. Partly for this work he received the Nobel Prize in 1961.

13.6 BARYONS

Nucleons are the lowest mass members of a family of particles called *baryons*. After searches inspired by theory, some full sets of baryons have been identified. All can be formed by means of collisions of protons, neutrons, and/or pions in any combination. In any given collision one cannot predict exactly what will be formed, beyond the constraints of the conservation laws. On the average, when many similar collisions are tried, however, nature follows very predictable statistics. Baryon number B is of course conserved, with the value $+1$ assigned to each baryon (-1 for antibaryons). All baryons are fermions with spin 1/2. All have antiparticles. Only the proton seems to be truly stable. The rest that live longer than about 10^{-22} second are *metastable*, surviving until well after the collision and therefore being held together until a weak decay can occur. Baryons are distinguished by their strangeness S and their isospin I, each having a unique set of values. Luis Alvarez, who received the Nobel Prize in 1968, used the bubble chamber to find new metastable and strange particles.

The next heavier particle is the *lambda*, named for its uppercase Greek symbol Λ^0 which resembles the tracks of the two particles that result when the lambda decays. It has a rest mass-energy of 1116 MeV, somewhat heavier than the proton. It decays into a nucleon and pion in about 10^{-10} second. Discovery of the lambda in 1947 required the introduction of the quantum number for strangeness. The lambda has $S = -1$. To conserve strangeness, some other particle with $S = +1$ must be created along with a lambda. S need not be conserved when the lambda decays. The lambda provides a test of time reversal symmetry. Its strange weak decay is certainly a candidate for suspicion:

$$\Lambda^0 \rightarrow p + \pi^-$$

Because the reversed reaction is so hard to produce, the test was done by examination of the polarization of the proton in comparison with the lambda. No detectable violation has been found.

Next heavier are the *sigma* particles (uppercase Greek letter Σ) which come in charged and neutral forms around 1190 MeV, a little more than the lambda. The three forms differ by only a few MeV in mass, but this is one of many unusual facts about particles that requires a sophisticated explanation. The charged forms decay in about 10^{-10} second to a nucleon and pion. The neutral form decays faster (about 10^{-19} second) to a lambda. The strangeness of the sigmas is $S = -1$.

A little heavier at 1315 MeV and 1321 MeV are the two *cascade* or *xi* particles (the symbol is the uppercase Greek Ξ). One is neutral and one is negative. They decay only to lambdas and pions in about 10^{-10} second and have $S = -2$. The name comes from the "cascade" of decays that follow their creation.

The set of eight baryons is topped off (for the moment) with the *omega* (Greek uppercase Ω), which has a mass-energy of 1672 MeV, not quite double the proton. It is negatively charged and usually decays to a lambda and a kaon or a cascade and a pion in 10^{-10} second. It is the only member of the set with spin 3/2. After the other baryons had been found experimentally, the existence of the omega was predicted to be a strange particle with $S = -3$ by a new theory (the "eightfold way," covered shortly). It was searched for at Brookhaven with the new liquid hydrogen bubble chamber in 1963 and found. A beam of 5 GeV K^- mesons when collided with protons causes the following reaction:

$$K^- + p \rightarrow K^0 + K^+ + \Omega^-$$

An analysis of the discovery photograph, sketched in Fig. 13-4, illustrates the sort of detective work the particle physicist is faced with. A series of short lived particles must be traced until identifiable or stable results are found. Neutral particles (invisible paths shown dashed) can only be inferred from what comes next. A total of about 3.2 GeV was available in the CM frame, to make the kaons and the omega, whose rest mass-energies total to about 2.6 GeV. Thus there was barely enough to ensure omega production, and the kaons and omega together carried away about 0.6 GeV of kinetic energy. The omega decayed to a pion (π^-) and cascade (Ξ^0). The cascade had to be identified by its decay products (two gamma rays and a lambda). The gamma rays could only be inferred from their change into electron-positron pairs, readily identified by their paths in the magnetic field superimposed on the chamber. The lambda decayed to a proton and pion (π^-). The

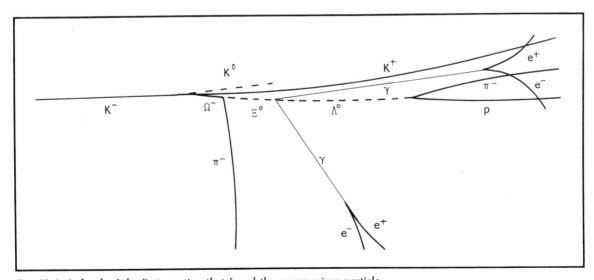

Fig. 13-4. A sketch of the first reaction that found the omega-minus particle.

Table 13-2. Baryons.

Name	Symbol	Rest mass (MeV)	Charge	Isospin	Strangeness	Life (seconds)
nucleon	p	938.2796	+1	1/2	0	stable?
	n	939.5731	0	1/2	0	898
lamba	Λ^0	1115.60	0	0	−1	$2.632(10)^{-10}$
sigma	Σ^+	1189.36	+1	1	−1	$8.00(10)^{-11}$
	Σ^0	1192.46	0	1	−1	$5.8(10)^{-20}$
	Σ^-	1197.34	−1	1	−1	$1.482(10)^{-10}$
cascade	Ξ^0	1314.9	0	1/2	−2	$2.90(10)^{-10}$
	Ξ^-	1321.32	−1	1/2	−2	$1.641(10)^{-10}$
omega	Ω^-	1672.45	−1	0	−3	$8.19(10)^{-11}$
delta	Δ^-	1230	−1	3/2	0	$<10^{-23}$
	Δ^0	1231	0	3/2	0	$<10^{-23}$
	Δ^+	1232	+1	3/2	0	$<10^{-23}$
	Δ^{++}	1234	+2	3/2	0	$<10^{-23}$

Notes: Each baryon has an antiparticle shown with a bar over the symbol. Each is a member of a multiplet of different charge states, with the number of members calculated from isopin I by 2I + I. All are fermions with parity " + " and spin 1/2, except the omega and delta with 3/2, which are excited states (resonances) of baryons for which no ground state exists. The delta is essentially unstable and has a unique isopin and strangeness different from the other five baryons.

gently curved tracks of the heavier charged particles allow their masses to be calculated, and conservation laws usually allow other invisible particles to be deduced.

Table 13-2 summarizes the properties of these baryons. Typical of collisions involving baryons, interaction of a pion and proton can produce any of the following seven results: pion and proton back again (an elastic collision), pion and neutron, two pions and a proton, three pions and a proton, a kaon and a lambda (Plate 10), a kaon and a sigma, or a pion, kaon, and lambda. "Atoms" can also be made with any of these baryons or mesons occupying a state normally held by an electron, providing new ways to study their behavior.

In addition to the above baryons and mesons, many more new ones seem to be made in high energy collisions. For a long time it was thought that new particles would proliferate without end as hundreds were entered in particle charts. Then it was realized that many of the heavier ones, which are so short-lived as to be called frankly "unstable," are related to the lighter ones already mentioned through having the same S and I. They have come to be recognized as *resonances* or unstable excited states of the stable and metastable

baryons and mesons. The first resonance (discovered by Fermi in 1952) was obtained by striking a proton with a pion to form a *delta* (uppercase Greek Δ). It is the lowest mass resonance, with a mass-energy of 1238 MeV and a spin of 3/2. When a positive pion is used the delta has the unusual charge of +2. The delta is a temporary association of pion and proton, lasting less than 10^{-22} second. Its lifetime can be predicted by applying the uncertainty principle to the "line" width of about 100 MeV.

Figure 13-5 is a sketch of the sort of resonance data the physicist uses to identify these excited energy states. The size of the resonance formed is characterized by the cross section of the reaction, which turns out to be much higher for the positive pion than the negative, although the same delta is formed. As the energy available in the CM frame is increased from 1000 MeV toward 2000 MeV, there is a range where energy is readily absorbed to form the delta, centered around 1232 MeV. As is typical, resonance peaks at higher energies are also formed, labeled N(1520) and so forth. N denotes that the resonance is an excited state of a nucleon. The proton has been boosted to an excited state with spin 3/2 for a brief time. The res-

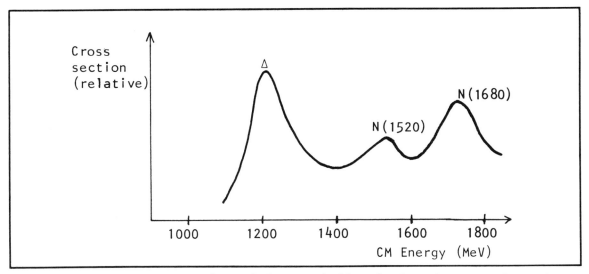

Fig. 13-5. A high energy resonance curve, showing the formation of a delta and nucleons when a pion and proton interact at various energies.

onance at 1680 MeV has spin 5/2. Each of these resonances will most likely decay to a nucleon and a pion but more rarely to a more exotic baryon and meson.

Each of the basic stable and metastable baryons is the basis of a set of higher states. Each one represents a unique combination of strangeness S and isospin I as shown in Table 13-2 and Fig. 13-6. Since the omega particle has spin 3/2, it is not the lowest state possible. However, no spin 1/2 state has been found at lower energy. Another unique combination of S and I, the delta, also has spin 3/2 and no lower state. It has a quadruple of charged forms: -1, 0, $+1$, and an unprecedented charge of $+2$. Unlike the omega, which is in the basic set because of its relatively long lifetime, the delta is a resonance that decays in less than 10^{-22} second. Some other "excited" baryon states are labeled "*". The baryons, like the mesons have a spectroscopy that will be explored further. Resonances are known as high as 5500 MeV.

The lowest mass resonance for a meson is called the *rho* (Greek letter ϱ), at 770 MeV (Fig. 13-3). It can be formed by colliding two pions, and almost always decays back to the same. It has spin 1, indicating the spinless pions form an angular

momentum state with $L = 1$. Some excited states of mesons have angular momentum zero. Hundreds of meson resonances are known with energies as high as 10 GeV. CM energies much higher than this are available, but this does not mean the limit to possible resonances has been found.

13.7 A THIRD SPECTROSCOPY-THE EIGHTFOLD WAY

The phenomenon of resonance—the formation of excited states—should remind the reader of energy levels encountered in other situations, the atom and the nucleus. Now we are in an energy regime a thousand times higher than that of the nucleus, and a "third spectroscopy" is revealed (attributable to Victor Weisskopf). The nucleon and other metastable baryons have a series of energy levels available to them; some lower ones were diagrammed in Fig. 13-6. The transitions from delta states to nucleon states are shown as an example in Fig. 13-7. Pions or gamma rays of very definite energy are emitted, corresponding to spectral lines. The observed (and predicted) patterns of energy levels are beginning to tell us much about the theory and structure of baryons and mesons. We are beginning to get explanations as to why there are so many particles and why they are distinct in accord with the various quantum numbers.

In 1961 Murray Gell-Mann brought together the mathematics of symmetry and the observation of the particle count and properties in Fig. 13-6 and proposed that the stable and metastable baryons form an *octet* (Fig. 13-8A), a set of states of a single entity, not necessarily one particle, whose identity is yet unknown. His theory was strengthened by the fact that some excited states of the baryons form a *decuplet* of ten states corresponding closely to the octet (Fig. 13-8B). One state of the decuplet had not been found experimentally, but this model predicted its properties well enough that it soon was (recall Fig. 13-4). The omega tops off the decuplet and is sufficiently

stable that it is listed as a metastable baryon.

When particles with the same S and I are arranged on an energy level diagram as in Fig. 13-6 or 13-8, it is apparent that some have nearly the same mass-energy. Each set so formed is called a *multiplet—singlet* for one member, *doublet* for two members, etc. The number in a multiplet can be calculated from $2I + 1$, just as was done for angular momentum earlier. Since the energies are not exactly the same, we observe distinct particles. But the closely spaced energy levels for a single type of particle or multiplet indicate that there may have been level splitting from an initially degenerate state. A close spacing such as about 3 MeV indicates that a non-strong force is involved. Indeed it is the electromagnetic interaction that distinguishes particles with different charge. Each has the same isospin I, yet a slightly different

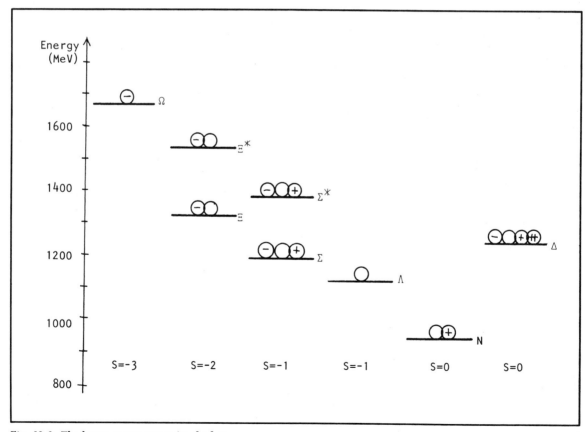

Fig. 13-6. The lower energy states for the baryons.

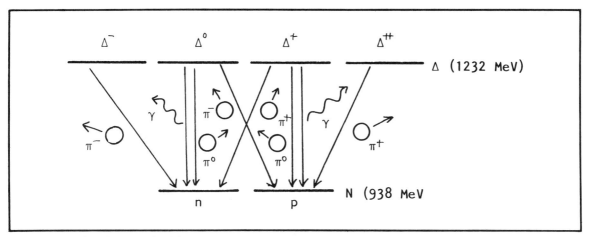

Fig. 13-7. *Possible decays of the delta particle, shown as transitions from delta energy states to nucleon energy states.*

energy. Isospin is conserved in strong interactions but not in electromagnetic.

On a larger scale, the baryon octet represents the splitting of one kind of entity, a basic or *generic* baryon, into a number of types of baryon whose energies differ typically by about 200 MeV. The nucleon is about 200 MeV below the next state, the lambda, and so forth. The quasi-uniform spacing is apparent in Fig. 13-6 and 13-8. If the generic baryon is split, something must not be conserved. To get 200 MeV splittings, something related to the strong force is needed. Strangeness is not conserved, and this forms the baryon octet. The same argument holds for the baryon decuplet. Not all states in the decuplet are higher than in the octet. We have seen the overlapping of sets of spectral lines before.

Another step in classifying particles was made by Tullio Regge in 1959. He worked on the relation between the energy of a particle or resonance and its spin state. More massive particles usually had higher spin states and simple proportions were found, called "Regge trajectories." One could predict the masses of higher resonances of both baryons and mesons. Some took this to mean that an "elementary" particle was no more elementary than its resonances were.

An explanation of "elementary" particles that was in vogue for some time was formulated by Geoffrey Chew and S. C. Frautschi. Called the "bootstrap hypothesis," it relied on the fact that each particle should have a swarm of virtual particles around it. Furthermore, it was supposed that the elementary particles were nothing more than composites of each other. Each particle is in existence by its own "bootstraps," lifting itself into existence by causing other particles which cause it. The search for the building blocks of matter would end with the hadrons, leptons, and photon. All observed physical quantities and symmetries are hypothesized to be the final and definite result of all this activity. This sort of theory is both philosophically appealing and unpleasant. Fortunately later developments seem to have laid it to rest for the time being.

In 1961 Murray Gell-Mann proposed a deeper classification of the baryons, based on the octet and decuplet. His scheme has been named the *eightfold way* because eight quantum numbers and symmetries are involved. There is also reference to an old Buddhist saying about the eight steps to the perfect life, possibly the first of many connections to be drawn between Eastern philosophy and Western physics. Explaining the eightfold way would lead us into mathematics of a different and possibly more complex sort than has already been discussed. Just briefly, the mathematics is *group theory*, which describes the various symmetry operations that can be done. A group called *SU*(3) has eight components,

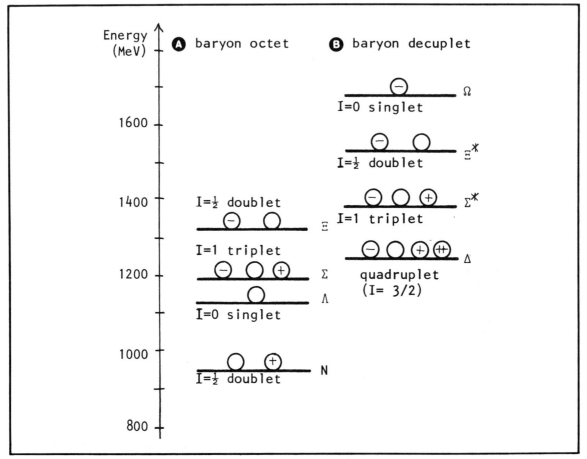

Fig. 13-8. The lower baryon states arranged as an octet and a decuplet in the eightfold way.

representing the eight different ways objects such as particles can be related.

An interesting connection to physics occurs when a certain "breaking" of the symmetry is done. Then the objects groups into exactly the same arrangement as has been observed for the baryons: a singlet, two doublets, and a triplet (see Fig. 13-8A). Gell-Mann must have been onto something, and he received the Nobel Prize in 1969. He appears at a 1966 lunch conference in Plate 11, along with McMillan, Fitch, Weisskopf, and other prominent physicists. And there is more

success with the eightfold way. Another representation of $SU(3)$ has 10 components, which split into a quartet, a triplet, a doublet, and a singlet, exactly matching the excited baryons in the decuplet (Fig. 13-8B). The fundamental symmetry of the baryons has been established in a very general and abstract way but a physical explanation that appeals to our common sense remains to be made. Group theory alone cannot predict exact masses, lifetimes, forces, and so forth. A better theory, using new entities called "quarks," is in the next chapter.

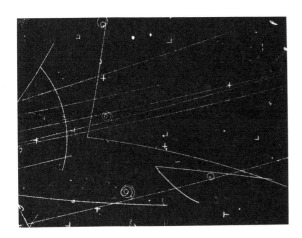

Chapter 14

The Inscrutable Quarks

P ROBABLY NO DEVELOPMENT IN POSTWAR THE-
oretical physics has tantalized the general
public more than the invention—or discovery?—
of quarks. Their introduction also seemed to cause
much controversy among physicists. Now quarks
are well established in the physicists' picture of
reality, despite their never having been directly
photographed as particle tracks. The quark model
of the structure of formerly "elementary" particles
is widely accepted because the model has predicted the existence of several new particles and
these have been found! Nobel Prizes based on work
with quarks are almost as abundant as the quarks
themselves and are being awarded quickly. Once
a wild idea, quarks now seem to be an inevitable
development in our understanding of the deepest
structures of nature.

14.1 QUARKS AND THE EIGHTFOLD WAY

In 1963 Murray Gell-Mann and George
Zweig independently arrived at a new model of
hadrons, inspired by Gell-Mann's discovery that

the eightfold way, based on the mathematics of
groups, fitted the known particles and predicted
more. If hadrons are considered to be made of
other, more "fundamental" particles, a physical
explanation of the eightfold way emerges. The new
constituents were named "quarks" by Gell-Mann,
adopting the word from "Finnegan's" "Wake",
by James Joyce. The beginning of the last chapter
of Part II says, "Three quarks for Muster Mark!"
Not only is this an instance of how physicists link
their work to the larger human culture, but Joyce
and Gell-Mann agreed on the correct number as
well.

The quarks introduce some new and unusual
properties for particles. These will first be discussed in general terms. Some of the evidence for
the actual existence of these properties will be
reviewed later.

1. Quarks are charged, but the amount of
charge is a fraction of a standard unit (one electron or proton). Some quarks have $+2/3$ and
others $-1/3$ of a unit of charge. This ought to

make them stand out in pictures of particle tracks, because the curve of a track in a magnetic field depends on the exact size of the particle's charge.

2. Quarks interact by means of a new version of the strong force. The interaction between quarks seems to be such that a single quark cannot be isolated from other quarks so that it may be "seen" more directly. The force is low when the quarks are close together, but seems to approach infinity if they are pulled apart—which means they cannot be separated. This should remind us of the fact that a hadron cannot be "ionized"—that is, have one constituent removed. The two or three bound quarks come in a *bag* about $0.5(10)^{-15}$ meter (0.5 fm) in diameter. The bag constitutes a rather impenetrable barrier at moderate energy. Thus the presence of quarks inside a nucleon is not apparent in the ordinary spectrum of nuclear states. There are theoretical reasons why quarks cannot be separated, and all experimental efforts to date have not revealed any individual quarks.

3. Quarks are generally thought to be truly fundamental particles, being pointlike, having no structure, and being indestructible. The different quarks seem to have widely differing masses, difficult to measure accurately.

4. All baryons are made from three quarks, and all mesons are made from two. Leptons, photons, and other bosons remain fundamental, having no constituents.

5. At this time six different kinds of quarks have been established, plus the corresponding antiquarks. Each kind of quark is called a *flavor*, and they have been named the *up*, *down*, *charmed*, *strange*, *top*, and *bottom* quarks (not your usual sort of flavors). For a while it was feared that quarks, too, would proliferate, as one after another was found necessary to explain experiments. But at the present the count seems relatively complete. Each flavor does come in three different *colors*, but this characteristic should not be thought of as differentiating them into more particles. The notion of color has nothing to do with our ordinary perception of color but is another of those whimsical concepts applied to unfamiliar aspects of the microworld.

6. It may be a coincidence of numbers, but the number of different kinds of quarks known happens to be the same number as the number of leptons known. Moreover, the charge of leptons is an exact multiple of quark charge.

Quarks have more usual properties, too. They have spin 1/2 and, being charged, have magnetic moments. Their calculated magnetic moments agree well with the measured magnetic moments of many particles containing them, especially for the lambda. The distribution of charge in a neutral particle such as the kaon has been found to agree with the quark model's prediction. The properties of quarks are shown in Table 14-1. Each quark has quantum numbers assigned to it that

Table 14-1. Quarks.

Type (flavor)	Symbol	Charge	Quantum Numbers: Isospin	Strangeness	Charm	Bottomness	Topness	Mass-energy (est., GeV)
down	d	− 1/3	− 1/2	0	0	0	0	0.1-0.2
up	u	+ 2/3	+ 1/2	0	0	0	0	0.1-0.2
strange	s	− 1/3	0	− 1	0	0	0	0.3-0.4
charmed	c	+ 2/3	0	0	+ 1	0	0	1.5?
bottom	b	− 1/3	0	0	0	− 1	0	5?
top	t	+ 2/3	0	0	0	0	+ 1	40?

Notes: The antiquarks (bar over symbol) have the opposite charge. All quarks have baryon number B = 1/3 and spin 1/2. Antiquarks have B = − 1/3 and, the opposite sign on the other quantum numbers.

characterize its flavor. These were found necessary to explain what is conserved in observed reactions. Each quark carries a baryon number $B = 1/3$ ($-1/3$ for antiquarks).

There is much much more to the quark model of nature. We will not be able to examine the mathematical side of the theory in any detail, but we will find many new physical facts and theories to explore. The quantum mechanics first developed for atoms, then found useful for nuclei, now fits the quark level just as well. The quarks are the particles which underlie the third spectroscopy and the energy states of hadrons. Whether quarks represent the end of the search for the building blocks of nature is yet to be proven, if it can ever be proven.

14.2 MESONS AND QUARKS

Research on mesons has been a fertile field for making discoveries about quarks, and it is appropriate to review the knowledge of meson states at this point. In 1961 only the pions and kaons were known. The three pions of isospin 1 form a triplet, the two kaons of isospin 1/2 and strangeness 1 form a doublet, and the two antikaons of isospin 1/2 and strangeness -1 form a doublet, as was shown in Fig. 13-3. Together this group of mesons would form an octet if one more particle were found. The eta was predicted by the eightfold way applied to mesons, and it has isospin 0. The meson energy level diagram consists of four multiplets, each of which has excited states as was shown. There is no stable ground state since even the pions are metastable. The upper (excited) states are resonances.

The new property of matter called *charm* was established in 1974 when a new particle (or resonance) called *J/psi* (*J* and the Greek letter ψ) was discovered. It has a rest mass-energy of about 3.1 GeV, an unusually heavy particle at that time, and a spin of 1, making it a boson. It is neutral and decays to nonstrange particles, having no strangeness itself. Except for its mass it behaves the same as particles already known. There seemed to be no need for it. It did have a longer lifetime than other unstable particles known,

10^{-20} second. Samuel Ting and Burton Richter, each leading a group independently, found the *J/ψ*, Ting using collisions of protons with beryllium nuclei and Richter (Plate 13) using electron-positron collisions. Ting called it *J* and Richter ψ, and its dual name remains. They both received the Nobel Prize in 1976 for this work.

The existence of charm was expected initially on the basis that the strange quark should have a companion. The leptons come in pairs, and so should the quarks. The name "charm" was given by James Bjorken and Sheldon Glashow. Charm, like strangeness, should not be conserved in weak decays. Charm was also expected to exist because charge is not always exchanged in strange decays. When it is not, the strangeness is carried from the initial particle to the final one by a "neutral weak current" (covered in the next chapter). These considerations led to the prediction of a series of new mesons and baryons involving charm. The *J/ψ* was expected to contain a new quark because it was so slow in decaying to particles which contain the known quarks.

The annihilation of electron and positron creates a virtual photon, which possesses only one quantum number, spin 1. This favors its decay to equal numbers of particles and their antiparticles, or perhaps to a single particle *J/ψ*, which has spin 1 and no other nonzero quantum numbers. Thus if it is made of quarks (just two), they must be quark and antiquark and rather heavy. Neither *J/ψ* nor the quark-antiquark pair can be observed more directly, but an analysis is made of the hadrons pouring out from the decay of *J/ψ*. Based on theory, it is possible to deduce the sum of the squares of the original quark charges from counts of the decay products. Since the quarks carry fractional charges, the fact that there were quarks was proven without seeing any.

The adjustment of input energy in an electron-positron collision has to be very precise. The resonance that is *J/ψ* is very sharp, and this new particle is not produced if the energy is wrong by more than a few MeV. In fact, this bound state of charm and anticharm has been called "charmonium" and likened to a tiny atom. It has ex-

209

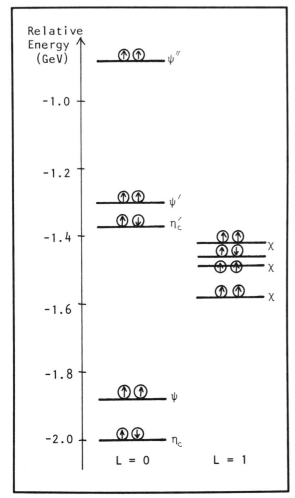

Fig. 14-1. Known and suspected energy states of a bound charmed quark and antiquark ("charmonium"), on an arbitrary energy scale. (The quarks occupy one of the states shown, forming the particle shown, depending on energy, spin alignment, and the angular momentum of the bound pair.)

cited states known as other mesons (Fig. 14-1), which were found as soon as the energy range was established by the finding of J/ψ. The charmonium atom can be made to jump between states, emitting photons in accord with this third spectroscopy of nature.

The J/ψ contains charm, yet is charmless because the charm and anticharm cancel. There should be mesons containing a charmed quark and some more common flavor. The first one found

with explicit charm was the D^0 meson, containing a charmed and an up antiquark. Also notable is the F meson, combining charm and strangeness. Many more have been found, as shown in Table 14-2. Combinations of quarks without charm are much lighter than those with charm. Charmed baryons have also been found, as shown in Plate 12 where a high energy photon collides with a proton.

A very heavy meson, the *upsilon* (upper case Greek ϒ), was found at Fermilab in 1977 by a large group of physicists (see Color Plate 27). Typically, experiments are done by large democratic groups now. A leader of the upsilon group was Leon Lederman. Collision of protons with copper nuclei were used, resulting in many known but unwanted particles. A broad resonance was found around 10 GeV, and it has a relatively long lifetime. The upsilon must contain some new and heavier quark so that it cannot change into the ordinary hadrons. Since it did decay to a virtual photon, it must consist of a quark-antiquark pair. The new quark flavor was named *bottom* (also occasionally called "beauty").

Actually, the upsilon was resolved into three resonances at 9.4, 10.0, and 10.4 GeV, indicating several excited states for the quark pair. Theoretical calculation of these states has been rather successful. The net bottomness of the upsilon is zero, but mesons (and baryons) exhibiting bottomness are expected to be found. More recently, electron-positron collisions have also produced the upsilon, which then decays to a bottom-

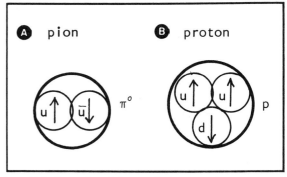

Fig. 14-2. Examples of the quark structure of a meson and baryon (simplified models showing the spins).

210

flavored meson (said to exhibit "naked bottomness." This B^0 meson has lower mass (5.2 GeV) since a heavy antibottom quark is paired with a light down quark. Since the upsilon resonance at 10.4 GeV has just enough mass to make two bottom mesons, they are produced nearly at rest with no other accompanying particles. The lifetime of the bottom quark can be estimated at 10^{-14} second. It then decays to lighter quarks, forming new mesons in the process.

If quarks come in pairs, maybe there exists a sixth one to be paired with the bottom quark. It was named *top* (sometimes "truth") in case its existence is verified. A giant group of 151 physicists at CERN, nominally headed by Carlo Rubbia, found evidence for it in 1983. It was expected to appear in particles of at least 30 GeV mass, and the proton-antiproton collisions available at 540 GeV should be more than sufficient to find "topness." The particle containing the top quark has not been identified, or named, but the evidence is that top-antibottom and bottom-antitop pairs were formed. These do not live long; the top quark can change into a bottom quark, releasing leptons at huge energy. The bottom does not stay around but annihilates with the antibottom quark to form bunches ("jets") of hadrons and a fast lepton. The mass of the top quark seems to be around 40 GeV. Work on the top quark continues.

14.3 COMPOSITION AND INTERACTIONS OF BARYONS AND MESONS--QUARK MODEL

Having seen some evidence for quarks from meson experiments, we are now in position to give the compositions of all known hadrons. Since there are many excited states, we shall restrict consideration to the familiar ones already discussed (which contain the up, down, and strange quarks), plus some new ones found to hold the less familiar quarks (charmed, bottom, and top). Using the symbols *u, d, c, s, b,* and *t* for the six flavors of quarks, Tables 14-2 and 14-3 give the structures of mesons and baryons, respectively. Since there are "only" twelve quarks and antiquarks, when taken in pairs it would seem that 66 distinct

mesons are possible. Many have not been found, particularly those with the heavy bottom and top quarks. Familiar mesons such as the pions have several alternate compositions which are equivalent because the *u* and *d* apparently have the same mass. Duplication also occurs because two quarks can have several states of angular momentum, and their spins can be parallel or antiparallel. For example, the A_2 meson has both quark spins parallel, and the quark-antiquark pair is bound in an angular momentum state of 1 unit.

A familiar meson, the pion, is made of a *u* and $\overline{u}$ as shown in Fig. 14-2A. Quark and antiquark form a neutral particle, and it should be clear why this pion is metastable. To obtain a charged pion, two different flavors are needed, always one an antiquark. Figure 14-2B shows the composition of the proton, *uud*. The baryons have been explored less than the mesons, but those having the *u, d,* and *s* quarks have been completely enumerated as well as observed. The theory for three quarks is more complex than for two. There are ten ways to combine three distinct quarks, and these form the 18 baryons in the octet and decuplet. The eight "duplicate" uses of three quarks form distinct baryons because of higher angular momentum states—for example, Λ^0 and Σ^0. Because of the charges of the three quarks, the quark model readily predicts that no baryon can have charge greater than $+2$ or less than -1.

The mathematical description of particle reactions in terms of quarks is done mainly with the wave/field model. But, as will be seen, quarks as particles provide a simple and clear picture that "explains" in its own way the reactions and decays of the hadrons. Thus with the usual cautions, the structures of a pion and a proton were shown graphically in Fig. 14-2, replacing abstract symbols with particles. The quark spins must be antiparallel to satisfy the zero spin requirement of the pion.

Interesting baryon states containing nonstrange quarks *u* and *d* are the *deltas* (Δ). The decay of the Δ^+ (*uud*) in Fig. 14-3A results in a nucleon and pion, by two different *channels*. In one channel the proton is composed of *uud*, so

Table 14-2. Quark Structure of Mesons.

Particle	Symbol	Mass-Energy	Quarks	Total Spin	Angular momentum
pion	π^+	0.140 GeV	$u\bar{d}$	0	0
	π^0	0.135	$u\bar{u}/d\bar{d}$	0	0
	π^-	0.140	$d\bar{u}$	0	0
kaon	K^+	0.494	$u\bar{s}$	0	0
	K^0	0.498	$d\bar{s}$	0	0
eta	η	0.549	$u\bar{u}/d\bar{d}/s\bar{s}$	0	0
rho	ϱ^+	0.77	$u\bar{d}$	1	0
	ϱ^0	0.77	$u\bar{u}/d\bar{d}$	1	0
	ϱ^-	0.78	$d\bar{u}$	1	0
omega	ω	0.78	$u\bar{u}/d\bar{d}$	0	0
kaon	K^{+*},K^{0*}	0.892	$u\bar{s}/d\bar{s}$	1	0
eta	η'	0.958	$u\bar{u}/d\bar{d}/s\bar{s}$	0	0
S	S	0.975	$u\bar{u}/d\bar{d}/s\bar{s}$	0	1
phi	ϕ	1.04	$s\bar{s}$	1	0
H	H	1.19	$u\bar{u}/d\bar{d}/s\bar{s}$	1	1
B	B^+,B^0	1.235	$u\bar{d}/u\bar{u}/d\bar{d}$	0	1
A	A^+,A^0	1.270	$u\bar{d}/u\bar{u}/d\bar{d}$	1	1
D	D	1.285	$u\bar{u}/d\bar{d}/s\bar{s}$	1	1
A_2	A_2^+,A_2^0	1.32	$u\bar{d}/u\bar{u}/d\bar{d}$	1	1
g	g^+	1.69	$u\bar{d}$	1	2
D	D^+	1.869	$c\bar{d}$	0	0
	D^0	1.865	$c\bar{u}$	0	0
F	F^+	1.97	$c\bar{s}$	0	0
D	D^{0*},D^{+*}	2.01	$c\bar{u}/c\bar{d}$	1	0
eta	η_c	2.98	$c\bar{c}$	0	0
J/psi	J/ψ	3.10	$c\bar{c}$	1	0
chi	χ	3.415	$c\bar{c}$	1	0
B	B^+	5.271	$u\bar{b}$	0	0
	B^0	5.274	$d\bar{b}$	0	0
upsilon	Υ	9.46	$b\bar{b}$	1	0

Notes: "/" denotes alternate quark compositions. Sometimes two or three charge states are listed on one line. Mesons are neutral unless shown otherwise. The omega meson is not to be confused with the omega baryon (different symbol). There are many more mesons (resonances) with angular momentum 1 and higher. Many pairings of quarks have not been found as mesons yet. All have the structure $q\bar{q}$. The order in which the quarks is written is not significant. The arrangements of the spins can be deduced from the total spin: one spin up and one down give 0; both spins up give 1.

nothing seems to have happened, yet an energetic pion is given off. Looking closer we see that all spins were up in the delta, to give the observed spin of 3/2. The proton has spin 1/2 and so must have one quark spin down. The decay must involve a spin flip, resulting in the emission of a virtual photon. The photon then most likely changes into a neutral pion ($u\bar{u}$ or $d\bar{d}$) but, rarely, remains a gamma ray. In the other channel, a u quark must change to a d quark to form the neutron. If the available energy materializes to form a pair $d\bar{d}$, then the d remains bound in the nucleon, and the $\bar{d}$ combines with a u to form the pion π^+ that

is emitted. No spin flips were needed.

Let us consider the strange decay of the lambda, as shown in Fig. 14-3B. It usually decays (slowly) to a proton and π^- {pi}, sometimes to a neutron and π^0. Not only are there spin flips in either channel of decay, but the strange quark s must change to a d quark. It takes a lot longer for a quark to change flavor than to flip its spin. The pion, of course, cannot carry away the strangeness; strangeness must decay away. The pion does form after a $u\bar{u}$ pair materializes.

What about the decay of the neutron to a proton and leptons? A d quark would have to change into a *u* quark, with no virtual pairs formed since there is too little surplus energy. This must be a very slow process since the neutron can live about 16 minutes. Whether quarks are behind this weak interaction will be discussed later.

In Fig. 14-4 a reaction is considered. When two protons collide, there are four *u* quarks and two *d* quarks to interact. Within the constraint that two baryons must be produced, many results are possible—virtually any combination of the quarks. Excess energy will materialize as one, two, three, or more, quark-antiquark pairs, $u\bar{u}$, $d\bar{d}$, and perhaps $s\bar{s}$ or $c\bar{c}$. Members of these pairs join in the interaction. A result such as shown in Fig. 14-4 might occur if the energy is high enough to form

the strange quark and antiquark. Additional pairs materialize until possible mesons can be formed. The meson table can be used to identify the results.

When two nucleons collide, each quark has a chance of hitting three other quarks, with nine possibilities. When two mesons collide, each quark can hit one of only two others, giving a total of four possibilities. When a meson and a nucleon collide, there are six possible interactions. The existence of quarks has been confirmed by examining the relative cross sections of these different collisions. The cross section for nucleon-nucleon collisions is indeed appropriately greater than for the others. Examination of the spin effects in proton-proton collisions has revealed the possible

Table 14-3. Quark Structure of Baryons.

Particle	Symbol	Mass-energy	Quarks	Total Spin	Isospin
Octet:					
nucleon	p	0.9383 GeV	uud	1/2	1/2
	n	0.9396	udd	1/2	
lambda	Λ^0	1.116	dus	1/2	0
sigma	Σ^+	1.189	uus	1/2	1
	Σ^0	1.192	dus	1/2	
	Σ^-	1.197	dds	1/2	
cascade	Ξ^0	1.315	uss	1/2	1/2
	Ξ^-	1.321	dss	1/2	
Decuplet:					
delta	Δ^{++}	1.234	uuu	3/2	3/2
	Δ^+	1.232	duu	3/2	
	Δ^0	1.231	ddu	3/2	
	Δ^-	1.230	ddd	3/2	
sigma	Σ^{+*}	1.382	uus	3/2	1
	Σ^{0*}	1.382	dus	3/2	
	Σ^{-*}	1.387	dds	3/2	
cascade	Ξ^{0*}	1.532	uss	3/2	1/2
	Ξ^{-*}	1.535	dss	3/2	
omega	Ω^-	1.672	sss	3/2	0
A charmed baryon:					
lambda	Λ_c^+	2.282	duc	1/2	

Notes: The octet and decuplet use all possible combinations of d, u, and s quarks. Only 10 combinations are possible, so some of the higher energy baryons consist of higher angular momentum states of the same combination that formed a lower energy particle (for example, dus). The order in which the quarks are written here is not significant. The octet has two spins up and one down; the decuplet has all spins parallel. Additional multiplets are being determined as additional quarks are added to the combinations. Each baryon has many higher energy states. Each multiplet of charge states is determined by the isospin for that baryon.

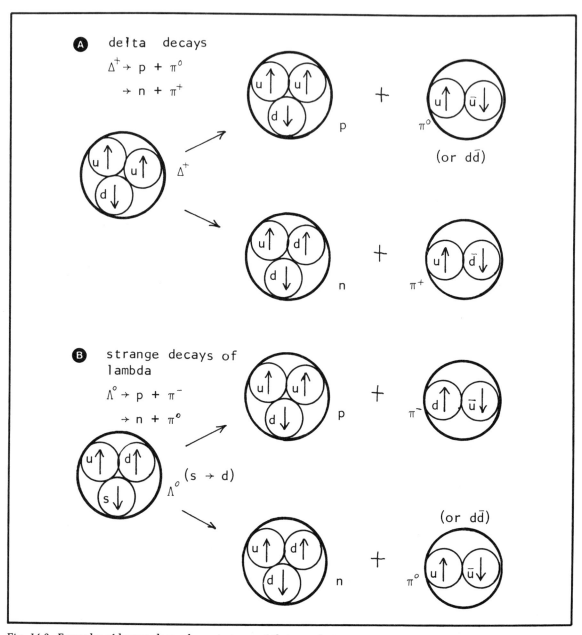

Fig. 14-3. Examples of baryon decay shown in terms of their quark structures.

existence of particles with spin inside the protons—more quark evidence.

The collision of electrons and protons at energies like 20 GeV is called "deep inelastic scattering." It is done at such high energy that structure about 1/3 the size of the proton is detected.

The structureless electrons provide a simple electromagnetic interaction and search out only the charge structure of the protons. The results are that the proton contains three tiny hard constituents.

Evidence for quarks is found in the appearance of *jets* of particles (mostly pions) from high energy collisions. In the lab frame one would expect all products to appear as a forward "jet." But jets instead of randomly diverging particles occur in the CM frame as well and are an indication of the inseparability of quarks. As shown in Fig. 14-5, the energy from an electron-positron collision materializes as a quark-antiquark pair. As soon as the pair separates a certain amount, the force between them increases so much that additional pairs materialize from the rising field energy. This process continues until a number of mesons (and baryons) form and appear as members of the jets.

14.4   TRYING TO FIND FREE QUARKS

Much effort has gone into the attempt to catch a "naked" quark on film or otherwise. Until theoretical reasons became compelling for quarks to remain bound to each other, physicists were certain that a few loose ones might lurk in ordinary matter. We might expect some left over from the "Big Bang" at the beginning of the universe or as high energy "cosmic rays" coming from elsewhere.

Free quarks cannot decay since they would violate conservation of charge. There is no known decay product. Therefore any free quarks would remain unless they encountered antiquarks of the

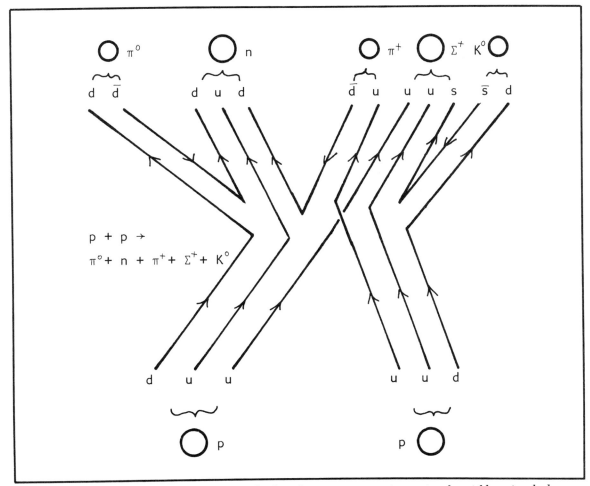

Fig. 14-4. A Feynman diagram tracking their quarks through a proton-proton interaction that yields various hadrons.

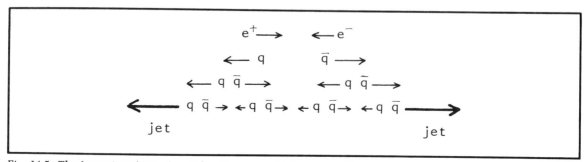

Fig. 14-5. *The formation of jets of particles after a very high energy collision as an indication of the materialization of quark-antiquark pairs formed as the first quarks become too far apart. (Shortly thereafter all pairs appear as mesons in the jets.)*

same flavor (a rare encounter). Also, the high energy accelerator provides enough energy to form at least one loose u or d, even if the event is very rare. The passage of a quarks would leave a different density of ions and therefore bubbles in a bubble chamber than normally charged particles would leave. The tracks would stand out.

Quarks arriving from across the galaxy would be trapped in the Earth's crust and could, if plentiful enough, be found in samples of matter. One might attach itself to an atom, acting the role of a very heavy electron or entering the nucleus. Such an atom would not be neutral and could be separated from other atoms by means of the mass spectrometer. It would also have unusual spectral lines and could be detected by heating the sample. Many searches have been done for quarks in matter. The supposed chemistry of atoms containing quarks has been studied, and searches have concentrated on sea water, minerals, and metals, especially iron and niobium. Despite a sensitivity that allows the detection of one unusual particle or atom in about 10^{22}, no quarks have been found yet. Nevertheless, attempts to modify quark theory continue so that free quarks could be explained if they were found. This is quite the reverse of early attitudes toward quarks!

14.5 COLORED QUARKS

Further complexity lies ahead. Quarks (with spin 1/2) are supposed to act like fermions, obeying Pauli exclusion and not all occupying the same states. Yet certain baryons (Δ^{++} and Ω^-) are made of three supposedly identical quarks with

all their spins up. Either Pauli's exclusion is violated, or there is some difference among the quarks. Many other baryons have two identical quarks with spins up in the same state (for example, in the nucleons). Some evidence from collisions also indicates that quarks of the same flavor are distinguishable and contribute to interactions. The decay of the neutral pion is best explained by "colored" quarks. A good explanation is also needed as to why only two or three quarks bind together.

The answer is traceable to Oscar Greenberg in 1964 but its acceptance was more recent. In addition to flavor, quarks have *color*. Each flavor can have one of three colors. Here is more physics based on the number three, quite novel in comparison to older philosophical approaches to nature. Now it seems likely that there are 18 distinguishable quarks (36, counting antiquarks), and the supposed simplification of elementary particles seems in danger of extinction.

The use of "color" is metaphorical only, and has nothing to do with ordinary colors of the rainbow. Nevertheless, the color names are used: red, green and blue. Another name that has been used is *color charge*, with the implication that there is a new kind of charge that comes in three varieties. Color is another unobservable, yet evidence demands that such a concept be defined. Each quark in a baryon (or meson) must have a color such that when three (or two) are combined (Color Plate 1), the colors blend or cancel to produce colorless particles (white). It so happens that red, green, and blue are the three "primary" colors of light that, when added properly, produce white light. Each color also has an anticolor or complementary color, which, when added to the primary color, produces white. Red and cyan (a light blue) are complementary, and so are green and magenta (a reddish blue), and blue and yellow. (The complementaries appear when a color negative is made.) By this model a meson made of red and cyan (antired) quarks (Color Plate 1b) will appear colorless (white). But what about a red and antigreen meson? Here the analogy to ordinary color breaks down. The color of any quark "cancels" the color of any other antiquark. We shall limit the color names to three and use "antired" instead of "cyan," but the Color Plates will show antired as cyan.

Supposing three colors, how is a baryon made of them? Does the proton possess, for example, a red u, a green u, and blue d? The answer seems to be that all three colors are present for each quark. The quarks are continually changing color in such a way that each experiences all three colors consecutively, while its two partners have the other appropriate colors to form "white." It is as if coordinated pointers rotate and set the color of each quark (Color Plate 2). There are actually only two independent colors in a baryon, since the specification of two implies the third. The color models shown in the Plates are representations only, since each quark is probably a point particle. There is some truth to their depiction as spheres because, as will be seen, each quark surrounds itself with more of the same color.

The introduction of color has been made compatible with the eightfold way and its underlying SU(3) group theory. It is necessary that all hadrons be color singlets; that is, one of each color is present, no more, no less. This "color symmetry" is compatible with a "color gauge theory," to be discussed in the next chapter when we return to consideration of the strong interaction among quarks. Eventually it should be clear why the colored quark theory, complex though it seems, must be this way.

Chapter 15

Field Theories
for Particles

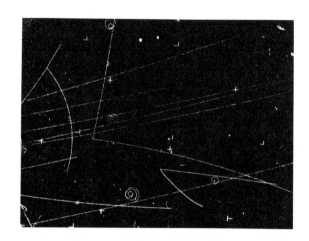

A N ELABORATE SET OF THEORIES USING FIELDS
has been constructed, not only for quarks, but
for weak interactions and for combining several
of the known forces. We are approaching the
climax of current particle physics: the unification
of three of the fundamental forces into Grand
Unified Theory. Gravity may also be added. We
have met field theories before, especially Maxwell's electromagnetism. Einstein pioneered attempts at further unification when he tackled
gravity and electromagnetism together. (No other
fundamental forces were known at that time!) Unfortunately little headway was made, as gravity
has been particularly resistive to being combined
with other forces. The basic elements of field
theories are reviewed here before discussing their
applications to quarks and to unification.

15.1 ELEMENTS OF
MODERN FIELD THEORY

As indicated for the theoretical success of the
quark description of the hadrons, symmetries play
a crucial role. One might expect nothing deeper
from nature than that all aspects have symmetry.
The symmetries important in particle physics are
symmetries of translation and rotation in space
and time, charge conjugation, handedness, rotations in an imaginary space of quantities such as
isospin, strangeness, charm, bottomness, topness,
and more to be discovered. Each symmetry operation leaves the corresponding quantity invariant.
Each symmetry has a corresponding conservation
law pertaining to the conservation of the quantity: momentum, angular momentum, energy,
charge, parity, isospin, lepton number, baryon
number, strangeness, charm, and so forth.

Each symmetry can be broken. As a simple
example, translation through space is not symmetric if a wall is in the way. Then momentum
is not conserved for a particle if it hits the wall.
The wall has broken the symmetry and made one
part of space different from another—namely, that
the space behind the wall is inaccessible to the particle. We saw that the weak interaction breaks

many symmetries—charge conjugation, handedness, time reversal, strangeness, charm, isospin, and more. The electromagnetic interaction breaks only the symmetry of isospin, so that the neutron and proton have different mass-energies. The strong interaction preserves all known symmetries. (Should not some interaction do so?) Any symmetry breaking is observed by physicists as a difference in energy—a splitting of what had been two identical (degenerate) cases—or as a difference in reaction rates.

A major ally of the theoretical physicist is the *PCT theorem*. This theorem simply states that when a parity transformation P, a charge conjugation C, and a time reversal T are all done at once to a particle or reaction, the result must be the same as before. Each is a binary operation; " + " is changed to " − " or vice versa. The PCT theorem was derived from fundamental considerations in quantum mechanics and relativity combined. If it is proven wrong experimentally, then quantum theory or relativity or both must also fall. So far we are safe. The notorious neutral kaon manages to violate P, C, and T in such a way as to preserve PCT.

Every force has a field, and every field is "carried" from one particle to another by quantized particles called, among other terms, *carriers*. They are always bosons. The photon, the carrier in QED, interacts with only one kind of particle at a time. It does not carry charge, nor does it carry something else that changes an electron into, for example, a muon. It can couple an electron and antielectron (positron), however. Since there is only one kind of photon, the symmetry group of QED is called U(1). Carriers of other interactions act more complexly and so the symmetry groups are larger. The carriers of the weak force carry a new kind of "charge" called *weak charge* from one particle to another. The transport of weak charge is called a *weak current*. The symmetry group of the weak theory is SU(2). The carriers of the color force carry *color charge* from one quark to another, changing the colors of the interacting quarks. We have learned that quarks fit the SU(3) symmetry group.

It has been shown that the electron (or any charged particle) has an electromagnetic self-field. Its existence can be explained in terms of particles with the Feynman diagrams: virtual photons, electrons, and positrons are continually flashing in and out of existence around each electron. They partially shield the electron charge and contribute to its mass-energy. The same picture applies to all particles and the fields around them. The electron and other participants in the weak interaction must have weak fields around them, punctuated with the appearance and disappearance of carriers of the weak force and of particles that so interact. We shall soon meet the carriers. At one time it was thought that the proton had a cloud of pions, carrier of the strong force, and virtual protons and antiprotons around it. Now with the advent of quarks, the focus has changed. There must be carriers of the strong force (now called the *color force*) between quarks, and a cloud of carriers and quarks of various colors around each quark.

Because of the self-field, the constant of interaction between two charged particles is no longer a constant (such as k or $\frac{1}{4\pi\epsilon_0}$ but instead increases at close range. The reverse occurs for the carriers of force between quarks. The color charge of the quarks is dispersed when the quarks are close together so that they interact more weakly than when farther apart. Most particles also have clouds of weak carriers and weakly-charged virtual particles around them. Some of the weak carriers possess weak charge, too, and the net weak charge is spread out around the central particle. The result is that the weak interaction decreases at small distances (but also at large distances, unlike the color force).

Some related terminology should be assimilated here. A photon is called a *vector* carrier because it has spin 1. Its electromagnetic field is a vector field (with three components). The photon is also called a *neutral current* when it passes between two electrons because it carries no charge. All carriers are bosons because there must be no limit on their number and no need to account for their states. There could exist spinless bosons, which would be *scalar* carriers, transporting no spin. Three vector bosons have recently been found for the weak interaction. Two carry both electric and weak charge. One carries only weak charge and so is called a "neutral weak current." If the distinctions between electric, weak, and other postulated kinds of charge are kept in mind, these terms will not seem self-contradictory. Eight carriers are postulated for the quark or color force, most of which carry a color charge.

Any particle or carrier which has spin can exist in right or left handed forms. This is called *chiral symmetry or chirality*. Its spin can point along its direction of travel or against it. This handedness should not be confused with the intrinsic parity of a particle, which is a fixed value for a particular kind of particle (" + " or " − "). The massless photon can never be stopped and turned around, since it always has lightspeed. Therefore despite having spin 1 it cannot be found to have three components, giving a transverse form that is neither right nor left handed. Nor can it change its handedness. Electrons have mass and spin 1/2 and can occur in right and left forms. The vector bosons for the weak force can also be found in the transverse form. Since it is not certain whether the neutrino mass is zero, it is not certain whether it can change handedness. Currently the neutrino is assumed massless as no definite

results to the contrary are available. Only left handed neutrinos and right handed antineutrinos have been observed. It will be seen that the weak force depends on handedness, as one would expect from observed parity violations in weak interactions.

Gauge symmetry can be a particularly elusive concept, yet a vital one, in modern field theories. Actually it has been known since Maxwell's unification of electricity and magnetism—the first unified field theory. Lorentz found the first gauge transformation in the late 19th century. Gauge symmetry applied to electric charge predicts the conservation of charge. It can be generalized to many other properties of matter. The energy in an electric field and the shape of the field should depend only on the difference in potential between one place and another in the field. Transformation of the reference point for the potential energy should have no effect on the field or energy. The same physics results for both electric potential fields shown in Fig. 15-1. When charges are in motion, gauge symmetry is preserved if magnetic potential is also considered. More generally it might be said that gauge symmetry not only removes the dependence of physics on the sizes of quantities—that is, the units of measure—but also any dependence on the zero point of the units.

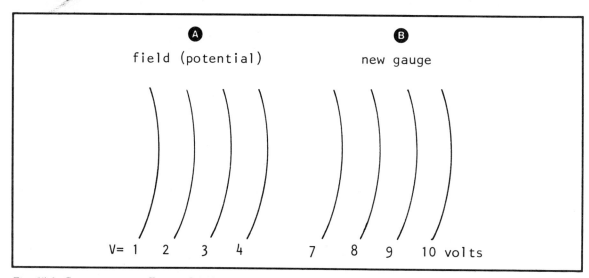

Fig. 15-1. *Gauge symmetry illustrated with a rescaling of potential without any change in the field.*

Hence charge can be measured in such a way as to total to zero, giving us the conservation of charge.

The additional phrase "non-Abelian" is often added to describe modern "gauge fields" which have gauge symmetry. This means that the algebra is noncommutative and the order in which operations is done matters (leading, as we have seen, to the uncertainty principle). It is now known that the zero mass of the photon is a consequence of the gauge symmetry of the electromagnetic field. Gauge symmetry results from scale invariance (proposed by Hermann Weyl in 1918). There is gauge symmetry when rotation of a variable in an imaginary space gives a physical invariance. The term "gauge" is traceable to an early but inaccurate analogy with the use of gage blocks by machinists to measure distances accurately.

The symmetries that apply to particles turn out to require a more stringent symmetry than has heretofore been implied. An operation such as rotating coordinates or C, P, or T is normally thought to apply to all of space at once. Symmetry under such a rotation is *global* or universal. The other form of symmetry is therefore *local*. The symmetry operation varies from place to place. It is as if the units of measure for a quantity could be varied from place to place without messing up the physics.

The astonishing result is that local non-Abelian gauge symmetry works in creating physical laws, and works very well indeed. With hindsight from using gauge symmetry in a local way, physicists can now deduce a gauge theory such as Maxwell's electromagnetism without knowing any experimental evidence about electricity and magnetism! Once it is decided that translational, rotational, charge, and a few other symmetries are to be locally preserved, both mathematics and nature automatically provide the correct electromagnetic field that makes this so. The gauge approach, used locally, is so powerful that working physical laws can be pulled out of a mathematical hat. This power has proven to be what was needed to make headway in really understanding the fundamental forces. The gauge approach has also relieved physicists of the embarrassing infinities that appeared in earlier approaches. QED was successful not because the infinities were faked out but because the underlying gauge symmetry was accidentally built in, thanks to Maxwell's equations. This is perhaps very subtle progress in physics, hardly dramatic to outsiders; yet this chapter will show how much can come from these breakthroughs. The Nobel Prize committees fortunately search deeply enough to reward such work.

15.2 THE THEORY OF WEAK INTERACTIONS

The weak force has been found to discriminate between left and right hand particles. In fact, only left hand particles (and right hand antiparticles) participate in weak interactions. Therefore only they possess weak charge. The range of the weak force is very small, about 10^{-17} meters or 0.01 fm. Weak charge is usually given the values $+1/2$ for left hand particles and $-1/2$ for right hand antiparticles. (At least there are only two values!) As far as weakness is concerned, particles come in doublets, hence the SU(2) group as found by S. Bludman in 1958. The electron and neutrino are members of one doublet. The d and u quarks form another doublet. The antiparticles of these form two corresponding doublets. The right hand particles (and left hand antiparticles) have no weak charge so each is a singlet state. Because a weak interaction could turn a left hand spin 1/2 particle into a right hand spin 1/2 particle, the weak carrier must be a spin 1 boson.

From the short range, the mass of the weak carriers can be estimated to be 100 GeV. It was a major triumph of theoretical physics to predict their properties and then have them found. Progress began in 1973 when "strange decays" were found, involving neutral weak currents. In Fig. 15-2 neutral hadrons of varying strangenesses decay to other particles with less strangeness via some mechanism which does not carry electric charge. Although quarks were not suspected then,

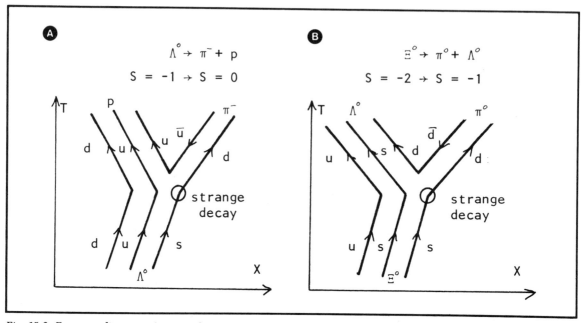

Fig. 15-2. Feynman diagrams of strange decays in which there is a neutral weak current. (Only the strangeness decays as weak charge is transferred; there is no electric charge transferred.)

the s quark decays to a d. The decays were known to be weak interactions by their slow time scales.

Surprisingly, success in understanding weak interactions was aided by the unification of electromagnetic and weak interactions, a story to be covered shortly. In 1982 and 1983, after many tries, data was collected with the Super Proton Synchrotron (SPS) at CERN, smashing protons against antiprotons at 540 GeV in the CM frame. Large elaborate detectors UA1 and UA2 surrounding the interaction region (Color Plates 28 and 29) helped to identify about 90 W and 10 Z^0 "intermediate vector bosons," aided by on-line data processing and computer reconstruction of events. The W, which comes in electrically positive and negative forms, has about 80 GeV of rest mass-energy, and the Z^0 has about 90 GeV. These are the most monstrous elementary particles yet isolated by scientists, although heavier ones are predicted. Success was due in part to Simon van der Meer's technique for obtaining a dense beam of antiprotons so that rare events could be observed, and in part to Carlo Rubbia's huge 2000 ton UA1 detector and huge team of 134 physicists.

Both received the Nobel Prize in 1984 for this work (Plate 14).

The W^-, W^+, and Z^0 live very briefly, then decay to electron and antineutrino, positron and neutrino, and electron-positron pair, respectively. The Z^0 has a role similar to the photon; it just acts on a different kind of charge. The W^+ is the antiparticle of the W^-. Both carry one unit of both electric and weak charge. The Z^0 only carries force, not any kind of charge and thus leaves particles unchanged. The existence of the weak bosons allows an explanation of the weak decay of the neutron to a proton. In Fig. 15-3 a neutron (containing quarks udd) changes to a proton (quarks uud) when a d emits a W^- to become a u quark. The charge of d is $-1/3$ and of u is $+2/3$, so removing -1 unit of electric charge is consistent. The W^- stays around only about 10^{-25} second and then decays to an electron and antineutrino, just as expected for beta decay of the neutron. The slowness of the process is related to the weakness of the force, which is almost the same as saying that one must wait a long time to see such a monstrously heavy virtual particle as

222

the W^- materialize to allow the decay of the d quark. Only the weak interaction can change quark flavor.

The behavior over distance of the weak force is now better known. It vanishes to zero at distances greater than about 10^{-17} meter. It increases to and remains at a value of about 10^{-5} of the strong force over the range from 10^{-17} to 10^{-25} meter. A single value of strength cannot be assigned to it. This description is for the "naked" weak force; shielding by virtual particles changes the behavior at small distances, further weakening it.

15.3 THE ELECTROWEAK THEORY

Sophisticated consideration of the role of electric and weak charge and boson carriers in the electromagnetic and weak interactions led to the hope, and then to the reality, of the unification of these two fundamental forces. The goal was achieved finally by Sheldon Glashow, Abdus Salam, and Steven Weinberg in the 1960s, and they received the Nobel Prize in 1979 before their predictions were confirmed! The new theory of nature is called the *electroweak* force.

Theories can be combined by multiplying the respective groups which represent them. It was expected that the electroweak force would be

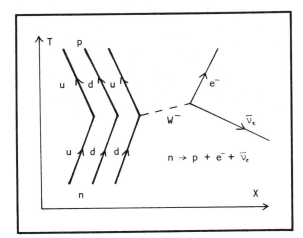

Fig. 15-3. *The Feynman diagram of the quarks and vector boson involved in the weak decay of the neutron.*

represented by U(1) for electromagnetism times SU(2) for weak theory. The symmetry of the combination is observed to be not as perfect as the symmetries of its parts. Weak charge is not conserved in the real world and this results in its carriers having masses. What has happened is that the ideal electroweak theory is perfectly symmetrical but its participation in nature results in some symmetry breaking. The major influence is the ability of the vacuum to generate virtual particles. At a very small scale, less than 10^{-18} m (1 attometer), representing energies much greater than the W and Z boson masses (about 1000 GeV), they interact as freely as photons, and the electroweak force is symmetrical. In fact, at such close quarters the weak bosons have the same energy as photons. They are indistinguishable! At much larger distances the symmetry between the weak bosons and the photon is broken, and nature splits the electroweak force into two rather distinct interactions. In the micro-microworld where the electroweak force reigns, weak charge is conserved and the handedness of particles is fixed because all particles have negligible mass compared to the energies involved.

Two problems remain with the present form of electroweak theory. The forces are still expressed differently with their own strength constants. Also, the quantization of electric charge is not fully dealt with. For practical purposes physicists may wish to preserve separate theories for electromagnetism and the weak force. The old theories remain useful, although the weak theory was not very successful until unification brought greater understanding.

15.4 QUARKONIUM

Yes, *quarkonium* is an "atom" made of a quark and an antiquark. It is useful for studying the force between quarks in its simplest manifestation. Quarkonium is assembled by colliding electrons and positrons at 10 GeV or higher. Virtual photons form and quickly (but rarely) materialize as quark-antiquark pairs. Quarkonium has energy levels with a hundred million times more energy than positronium, the simplest "atom" previously

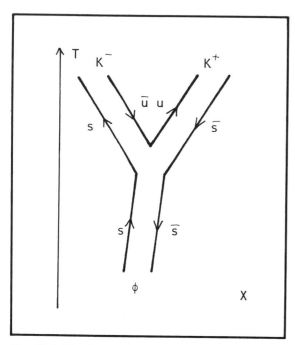

Fig. 15-4. The Feynman diagram of the decay of the phi meson ("strangonium").

discussed. It is also a hundred thousand times smaller. The existence of a quarkonium atom is verified by surrounding the collision region with the Crystal Ball detector, a sphere of 732 solid state detectors (Plate 7). Sometimes the two quarks have no other choice but to annihilate each other, a relatively slow process. Meanwhile, the "atom" may change state, emitting a high energy photon. Ultimately other quark-antiquark pairs materialize nearby, combine with the quarkonium constituents, and form recognizable hadrons. An example is shown in Fig. 15-4 for the phi meson which decays to kaons.

We have met some forms of quarkonium already. "Charmonium" is the brief combination of the c and $\bar{c}$ quarks to form the J/ψ and related particles. Eight energy states of it have been measured and more are predicted (recall Fig. 14-1). The lowest state is the charmed eta meson. Another set of states belong to the chi meson. "Bottomonium" is the combination of the b and $\bar{b}$ quarks to form the upsilon particle. Four energy states have been measured for it. "Toponium" is

expected to be made at greater than 40 GeV. The lowest state of "uponium" and "downonium" is the neutral pion, and "strangonium" in its lowest state is the eta meson. The light quarks in these move at relativistic speeds and do not provide the simplest possible "atoms" for study.

The color force is of great interest. It is possible to deduce from the measured energy levels of charmonium and bottomonium that the color force decreases with the square of the distance for distances less than about 0.5 fm and levels off to a constant value at greater distances. The constant value has been found to be about 16 tons. That is a lot of force between just two elementary particles! No wonder free quarks cannot be separated far enough to take their picture. At smaller scales the quarks become relatively free to move about, a condition called "asymptotic freedom" invented by David Gross, Frank Wilczek, and H. D. Politzer in 1973.

15.5 QUANTUM CHROMODYNAMICS

The theory for the color force and the behavior of quarks must be a gauge field theory that is non-Abelian. Since only the strong/color force preserves isospin symmetry, this must be compatible with the gauge theory. C. N. Yang and Robert Mills invented such a gauge theory in 1954 and it is often referred to with their names (Yang-Mills). It has had to be modified substantially since then. J. J. Sakurai first suggested in 1960 that it could be applied to the strong force. Another ingredient for a new theory of quark interactions is the abundant use of Feynman diagrams to aid calculations when clouds of virtual particles are possible. Computer programs have been developed to do the enormous amount of algebra in the calculations.

The concept of a vacuum field or Higgs field was introduced by F. E. Brout, Robert Brout, and Peter Higgs. It is necessary to give some carriers mass in the Yang-Mills gauge field. This field is defined to have the lowest possible energy, but it cannot be zero. It would cost energy to cancel this field to nothingness. The Higgs field provides a direction against which to measure the orientation of isospin. The use of a Higgs field provides "spontaneous symmetry breaking," so that the theory can be symmetric but the observable effects—for example, that a proton and neutron are not the same—can be accommodated. We ultimately require symmetric theories in every imaginable respect if they are to be universal. The Higgs field has spinless scalar carriers, and these "Higgs bosons" are expected to be very massive and the subject of much searching. Gerard 't Hooft carried out the first successful calculations of this modified renormalized Yang-Mills theory in 1971. Weinberg, Salam, and John Ward adapted it to the weak interaction and moved toward the electroweak theory as discussed earlier. The theory predicts the existence and properties of the weak bosons, the photon, and the Higgs carrier.

Quantum chromodynamics (QCD) is a non-Abelian gauge theory that sprang from the work on the weak force. Theoretical difficulties could be repaired if a quark with a property called charm were found, and it was found. QCD is modeled after quantum electrodynamics (QED), with color charge replacing electric charge. Its gauge symmetry is based on an invariance when colors are changed. Therefore "white" is a result of the gauge symmetry. Selection of a color is handled as isospin was handled; an imaginary pointer can point in one of three directions in an imaginary "color space." To keep a baryon colorless, the pointer in each of its three quarks must, at any time, be at a different color (Color Plate 2). The three pointers rotate in step (in phase) to preserve "whiteness." At any given time, each quark has the color indicated by the pointer. The three color charges require the symmetry group SU(3).

Unfortunately, it is necessary to have more fields in order to preserve color invariance. The QCD fields have eight components. Therefore they need eight carriers of zero mass and spin 1 called *gluons* (obviously named for the way they glue matter together). The gluons carry no charge but each must carry two colors. It takes eight different gluons to be able to convert any color quark into any other color. They are represented as having size in Color Plate 3, but actually gluons are point

particles. Each gluon carries one color away from a quark and an anticolor complementary to the new color it had to leave behind on the quark. For example, in the Feynman diagram of Color Plate 4a, the green quark emits a green-antiblue gluon (symbol $G_{G\bar{B}}$). The gluon carries away the green and it also carries the antiblue color yellow, having left blue with the quark. A quark can also emit a gluon without changing color. There are only two independent gluons, G_1 and G_2 which can carry a color and its anticolor, and they are shown grey in Color Plate 3 to avoid an incorrect representation. As an example, a red quark would emit a red-antired gluon as a mixture of G_1 and G_2 shown with the appropriate color in Color Plate 4b; the antired color cyan returns red to the quark.

The quark-gluon equivalent of the exchange of a pion between nucleons is illustrated in Color Plate 5. Here we see the details behind what was originally called the strong nuclear force. It should be kept in mind that in this Feynman diagram, any particle (or color) moving downwards, against the flow of time, is an antiparticle (or anticolor). To remind us that the quark colors of the proton and neutron are always changing, some gluons are shown which bring about the changes. The quarks in the pion also exchange colors periodically. The nuclear force is a precise leakage of QCD when hadrons approach to within 10^{-15} m and the quarks "feel" each other.

So far nothing about QCD explains why quarks cannot be freed, nor why gluons are not observed. The effects of asymptotic freedom of quarks and of the existence of gluons have been observed in experiments at DESY. After an electron-positron collision, quark-antiquark pairs are formed whose decays usually result in two jets of hadrons instead of randomly diverging particles (recall Fig. 14-5). As the original two quarks draw apart, they begin to feel the color force and cannot move farther apart. This causes more pairs of quarks to materialize adding to the showers in two singular directions and creating jets. Occasionally, a quark emits a gluon first, which is also free of the color force near the center of interaction.

When it reaches the limit allowed by the color force, it too forms a stream of quark pairs in a third singular direction, giving a third jet as observed.

The explanation for the decrease in color force as quarks approach and the increase as they separate is based on a "camouflage" effect, opposite to the way virtual charges shield an electron. Because virtual gluons and quarks can materialize around a quark, their colors affect the color "seen" for the originally bare quark (see Color Plate 6). Like colors repel, and any two unlike colors attract, with a force proportional to the product of the color charges. Since the gluons are colored, "half" with the original quark's color and "half" with the anticolor (so to speak), and since the gluon colors cannot separate, they do not shield but instead enhance the bare quark's color. For example, each gluon with green, near a green quark, adds to the total green charge. From a distance the green quark seems to have a stronger color field. The virtual quarks do not contribute as strongly to the enhancement of color because they move apart. Quarks and gluons with other colors also materialize around each quark but do not enhance its particular color until it changes color.

Another analogy is that the color force acts like a rubber band (or spring); it is stronger at large distances and weak close by. The "bag" model is also used. Quark confinement has been modeled by using a lattice model of space for calculations. This was developed by Kenneth Wilson, who received a Nobel Prize in 1982 for applications to statistical physics. Space is not assumed continuous but rather composed of discrete, separated points. Since the quarks have spin and color charge, they also have a "color magnetic field." This leads to an energy difference as to whether quark spins are parallel or antiparallel. Mesons with parallel quark spins are heavier than those with antiparallel spins because the color magnetic fields repel in the former case and attract in the latter.

The masses of the quarks themselves have been difficult to ascertain. Values given, such as in Table 14-1, are estimated using the assumption that a particle such as a proton consists of about equal parts quarks and binding energy. The u and d quarks are known to have nearly, perhaps exactly, the same mass. This can be seen by comparing the neutron and proton masses. Three quarks of about 100 to 200 MeV must make up about half of the proton's mass-energy leaving the rest as binding energy. Particles constructed from strange, charmed, and other quarks are much heavier, and it is assumed, with experimental justification, that the corresponding quarks are similarly heavier.

Another prediction of QCD is that two or more gluons should combine to form a "glueball." Evidence for them has been found at SLAC. The reader would be justified in being dazzled by the structure of QCD. It is an enormously complex theory, hardly understood fully by anyone. It would be difficult to display the mathematical form. Calculations, from which predictions can be made, are very difficult. Moreover, the theory cannot predict certain basic facts such as the strength of the color force. Some constants must be measured experimentally first. It may seem uncanny that the precise properties of the proton depend on an evanescent cloud of virtual gluons and quarks around each quark. The exact value of the nuclear force for protons is determined extremely accurately by this seemingly indeterminant color interaction.

15.6 GRAND UNIFIED FIELD THEORY

A *grand unified theory* (GUT) of elementary particles unites three forces of Table 15-1 (excepting gravity) in one large field theory. GUTs are at the frontier of theoretical physics and there are several competing versions. Each makes predictions about nature which could be verified experimentally, allowing the correct theory to be chosen. But the energy and time scales involved are so large that experiments are difficult and some may be impossible. There is a *unification energy* and a *unification scale* at which the three forces act as one unified force. The energy is

Table 15-1. Fundamental Forces, Carriers, Symmetry, and Unification.

Force	Source	Carrier	Symmetry	Unification
electromagnetism	electric charges	photon	U(1)	electroweak SU(2) × U(1)
weak	weak charges	W^+, W^-, Z^0	SU(2)	grand unified SU(5)
strong/color	color charges	8 gluons	SU(3)	supersymmetry
gravity	mass/energy	graviton		

Notes: See also Table 2-1 for other basic properties of the forces. The carriers are massless spin 1 bosons, with the exception of the heavy intermediate vector bosons for the weak force. Supersymmetry is covered in Ch. 16.

estimated to be 10^{15} GeV, about 10^{11} times larger than the largest proposed accelerator. This energy corresponds to about 10^5 joules for a single particle or reaction, comparable to the energy needed to boil 10 grams of water.

The unification scale is about 10^{-31} meter, such an incredibly small size inside a proton that it is as if a whole new universe of physics exists within protons. We may have been thinking of the three quarks in a 10^{-15} meter proton as each occupying about one-third of the proton, but any two of them must approach within 10^{-31} meter in order to interact with the unified force. (Remember that the quarks are supposed to be point particles, so they could approach as closely as wanted.) We shall see shortly that one must wait a long time for two quarks to approach so closely. Instead of the picture of three substantial colored balls in a bag, one should think of the proton as a vast empty chamber in which three tiny specks roam freely in an otherwise empty universe.

Grand unification implies new connections between leptons and quarks. They should be able to turn into each other under the right conditions, and it can be no accident that the lepton charge is an exact multiple of quark charge. We have already observed that there seem to be equal numbers (six, plus the antiparticles) of leptons and quarks. All familiar matter is made of just one "family" of these, the up and down quarks and the electron and its antineutrino. The other two families contain the other heavier quarks and leptons, which happen to be in the order discovered (see Table 15-2). We must count the colors of the quarks, so that there are actually 8 particles in each family. Each particle has a unique electric and color charge (leptons having zero color). Color charge is measured in units of 1/2.

The symmetry group for each force can be combined in a grand multiplication. The new symmetry group could be U(1) × SU(2) × SU(3). It contains each group for separate use, too. The symmetry group SU(5) was found suitable in 1973 by Howard Georgi and Sheldon Glashow. An SU(5) group requires five particles to represent it, and the *d* quark (in three colors), the positron, and the antineutrino were chosen. All must be right handed. Each particle now carries four kinds of charge: electric, weak, and two independent color charges. (The leptons have zero color charge, as usual.) The positron has weak charge + 1/2, and the antineutrino − 1/2. The *d* quark has electric charge − 1/3.

Twenty-four carriers of the grand unified force are needed in one version of GUT to intermediate between each pair of these elementary particles. We know about twelve of them: photon, Z^0, the two Ws, the two neutral gluons, and the six colored gluons. The other twelve have been given the legendary symbol *X*. Each *X* can carry the three kinds of charge needed to change a lepton to a quark or vice-versa. When an *X* is involved, clearly the conservation of baryon number is violated. The interaction carried by *X* is sometimes called the "hyperweak force." Each *X* has an electric charge of 1/3 or 4/3, a weak charge, and one of the six color or anticolor charges. An antineutrino, for example, can change into a blue

d quark by emitting an X carrying antiblue (yellow) and an electric charge of $+1/3$ (Color Plate 7). Because all electric charge is now measured in thirds, there is no other possible way to divide charge. A particle with charge $3/2$ could not participate in the unification because it could not change into other particles. The SU(5) theory also provides an explanation as to why the d quark charge is $1/3$ of the positron's and opposite in sign.

Other particles such as u quarks and left handed particles can be made from combinations of pairs of the basic family just described in the GUT. The first family has a total of thirty distinguishable particles. It may not sound, at this stage, as if the physics of elementary particles has been simplified, since three families of thirty particles are possible, along with twenty four carriers. However, observed particles have been found to have several complex properties, and if symmetry is to prevail overall, these are the minimum numbers of building blocks required. Clearly many new particles are predicted to exist, notably the Xs. Also, SU(5) may be completely symmetric, but nature is not. The symmetry is quite broken. This has to occur at the tremendously high energy of 10^{15} GeV (10^5 joule) and extremely small scale (10^{-31} meter) of unification. The masses of the Xs must be about 10^5 joules; imagine those popping in and out of existence! Even if the Xs have huge masses, all the carriers would act as if they had negligible mass-energy compared to the huge energy of reactions at scales smaller than the unification one. In the range between 10^{-31} and 10^{-18} meter, the SU(5) symmetry is broken but the SU(2) $\times$ U(1) electroweak symmetry remains. Then the color force is distinguished from the electroweak force.

Physicists' confidence in the GUT based on SU(5) is based on several developments. One is that a calculation of the range at which all three forces approach the same value confirms that the unification scale is 10^{-31} meter. At that scale the color force is much reduced, and the electromagnetic and weak forces increase to the same value so that all three merge into one force, the hyperweak. Another is that calculations in GUT avoid all the unpleasant infinities that appeared in the preceding theories of QED and electroweak interactions. Also, estimates are being made of various particle masses, using basic calculations with GUT, and results are coming out reasonably well. The shortage of antimatter in the universe is another clue that favors GUT. And several startling experimental predictions are made by GUT, as will be discussed next. However, there is no hope of producing particles like the Xs. On the unfavorable side, GUT makes some erroneous predictions such as the value of the ratio of baryons to photons in the universe. Also it involves many constants that must be measured experimentally rather than calculated from basic principles.

15.7 FRONTIERS AT THE MICROLEVEL

The universe within the proton is a frontier spread over a broad scale of at least sixteen powers of ten in distance (if length has any meaning so small). But this micro-frontier cannot be discussed

Table 15-2. The Three Families of Leptons and Quarks, for the Grand Unified Theory.

Family	Leptons	Electric Charge	Quarks	Colors	Electric Charge
1	e^-	-1	d	R,G,B	$-1/3$
1	ν_e	0	u	R,G,B	$+2/3$
2	μ^-	-1	s	R,G,B	$-1/3$
2	ν_μ	0	c	R,G,B	$+2/3$
3	τ^-	-1	b	R,G,B	$-1/3$
3	ν_τ	0	t	R,G,B	$+2/3$

much further without involving theories and events at extremely large scales—the cosmological and astrophysical. The next chapter will examine developments at the largest scales.

The nonconservation of baryon number predicts, among other things, that the proton is not prevented from decaying. Andrei Sakharov first suggested in 1967 that baryon number might not be conserved. Baryon number is the only conservation law requiring stable protons. Proton decay would not violate any other law. The GUT predicts that protons can decay, or at least does not prohibit this faintly alarming possibility. We might wonder whether anything in the universe is stable, or why we are even here. However, we are rather safe from immediate dissolution because of the long time needed for a proton to decay. In order for two quarks within a proton to exchange the dreaded X particle, they must approach to within 10^{-31} meter of each other. This is much less likely than two cosmic grains of dust meeting in the universe. The probability of proton decay has been estimated to be such that one must wait at least 10^{31} years to witness it. Thus one proton in about 10 tons of material (10^{31} nucleons) is expected to decay every year. Clearly the stars will all die before protons start decaying in droves.

Color Plate 8 illustrates a mechanism of proton decay in a Feynman diagram. An X emitted by a d quark in the proton can change one of the u quarks into an anti-u quark. The combination $u\bar{u}$ is a pion which soon annihilates, leaving two gamma rays. Not only does the proton fall apart, but the positron emitted in the process wipes out the electron that may have accompanied the proton in a hydrogen atom. The whole atom is gone, leaving gamma rays. For those dreaming of huge energy sources, this is truly "total conversion" (TC) of mass to energy. And no catalyst or fancy equipment is needed, just a little waiting. At least 32 modes of proton decay are possible, most with even longer waiting times.

Following a sound prediction of proton instability, the experimental search was on. For example, a tank of 1000 tons of water should show nearly 100 decays per year, the minimum to be sure the process is reliably detected. Experiments are being done in deep abandoned mines where interference from cosmic rays is minimal. Any proton decay is expected to result in high energy leptons and pions, which will emit Cerenkov radiation in the water. Banks of photodetectors would sense this light. The results to date have been nil.

Protons seem to live a little longer, perhaps a lot longer, than they should. This has shaken but not destroyed physicists' confidence in GUT and all the ingredients that precede it. We may have to wait a great deal longer to see a proton decay, or examine really prodigious amounts of material. Although the average lifetime of a proton is 10^{31} years, they have been around only about 10^{10} years. At this age they are practically freshly minted. Protons might not decay steadily since Day 1 but rather wait a bit, give or take a few powers of ten. Physicists have practical experience with lifetimes only up to about 10^9 years (uranium isotopes).

Other evidence that there is nothing sacred about baryon number is that no difference in force has been found between neutrons and protons. Different elements differ in the proportion of neutrons and protons they contain. Roland von Eotvos in 1889 found no measurable difference in the way substances are attracted by the Earth. This experiment has been replicated more recently and many times.

Another object predicted by GUT is the magnetic monopole. Recalling that magnetic fields are known to come only from current loops (magnetic dipoles) we see that there is incomplete symmetry between magnetic and electric fields. If electric fields have point sources (electrons, for example), why not similar sources for magnetic fields? Dirac predicted the monopole from basic considerations in quantum mechanics long before GUT and without any inkling of the mass that could be involved. The "magnetic charge" of the monopole would be so large that the passage of a monopole would have a much stronger ionizing effect than the passage of an electric charge.

It could be accelerated thousands of times more easily in an accelerator, and it would slow down very quickly in matter so that any monopoles present on the Earth are likely to be moving very slowly. Another predicted property is that the monopole's field must violate time reversal. Most unusual is that monopoles may consist of layers imitating the evolution of the universe. GUT may reign supreme in the core of a monopole and symmetry is more and more broken in outer layers. The monopole may interact violently with quarks, re-emitting them as antiquarks and thus acting as a catalyst for the decay of protons and other particles. This is all speculation at this point with much theoretical work remaining and experimental evidence lacking.

If it exists, the magnetic monopole is very heavy, about 10^{16} GeV (more than the Xs!). Physicists are unlikely ever to be able to make one. Instead natural searches are being carried out. The monopole would have a quantized magnetic charge (an N pole), and also exist in antimonopole form (an S pole). Some, perhaps many, may have been formed at the beginning of the universe and still be around, raining on the Earth. Some may be detected in the proton decay detectors. Searches have found none so far, so that the upper limit of their abundance is about 1 per 10^{15} protons. Perhaps they have sunk deep into the Earth. If a monopole and its antipole were to meet, they would annihilate in quite a flash of energy—a million joules! From knowledge of the heat in the Earth's interior, it can be estimated that there are again very few, if any, monopoles present.

We frequently see that the frontiers of particle physics are bounded by available energy (in the CM frame) for making new predicted particles. After Fermilab comes on line with 2 TeV, how much farther can physicists expect to go? The perfection of superconducting magnets for increasing the magnetic field at virtually no cost in electric power is only a temporary solution. The special metals that are formed into the wire for superconduction have inherent limits on how much field they can stand before the magnetic field strength itself eliminates their superconductivity—sometimes catastrophically. The maximum fields may be only about five times what can be achieved with iron cores in conventional electromagnets, which is not much more than achieved at Fermilab. Therefore we can expect to scale up from Fermilab at the rate of about 2 TeV per kilometer of radius of beam path.

The largest accelerator on the drawing board is the so-called Desertron, named because only in the American southwest might enough land be found for it. More technically, it is the Superconducting Supercollider (SSC). As with most planned accelerators, it would use colliding beams of protons and antiprotons. With 20 TeV per beam, it would have a total of 40 TeV in the CM frame. The beam radius would be more than 10 kilometers, and the cost is expected to approach two billion dollars. The size, complexity, and cost of necessary detectors may make these the largest part of the project. Much new physics is likely in this range, as well as confirmation of existing QCD and electroweak theories. There are many particles in the spectrum of the heavier quarks still to be found. If, with unlimited funds, physicists could move into space and construct an accelerator with superconducting magnets in a ring a thousand kilometers in radius, the largest energy achieved would be 4000 TeV, or 4 PeV (Peta-electron volts). This is about 10^7 GeV, far short of that needed to reach the symmetries of GUT. Monopoles, Xs, and more, may remain forever out of our control.

Some new physics may turn up at energies we can reach. It may be in contradiction to QCD or GUT. Already new particles such as the zeta and the "squark" may have been detected. In 1984 at DESY, using the Crystal Ball detector, a particle at 8.32 GeV seems to have been found, although later analysis raises more doubt. The zeta does not fit any of the "standard" theories unless it is the elusive Higgs particle. It comes from a rare decay of the new upsilon meson, but not nearly rare enough to fit theory. Another more mysterious meson has been found at 2.22 GeV and named the xi. Efforts continue to measure the neutron electric (not magnetic) dipole moment. If it has one, it violates T, CP, and probably baryon conservation.

New mysteries are not confined to accelerator experiments. Underground detectors, including ones intended for proton decay in Minnesota, Ohio, and France, have found since 1983 a substantial number of unexplained events coming from the general direction of an unusual x-ray source in the galaxy, Cygnus X-3. The particles, if that is what they are, seem to be moving near lightspeed, are neutral, may live at least 30,000 years, are absorbed by the earth, and produce far too many muons in our atmosphere. All known particles are ruled out. Will these observations lead to new fundamental physics or eventually be found to be incorrect?

Speculation also continues to ever smaller scales. Possibly quarks are made of finer particles called "preons" or "prequarks" which come as "flavons," "chromons," and "somons." As the names indicate, these are supposed to determine the properties of the quarks and leptons, including their families. Another theory proposes constitutents called "rishons" in an effort to limit the number of truly "elementary" particles. Needless to say, the energy scale is large and the size scale is small, but not as extreme as the scales demanded for unification in GUT. In these new theories electrons have constituents, and our rather exact knowledge about the electron limits any such consituents to sizes smaller than about 10^{-18} meter. Therefore the energy scale starts at 100 GeV, within the reach of current accelerators. However, there is an energy mismatch. Quarks have energies much less than this and so cannot contain such constituents, it seems. Such contradictions may indicate we have indeed reached the end of the line!

On the theoretical side, perhaps someday the mathematical form of GUT and any larger theories will be simple enough to display to more of the educated public. Advances are made steadily both in understanding and in translating the understanding to simpler models. Certainly GUT is not the end, even if proven fully correct, because gravity remains to be dealt with. Matter (made of fermions) and force (carried by bosons) are still distinct. New theoretical constructs appear, such as *solitons* or "solitary waves." At ordinary scales waves that consist of a single packet of energy in motion have been found, and such solitons might be ideal models for elementary particles. The magnetic monopole could be a soliton, and solitons may provide for interconversion of fermions and bosons.

On the practical side, some of the many technological spin-offs from particle physics have been mentioned. These include medical instrumentation and treatment, superconducting technology, and computer hardware and software. Work in particle physics benefits other sciences through applications of mathematics first developed for particle theories. There is also the use of accelerators to produce beams of ions for implantation in materials to investigate them and modify them and beams of intense radiation for the study of materials. Imaging of the interior of the Earth may be possible by detection of neutrinos passing through. Unfortunately particle beam weapons are also proposed, and if it becomes possible to release more energy from matter than has been achieved to date, there is no indication that governments will not seize the opportunity to make more destructive weapons than those using fission and fusion of nuclei. For peaceful purposes new ways to obtain useful energy may be found with particle physics. For example, high energy collisions of nuclei that result in nuclear collapse could release energy, or proton decay could be catalyzed and harnessed on a large scale. Perhaps muons could be used to catalyze fusion reactions.

Chapter 16

Particles and Gravity, Astrophysics, and Cosmology

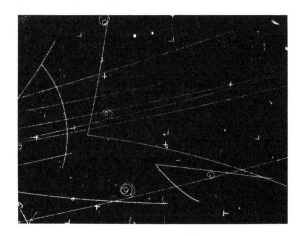

RECENT PROGRESS IN THE STUDY OF STARS, GAL-axies, and the universe at large has influenced progress in particle physics. Also, knowledge of the properties and interactions of elementary particles has aided the understanding of stars (in astrophysics) and of the history and structure of the universe (in cosmology). To conclude this book, we shall be interested in what can be learned about particles from exploring the universe. Gravity dominates the universe, so the time is at hand for the modern theory of this fourth fundamental force and a consideration of whether it can be united with the other forces.

16.1 GRAVITY IN GENERAL RELATIVITY

Einstein generalized his own theory of special relativity in 1915 with the introduction of *general relativity*. This theory combines several elements into one package. It deals directly with accelerating reference frames, it provides a mathematical theory for gravity, it again changes our conception of spacetime by introducing

curved spacetime, and it makes several predictions which were confirmed by observation and experiment. It is one of the greatest examples of a physical theory. However, Einstein did not provide a quantized or gauge field theory for gravity, and this has become the subject of recent study.

In addition to the postulate from special relativity that light travels at a constant and ultimate speed, general relativity uses the principle of *equivalence*. In this context equivalence refers to the indistinguishability of gravity and acceleration. A person within a sealed chamber cannot tell if it is an elevator accelerating upward (as shown in Fig. 16-1A) or if it is a room immersed in a gravitational field (Fig. 16-1B). The chamber might even be in orbit around the Earth, within its gravitational field, yet no net forces are felt inside (Fig. 16-1C). The classic test of equivalence is to see if more massive objects fall faster than lighter ones. This was done by Dutch scientists Simon Stevin and Jan Cornets de Groot in the 17th century some years before Gilileo, who has principle credit for this experiment. To high accuracy

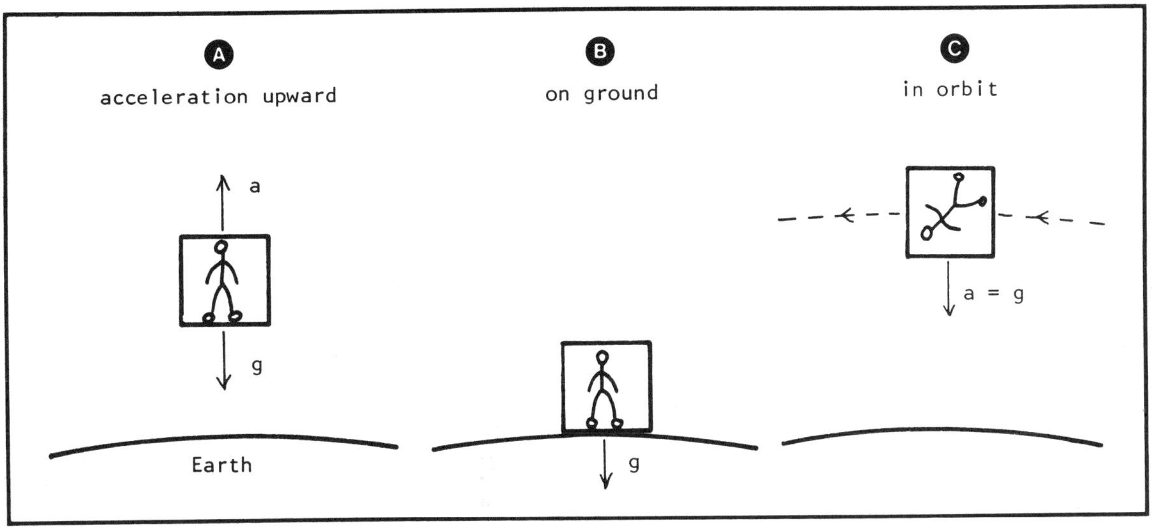

Fig. 16-1. The equivalence principle for general relativity: gravity and acceleration are indistinguishable.

the effect of gravity on mass cannot be distinguished from the acceleration of mass, as was confirmed in the Eotvos-Dicke experiment (originated by Roland von Eotvos in 1922, and repeated by P. G. Roll, R. Krotkov, and R. H. Dicke in 1964). Inertial mass and mass as a source of gravitational field seem to be the same, to an accuracy of one part in 10^{11}. We can no longer have inertial frames in which physical laws are the same, but we can have the new equivalence principle whereby the same physics occurs in different places regardless of whether gravity or acceleration is involved.

Since gravity and acceleration cannot be distinguished, a new way of describing how particles move in space is called for. Earlier, in classical mechanics, we assumed a Euclidean space with three dimensions at right angles. A particle would travel in a straight line if there are no outside influences. But what if the particle's path curves for no obvious reason? It is just as possible to describe the path in curved coordinates that curve along with it. In terms of curved coordinates, the particle could still move "straight;" it would follow a specific line called a *geodesic*. Since we cannot step outside of our own space to see if it is curved, we may be deluded. Particles may be following curved paths when we see them

as straight. Our senses are governed by whatever the local coordinate system is, curved or not. In fact, the light we use to see may follow curved paths and we would never know it. By the least action principle we will be safe in simply assuming that light follows geodesics in space and therefore light tells us where the geodesics are.

A common analogy is to use the curved surface of Earth. In some areas one can drive for hours in a straight line, or so it seems. Yet the actual path to an outside observer is a curve that follows the Earth's surface. Our only clue to the curvature is that the horizon seems close and never becomes closer. As one tries to travel "straight," objects appear at one horizon, are passed, and disappear at the other. This effect occurs only because the light we see by does not, in this case, curve around our planet. In much stronger gravity it would. On the planetary scale, the theory of general relativity says that the Earth simply follows a "straight" path in its orbit around the sun. It is the space that curves, both around the sun and toward it. The gravitational potential we have tried to imagine earlier seems to have a certain reality. A gravitating body such as the sun is at the bottom of a deep energy well. Space curves "downward" toward it, more and more steeply as the sun is approached.

A theory we may no longer need is gravity as action at a distance. The effects of gravity are embedded in the curved space. Every particle and piece of matter, however large or small, has the spatial coordinate system dimpled around it, as represented in two dimensions in Fig. 16-2. The mass is the source of curvature. This applies not only for mass but energy, since the two are the same. A hot brick curves space more than a cold one and therefore creates a stronger gravitational field. When the small curvature of space around each of a large number of particles is combined by superposition, the result is a deep curvature around a large body. The curved space theory nicely fits the way gravitational fields are combined. It also explains how gravitational attraction (and never repulsion) occurs. One particle near another seeks the lower energy at the bottom of the well (as in Fig. 16-2), unless it has a speed which allows it to orbit around the sides, much as a motorcyclist would travel around the inside of a cylinder.

Light follows paths such as the curved coordinate lines or geodesics of Fig. 16-2. The direction in which it leaves the vicinity of the gravitating body is the same as the direction at which it approached, if one observes the light's path from afar. But to a nearby observer, the light seems to come from a different direction. We are located near the sun, gravitationally speaking, and we might expect to see light from distant stars arrive bent from its original direction. The positions of stars are displaced outward from the sun. This is a prediction of general relativity, and the amount of deflection can be calculated. The sun is too bright to observe stars near it, but during the solar eclipse of 1919 an expedition was sent to measure whether stars near the sun are out of place compared to other times when the sun appears elsewhere. The effect was found, and the small measured deflections of about 1.5 seconds of arc agreed with calculations. Einstein triumphed again!

Einstein's equations that describe gravity are perhaps simple enough to display, although too complex for much explanation:

$$\frac{8\pi G}{c^2} \ T^{\mu\nu} \ = \ R^{\mu\nu} \ - \ \frac{1}{2} \ g^{\mu\nu}R$$

The four coordinates of spacetime are put on equal footing and numbered with the symbols μ and ν, which can have values from 1 to 4. There

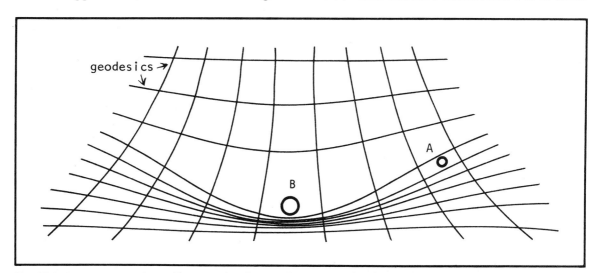

Fig. 16-2. A representation (in two dimensions) of the curved space around a mass which is a source of gravity. (The particle or planet A entering the "gravity well" of B should also form its own well. The grid lines shown are some of the geodesics, the paths light would travel.)

are ten independent ways this can be done with two sets of four numbers, so there are ten field equations contained in this one. The quantities with double subscripts are second order *tensors*, a higher sort of vector with two sets of components. Each direction has an effect on every other direction. $T^{\mu\nu}$ is the mass-energy tensor telling how much mass and energy is at each point in space. The right side of the equation is the curvature tensor, telling how space curves at a given point due to the presence of mass-energy there. The curvature is described by $g^{\mu\nu}$, the "Riemann metric."

The equations are local and must be solved over a region to find out how spacetime curves due to a given arrangement of mass and energy. It has been solved for a number of situations ranging from dense stars to simple models of the whole universe. The constants G/c^2 have the approximate value $7.4(10)^{-28}$ meters per kilogram, and the whole equation is expressed in terms of meters if $T^{\mu\nu}$ is in kilograms. For example, the mass of the sun when converted to length in this way gives 1.477 km. This is the radius to which space is curved around a mass the size of the sun, if all the mass is at one point. Since the sun is spread over a million kilometers, space curves much more gradually around it. It should be emphasized that Einstein's theory does not define or provide a way to calculate gravity but instead gives the shape of spacetime. It is a geometric theory that replaces older concepts of gravity.

A prediction that follows from general relativity and curved space is that the elliptical orbits of planets do not stay in one orientation but slowly shift around the sun. The orbit is not, in fact, closed, but when the shift is small during each orbit we say that the perihelion (the points of closest approach to the sun) precesses. This precession was seen for certain classical models of the atom (recall Fig. 9-3B). The effect is small, even for the orbit of Mercury, but it had been known to be about 43 seconds of arc long before Einstein. By calculating what it should be, he knew that his theory was correct.

Because spacetime, not just space, is curved,

one might expect some effects of gravity on time. Einstein's theory provides for the calculation of the change in intervals of time in a gravitational field. In particular, light emitted by a massive star will have its frequency shifted downward; the effect is called the *gravitational red shift* and has been repeatedly observed. (This red shift should not be confused with the Doppler shift due to motion of the source of light.) The increase in a time interval Δt is given by $\Delta t \Phi/c^2$ where Φ is the change in gravitational potential the light or an object experiences as it moves away from the gravitating body.

The gravitational effect on time has also been measured with the experiment that sent accurate atomic clocks in jetliners around the world (moving up to regions of slightly less gravity), with clocks aboard satellites, and with the Mossbauer effect. In the latter case, nuclei can be made to emit radiation at such a precise frequency that the small changes due to the height of the nuclear sample above the earth can be measured. The precision is better than one part in 10^{14}. The recent discovery that the gravitational fields of galaxies focus the radiation from objects behind them promises to provide more tests of general relativity. These are the so-called "galactic lenses."

We shall return later to consider some of the extreme cases of curved spacetime. The most extreme effect occurs when a body is so massive that it curves spacetime completely around itself. It disappears from all possible direct observation since light from it cannot leave, being confined to a closed spacetime. The object is called a *black hole* since it ordinarily cannot emit light and would absorb all light and other material that approach it. General relativity predicts the radius of the closed space to be:

$$R = 2GM/c^2$$

This is called the *Schwarzschild radius* after general relativist Martin Schwarzschild. The Earth would have to collapse to less than a meter in diameter in order to curve space so strongly that it closes off. The ordinary Earth curves spacetime so gradually that the radius of curvature is about 10^{16} meters (a light year).

There have been many other theories of gravity of varying degrees of consistency and agreement with observation. A good theory is expected to be complete, consistent, include special relativity, and agree with classical physics at low speeds or for small masses. Some theories competing with Einstein's differ substantially in their mathematics but make predictions almost indistinguishable from Einstein's. Activity in this area has slowed for lack of measurements that make clear distinctions. The emphasis has changed to the serious attempt to combine gravity with GUT, as will be discussed shortly.

Gravity can still be described as a field, not a stationary one but rather one with waves that propagate in space, in analogy to electromagnetism. The theory has been difficult to formulate, but that has not stopped a search for the waves, or for their quantized form, *gravitons*. The curved space equivalent would be ripples that propagate in the "fabric" of spacetime. A star sitting still would emit no waves, just as a stationary charge emits no radiation. But all stars are in motion, and some are in rapid motion around each other. They are expected to emit gravity waves with components more complex than those of electromagnetism, since gravity is a tensor field.

During the 1960s George Weber constructed an apparatus consisting of a suspended aluminum cylinder of almost four tons. If gravity waves of a certain frequency shook it, it was supposed to vibrate in resonance and the weak effect could be amplified and detected. Despite apparent initial success, other experimenters have not been able to consistently confirm the results and theoretical developments have continually pushed down the expected size of the effect. Gravity waves and gravitons are yet to be detected directly, but work continues. It is still hoped that such detectors might tell the form of the waves and either confirm Einstein's theory or a modification of it. In 1981 the existence of gravitational waves was verified indirectly by measuring the change in orbit of a binary star system.

16.2 SUPERSYMMETRY AND SUPERGRAVITY

The modern attempt to quantize the gravitational field has evolved to the notion of *supersymmetry* (SUSY). We have repeatedly shown how bringing more and more symmetries together allows the explanation of, then the unification of, the other forces. The same is now expected to hold for gravity. A symmetry is needed that contain previously established symmetries such as that of SU(5) for GUT. Hence the name of the unified theory as supersymmetry. Of course it must be a gauge theory. The work began about 1918 with Theodor Kaluza working on a unification of electromagnetism and gravity. He needed a fifth dimension for his geometric theory. A quantized version using wave mechanics was created by Oskar Klein in 1926. Progress was essentially dormant until the last decade.

An important requirement is to unify fermions and bosons, which are very incompatible partners. An imaginary "spin" orientation in an imaginary dimension is used to distinguish fermions and bosons, similar to the use of isospin. Under complete symmetry, rotation of the fermion-boson pointer would lead to an invariance. A supersymmetry mirror was also invented. A boson "reflected" in it becomes a fermion, and vice versa. Remarkably, this supersymmetry transformation also changes the spacetime position of the boson and provides the basis for gravitation. The Pauli exclusion principle is left intact. The lack of observed symmetry between bosons and fermions is explained by the action of symmetry breaking. Only at the beginning of the universe could conditions have been extreme enough to make bosons and fermions indistinguishable (total symmetry). After symmetry breaking there are particles with low masses and related "superparticles" with high masses.

Supersymmetry may proliferate particles. In some versions the total number of different particles is limited to a few hundred, but in other versions it is infinite. Almost all are inaccessible due to their high mass-energies, which start at approximately 10^{19} GeV (a billion joules). It would be

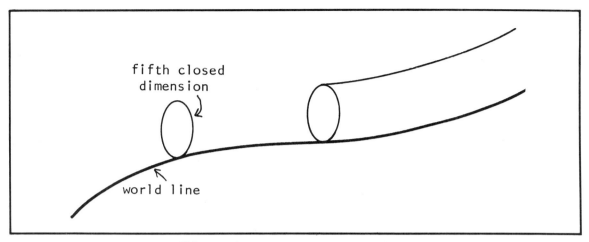

Fig. 16-3. A representation of a world line (geodesic) in spacetime with a closed circular dimension associated with every point, forming a tiny cylinder in five dimensions.

hopeless to consider producing them in accelerators. Scaling an accelerator up to the required size results in one that is about 1000 light years in radius! A few new particles are expected in the TeV range, within reach of the Supercollider. Each superparticle is a partner of a more common particle and may have its name taken from its partner with "-ino" suffixed or with "s-" prefixed (for example, the "selectron").

The supersymmetry fields require eight or more dimensions, perhaps up to eleven or twelve, to express. This is at least four more dimensions than ordinary spacetime. The dimensions are not supposed to be imaginary, but real ones all at right angles. The proposed reason why the extra ones are not apparent to us is that they are closed and rolled to a very small size, about 10^{-32} meter. Figure 16-3 illustrates how a tiny circle at each point along a world line adds a new dimension. The extension of the circle constitutes a cylinder. More dimensions can be added by rolling space into spheres or hyperspheres at each point. The most likely total dimensionality for supersymmetry is eleven, including seven such "hidden" dimensions. There are proofs that seven are needed to incorporate fields for the weak and strong interactions. The mathematical form of supersymmetry is quite the most complicated ever invented, comprehensible to very few physicists at this stage.

It is a geometric theory and contains general relativity.

The ordinary carrier of gravity over large distances, the massless spin 2 graviton, is kept in supersymmetry. The new supercarrier, the *gravitino*, only carries gravity over very short distances and has the unusual spin $\frac{3}{2}$. As a fermion, it must be exchanged in pairs. When Feynman diagrams are constructed with these carriers, convergence occurs readily and no embarassing infinities occur in calculations. Trouble had been expected because each of an unending series of virtual particles has mass-energy and would contribute to the curvature of space and therefore to local gravity. The breaking of supersymmetry is required to obtain calculated results in agreement with current observations about particles and the universe.

There have been occasional signs of particles that may be part of supersymmetry. Each particle is supposed to have a "superpartner" with a different spin. The partner to the quark is the "squark" (no kidding!). If produced, it will decay to a quark and a "photino", the partner of the photon. If a squark-antiquark pair is produced, two unseen photinos are expected, and the two quarks decay into unusual jets such as the ones seen recently at CERN. In addition, the lightest superparticles may not be able to decay and could remain from the beginning of the universe. Unfortunately, the results of early searches for the superpartners of the W and Z^0 bosons ("winos" and "zinos," expected to be as light as 35 GeV) are unfruitful.

If gravity is unified with GUT the result is called *Totally Unified Theory* (TUT) by some and superGUT by others. The first symmetry breaking produces *supergravity* and the Supersymmetric GUT. The part discussed in the previous chapter is GUT; the part we may never see unless we could look in the supersymmetry mirror is Super GUT, a whole other universe of "particlinos." It is hoped that a correct TUT will explain the energy scale at which its symmetries are broken, among other things.

More unusual effects involving spacetime have been proposed, based on quantized gravity theories, not necessarily supersymmetry. At the scale of 10^{-35} meter space itself is seething with quantum fluctuations and can no longer be considered smooth. These fluctuations last only about 10^{-43} second, as will be seen later for the Big Bang. They are a result of the energy of space itself, which materializes in chunks of about 10^{-8} kilogram, and they result in violent curvature of space. Holes develop, and even tunnels called "wormholes" from one part of space to another, as proposed by John Wheeler in 1957 in his geometrodynamics. Charges are described as wormholes into another part of spaces where anticharges reside. Space could have the structure of foam rubber, which would give it more than three spatial dimensions (as occurs with fractals). Since this is contrary to observation, quantum fluctuations in space are in doubt.

Another attempt at unification is called *superstring* theory. It requires ten or more dimensions, and particles consist of vibrations of "string" in the higher dimensions. The basic building blocks are no longer zero-dimensional point particles but one-dimensional pieces or loops of abstract string 10^{-15} m or smaller in size. String theory was invented in 1970 by Yoshiro Nambu and has grown rapidly in popularity in the past year. It could conceivably sweep away gauge theories since it seems to allow for fermions, bosons,and gravity much more readily. String sounds like a simple model, but with concepts such as the "heterotic" string that ripples one way in 10 dimensions while rippling the other way in 26 dimensions, mathematical physicists have a new conceptual world to explore. Whether a theory such as this comes to account for all properties of the observed universe remains to be seen.

16.3 EARLY HISTORY OF THE UNIVERSE (BRIEFLY TOLD)

The beginning and the evolution of the universe have become intricately linked with our understanding of elementary particles and fundamental forces. At the beginning was the only environment in which certain unifications of the four forces can occur. The beginning is commonly called the *Big Bang* because all the evidence is that the universe started in a titanic explosion. At the very earliest stage all matter and energy, and therefore all space, was compressed into a very tiny place, perhaps only a point.

After that, events occurred very rapidly for a while, at least on our scale of time. The velocity of expansion was in fact so near the speed of light that time dilation stretched a tiny fraction of a second out to the equivalent of years. As far as the many changes in the state of matter and energy are concerned, the first one second of the universe was an eternity. Figure 16-4 illustrates the history of particles, forces, energy, and effective temperature over time, on a logarithmic scale. Each increase in time by a factor of ten is shown on a uniformly-spaced scale. Our sketch of the history is almost the sequence of topics in this book and the history of the fields of physics in reverse, starting with supersymmetry and ending with atoms.

Near the beginning, times shorter than 10^{-43} second (the *Planck time*) are important, but time as such may be undefinable. It would probably be meaningless to talk about time t = 0. In some sense the beginning is infinitely far back in the past, despite the small numbers. Only the stages

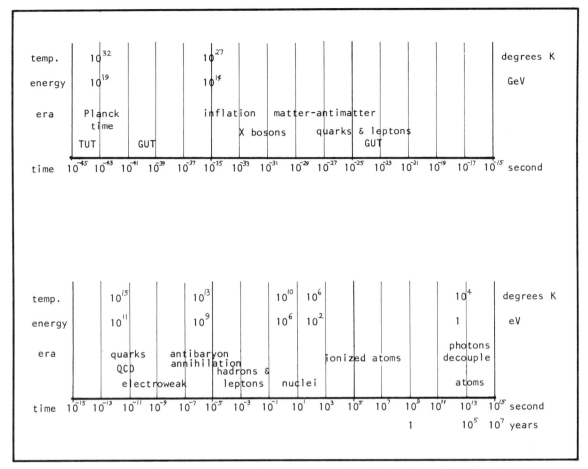

Fig. 16-4. The history of particles, forces, energy, and temperature on a logarithmic time scale for the universe, from near the Big Bang to the formation of atomic matter. (The early history of the universe is also the history of the breaking of symmetries—the unification of particle theories in reverse.)

of development, however rapid or slow, will have meaning. Physicists and astronomers are gradually pushing our understanding back to earlier and earlier times, although 10^{-43} second is a major stumbling block. Our present knowledge back to that time is mostly theoretical, with a good deal of speculation. At the Planck time each "particle" (if it could be called such) had so much energy that the gravitational field from its own energy made it an individual black hole. The effective temperature was 10^{32} K. Every particle was in its own universe. The radius of its black hole was the same as its quantum size. The amount of mass which does this can be calculated from:

$$M_{PL} = \sqrt{\frac{hc}{2\pi G}} = 2.2 (10)^{-8} \text{ kg}$$

The equivalent energy is about $1.2(10)^{19}$ GeV and is called the *Planck energy*. The Planck size, about 10^{-35} meter, is the size of quantum fluctuations in space itself due to gravity. As supersymmetry, TUT, or some other quantized gravitational theory is better understood, physicists may learn more about earlier times and smaller sizes than the Planck limits.

A Totally Unified Theory (TUT) reigned after the beginning until 10^{-43} second, when the average energy dropped below the mass-energy of superparticles. Then supersymmetry was broken and GUT reigned. All the forces except gravity were equivalent in a soup of particles, including X bosons, with typical energies of 10^5 joules. We observe now that matter as a whole is uniformly distributed around the universe and must assume that this is due to a very uniform expansion of matter long before galaxies were formed. The uniformity is traced back to 10^{-35} second when all of the universe was within "contact"—that is, light could travel from any part to any other part in the time available. (Thus the universe was only about 10^{-23} meter across.) There is also a "smoothness problem." Matter is nonuniformly clustered in galaxies, galaxies are nonuniformly clustered, and there may be nonuniform superclusters of clusters. This structure also had to originate in the very early universe.

Explaining the uniformity—the "horizon problem"—had been frustrating until a theory by Alan Guth in 1980 called the *inflationary scenario*. The analogy to symmetry breaking of GUT is the supercooling and sudden crystallization of a liquid. The matter undergoes a "phase change." Most of the energy is conjectured to have gone into Higgs fields in empty space (an excited or "false" vacuum). Then it suddenly "condensed" as more familiar particles—quarks and leptons. In doing so, the size expanded very rapidly, about 10^{50} times from 10^{-35} to 10^{-32} second and the temperature shot back up to 10^{27} K. These results are obtained in a superseding "new inflationary model" developed by A. D. Linde and independently by Paul Steinhardt and Andreas Albrecht. The high temperature returned the universe to a GUT era, where X bosons intermingled freely.

The inflationary models explain the "flatness problem"—the question of whether the universe has enough mass to make it contract at some future time, or whether it will expand forever. The mass density that is observed today indicates we may have a "flat" universe which will eventually cease expansion but may never contract. Such a special result must be built in from the beginning, but physicists find it hard to believe that the very early universe had this particular extremely accurate value for its initial mass density. The rapid inflation of size allows the universe to reach a "flat" mass density compatible with current observation without a special initial condition. The process is analagous to blowing up a balloon vastly until all its surface appears flat. The lack of smoothness in the universe is explained by quantum fluctuations during inflation, which are very small until expansion has blown them up to the size of galaxies.

The period of high temperatures after inflation was the last time very heavy particles such as monopoles and Higgs bosons could be formed, as well as very unusual objects such as small black holes. The fact that no monopoles seem to be around is also explained by inflation. Monopoles in quantity would have major effects, but astronomical observation does not find evidence of any. GUT would have them form as defects at the boundaries between each "bubble" or "domain," a tiny universe representing a quantum of space expanding during symmetry breaking. The latest inflationary models have our universe within a single domain. Countless other disconnected universes were discarded or just removed from all contact with ours during the Big Bang. Inflation is at the forefront of current theoretical work, and we may expect further improvements of the model. Physicists are pleased that a theory has been found which eliminates need to consider particular initial conditions for the universe. Possibly all the observed structure of the universe will follow from inflation applied to supersymmetry. It may even explain how the universe arrived at a state of order and is now tending to disorder, defining the direction of time flow.

A "long" time after quarks became distinct with the breaking of GUT, the universe cooled enough (to 10^{13} K) that they could coalesce into hadrons (baryons and mesons). This occurred at about 10^{-6} second and resulted in equal amounts of matter and antimatter (net baryon number zero). Physicists have been very concerned about explaining what became of the antimatter, since we seem to have only matter left in the universe (about 10^{78} baryons). Recent satellite gamma ray observatories have not found any blasts of gamma rays in space, signaling wholesale contact of matter and antimatter. A distinction between matter and antimatter is observed only in the strange decay of the neutral kaon. The matter-antimatter asymmetry can be traced back to the GUT era. One possible mechanism is that a tiny asymmetry in the decays of Xs and anti-Xs led to a small difference in the numbers of quarks and antiquarks. When baryons and antibaryons formed, there were similarly unequal numbers. Then all the matter and antimatter that could annihilate itself did so, leaving a small remainder, perhaps one part in a billion. A tremendous burst of gamma rays was produced, at least two for every two particles annihilated, and it has been measured that about a billion photons are still around for every proton. Henceforth the primordial universe was bathed in these gradually cooling photons.

When the temperature dropped to the level at which the electroweak symmetry breaks (about 1 second), electromagnetism became distinct from the weak interaction. Neutrinos decoupled from the rest of matter and no longer interacted significantly. There are about as many neutrinos left from this era as photons. Nuclear physics began at this temperature of about 10^9 K. Protons and neutrons combined to form helium, deuterium, and lithium nuclei, the first *nucleosynthesis*. The universe expanded too fast to form many of the light nuclei, or any of the heavy ones. The process ended at about 100 seconds. Hydrogen is the most abundant element, and one quarter of matter is helium (^{4}He). One part in 10^4 is deuterium and ^{3}He, and ^{7}Li is far down at one part in 10^9. The origin of heavier nuclei must wait until stars form, as discussed later.

After a hundred thousand years (about 10^{13} seconds), the universe had cooled enough that hydrogen and the other nuclei were no longer ionized. Instead of the brilliance of radiation from electrons jumping back to atoms after being knocked off by collisions, there was darkness. The abundant photons were "decoupled" from matter. Due to the expansion, these photons are now much cooler, red-shifted to a background of weak microwave radiation. It is called the 3 K background because its effective temperature has cooled to just 3 degrees above absolute zero. It is extremely uniformly distributed and was discovered in 1965 by Arno Penzias and Robert Wilson, for which they received the Nobel Prize in 1978. (It was predicted to exist by George Gamow, Ralph Alpher, and Robert Herman much earlier.)

16.4 LATER HISTORY OF THE UNIVERSE– ATOMS, STARS, AND GALAXIES

The development of the expanding universe after it had cooled enough to form atoms takes us away from particle physics and into astronomy and astrophysics. We may think of the current age of the Universe (about fifteen billion years, give or take a few billion) as being an enormously long time. But physicists now think of the first fraction of a second as containing most of the action. The freshly formed hydrogen and helium present everywhere as a hot gas collected into clouds of cool gas. These clouds gradually drew together under the influence of self-gravity and condensed to form stars after about a billion years (10^{16} seconds). As the stars collapsed, they were heated from the released gravitational potential energy until they reached the temperature at which fusion spontaneously begins.

We have discussed the details of how substantial amounts of nuclear energy are released when hydrogen, deuterium, and helium fuse to form heavier elements. Large stars became very hot, quickly used up their hydrogen, and expanded to red giants as they fused helium, then carbon, neon, oxygen, and silicon. They formed large amounts of all nuclei up to iron by nucleosynthesis, as first proposed by Fred Hoyle in 1946. William Fowler received the Nobel Prize in 1983 for his work on stellar nucleosynthesis.

During this nuclear "burning" copious amounts of neutrinos were created and spread through all space. These elusive particles carry away so much energy from large stars that eventually they cannot maintain enough pressure to hold their sizes and they collapse. Neutrinos contribute directly to supernova explosions. At first the protons in a heavy collapsing star capture electrons to produce neutrons and neutrinos, the latter escaping. Then the density rises so high that neutrinos are temporarily trapped. When the density exceeds that of nuclear matter, a shock wave blows up the star. The weak interaction is responsible not only for neutrino production but also for its role in beta emission and absorption, which sets the rate at which fusion reactions can occur in stars. The stars have not changed substantially the proportions of hydrogen and helium in the universe, despite all their activity. So it is possible to predict the number of different types of neutrinos that exist from the observed proportion of ^{4}He in the universe, a fact we shall find useful later.

Stars about ten times more massive than our sun are guaranteed to blow up as supernova, strewing their elements through the galaxy. As their cores collapse, further nuclear reactions occur. The abundant neutrons help form the heavy elements which have lower binding energy per nucleon than iron has. These include the unstable radioactive elements, also strewn across the galaxy. Second or third generation stars such as our sun, together with our planetary system, formed from this cosmic debris. Analysis of the long-lived isotopes found on the Earth and in meteorites gives us further knowledge of the age of the universe.

Meanwhile supernova cores live on as either neutron stars or black holes, depending on how much mass was shed. The former are like giant nuclei, a dense fluid of nucleons packed tightly together. All the electrons have combined with the protons to form neutrons, except in a crust. With a diameter of only about ten kilometers and containing several solar masses, a neutron star has the most dense substance known. The gravitational field at its surface is a trillion times that of our sun, which is already 28 times stronger than the Earth's. It is also capable of a magnetic field of up to 10^9 w/m^2, 10^{13} stronger than the Earth's. The neutron star can collapse no farther because

of Pauli exclusion. The neutrons cannot find states in a more compact arrangement. If the mass of the core were a little larger, even Pauli exclusion could not stop the relentless pull of gravity. Astronomers are beginning to detect radiation (gamma ray bursts) from neutron stars. For example, gamma rays from positron-electron annihilation are detected, red shifted just as if they had escaped from the enormous gravitational field of a neutron star.

Smaller stars, after a red giant stage where excess matter is gently blown away, will form "white dwarfs." They have a cooling core of 1.4 solar masses or less and a size similar to the Earth. They represent degenerate matter of another form, dominated by electrons and held up by electron pressure due to Pauli exclusion.

The expansion of the universe observed today is called the *Hubble expansion*; it is left over from the original Big Bang and was discovered in the 1920s by Edwin Hubble from studies of the Doppler red shifts of distant galaxies. Volume is being added to the universe at the rate of over a trillion cubic light years per second. Fortunately the space is being added between clusters of galaxies, and the neighboring stars are not moving farther away due to the general expansion.

The expansion as a whole can be visualized by considering the popular analogy with spots on a balloon. As the balloon's two-dimensional space is expanded by inflating with air, each spot draws apart from all others at the same rate. On the cosmic scale stars, galaxies, and even clusters of galaxies all act as particles. They share the motion of expansion, and they have smaller random motions as if they were a cosmic gas. They also have regular local motions, such as stars orbiting each other, the general orbiting of stars around the galaxy, and the orbiting of galaxies in a cluster. Although galaxies do not have a single concentration of mass, their stars act as a whole as if all the mass were at the center. As we go up the hierarchy of astronomical objects, we find higher and higher speeds involved, approaching 1000 km/s for galaxies. This is nothing, however, compared to the general expansion. Some distant galaxies and objects called "quasars" are moving so near lightspeed that the spectral lines in their light are shifted from ultraviolet to infrared. At about 10 billion light years, which happens to give us a view of objects shortly after they were formed, the limit of visibility is reached. Objects are not only faint, but their visible radiation is shifted too far to detect.

16.5 NEUTRINOS AND OTHER COSMIC "RAYS"

Despite their low cross section, it is now proposed that neutrinos from space be detected with a neutrino "telescope." Supernovas and neutron stars should emit them intensely enough to locate them by direction. A deep-sea instrument called DUMAND (Deep Underwater Muon And Neutrino Detector) is planned. It would be 5 kilometers underwater and detect the Cerenkov radiation resulting from the particles that are created when a neutrino hits a nucleus in sea water. In a cubic kilometer of water 1,331 detectors will be placed. Neutrinos over 10^{11} eV will be detected, and antineutrinos could be distinguished. DUMAND is preceded by other attempts at neutrino astronomy. A tank of cleaning fluid in a deep mine has been used by Raymond Davis for decades to detect solar neutrinos by their conversion of a chlorine isotope to radioactive argon. Controversy continues over why the sun emits fewer neutrinos that it should, by a factor of 3. Possibly the electron neutrino from fusion changes into the others, which cannot be detected. A new and expensive detector based on converting a gallium isotope to germanium is. being planned to resolve the issue. Superparticles such as 5 GeV WIMPs (weakly interacting massive particles) or "cosmions" have even been postulated as interfering with energy production in the solar core and affecting the predicted neutrino production.

Despite their name, *cosmic rays* are particles from space which reach the Earth's atmosphere. Their designation as "rays" is outdated but the name seems too entrenched to change. There is no longer any mystery about these "rays," and

photons are not normally included in the category. Cosmic rays are primarily protons, but electrons, positrons, alphas, and assorted common and heavy nuclei have been detected. The nuclei arrive without many if any electrons. Neutrinos, strictly speaking, are cosmic rays. Virtually all are stopped high in the atmosphere by collisions with nuclei of gas molecules. A collision results in a shower of tens to possibly millions of protons, neutrons, nuclei, muons, pions, electrons, gammas, and more. Many of the shorter-lived hadrons are produced but despite the huge time dilation, do not live long enough to reach the ground. A large shower can be seen with special cameras as a faint flash of light, and it strikes as much as a square kilometer of land.

The sources of cosmic rays have been speculated upon for a long time. The sun is known to emit ones of modest energy, mostly protons. The most common energy is about 10^9 eV, and there are about 10 of those per minute per square centimeter in space. Supernovas are capable of ejecting high energy particles and nuclei. Black holes and quasars may be able to generate cosmic rays. The highest energy cosmic rays observed are carrying about 10^{20} eV, which is more than ten joules per particle. They are very rare. Cosmic accelerators involving some fortuitous combination of magnetic and electric fields are postulated to boost fast particles to these ultrarelativistic speeds. Magnetic fields across the galaxy, and that of the Earth, affect the paths of charged particles strongly, and there is little hope of identifying sources on a line-of-sight basis. Low energy cosmic rays tend to strike near the Earth's magnetic poles, where they cannot be deflected away by the magnetic field.

Recently a bizarre double star called Cygnus X-3, with a neutron star orbiting a more normal star, has been found to be a major source of cosmic rays, if indirectly. Matter is sucked from the normal star and somehow accelerated to tremendous energy by the neutron star and its powerful magnetic field, emitting 10^{17} eV gamma rays. These are not affected by magnetic fields, and we can tell where they came from. When the gammas reach the Earth's atmosphere, they produce a shower of particles in this energy range. Enough radiation has been detected from this one source to account for most of the cosmic rays in our galaxy.

One might think that cosmic rays are our best bet for studying the latest particle physics, because of the high energy available. Indeed, cosmic rays were the only experimental method for earlier physicists, starting with Victor Hess who discovered them in 1912 and received the Nobel Prize in 1936. Major discoveries were made with cloud chambers that detected the particles that reach the ground from cosmic ray collisions high in the atmosphere. However, cosmic rays must strike essentially fixed targets, whether gas molecules or those of experimenters. We have learned that the energy available in the CM frame increases only by the square root of the particle energy in this situation. Thus a 10^{20} eV proton, the same as 10^{11} GeV, can yield only about 10^5 GeV (100 TeV) in the CM frame. This is only a factor of about two greater than the CM energy planned for the world's most powerful new colliding beam machine. Physicists will continue to rely on accelerators because of the complete control and high intensity available.

Cosmic rays provide additional evidence of the scarcity of antimatter. Only very rarely are antiprotons observed in cosmic rays, and these are probably produced in space when normal high energy protons strike other nuclei or dust grains. Cosmic rays contribute to nucleosynthesis. Beryllium and boron nuclei cannot be formed in fusion reactions, whether during the Big Bang or in stars, because there are no intermediate stable nuclei of mass 5 and 8. They can only be produced by collisions of cosmic rays with interstellar matter ("spallation"). Later the matter is gathered into new stars and planetary systems. These elements are not plentiful on the Earth, but we have enough of them from such an ephemeral source to find them useful.

Cosmic rays have a cosmic, if uncertain, effect on the evolution of life on Earth. They constitute sufficient radiation, especially at higher altitudes, to contribute to genetic mutation and other genetic damage. This might be said to be one more way human life in particular is tied to events across our galaxy. Evolution is probably affected by variations in cosmic ray flux. Cosmic rays have also been used by Alvarez to image the interior of Egyptian pyramids.

16.6 BLACK HOLES

We can do no more here than touch on the particle aspects of black holes. Much has been written to satisfy the public interest in them, not to mention the intense theoretical and astronomical interest. A black hole is the result of the concentration of an amount of matter into a single point (a *singularity*). At some radius around it (the Schwarzschild radius or "event horizon") there is an imaginary spherical boundary marking the limit of knowledge of the black hole. Supposedly no light or matter can return after venturing this close to the black hole. An exploding star about ten times heavier than the sun can form a black hole with a radius of a few kilometers. The concept, but not the name, was developed by Robert Oppenheimer and Hartland Snyder.

Black holes have very elementary particle properties. All identity of the matter that falls in is lost, and the black hole is characterized only by its mass-energy (how much fell in), its angular momentum (matter is likely to have arrived with some orbital speed), and its net charge. The black hole thus can violate most conservation laws, including baryon number conservation, another argument for the validity of GUT and proton decay. A black hole can be rotating. It also has a peculiar entropy arising from the almost total loss of information when matter falls into it. It does not violate the laws of thermodyamics, but it can be made to generate energy from matter.

The fate of matter entering the black hole is most peculiar. Aside from the destruction caused to material objects by the tremendous gravity near the hole (the sideways components of the force are enough to rip anything apart long before the hole is reached), there are effects on time. According to general relativity, time slows down in high gravitational fields. Any object nearing the hole would have its time so slowed that it would "think" it had never reached the singularity (in its own frame). But outside observers would see it torn apart into elementary particles by the all-dominating gravity (or the extreme curvature of space, to use the more modern view). And reaching the singularity is probably similar to experiencing a Big Bang in reverse.

Stephen Hawking proved in 1974 that quantum mechanics predicts bizarre behavior for black holes. The materialization of virtual photons and particles in pairs that takes place everywhere in space can occur at the horizon of a black hole. Figure 16-5 shows a Feynman diagram of such an occurrence. In a Feynman diagram, the hole has a fixed size in space and extends indefinitely in time; it is represented as a cylinder. One unfortunate virtual particle, for example, a positron, might be so aimed as to enter the black hole, and the other, an electron, is traveling in the other direction and escapes. Since a positron is the same as an electron moving backwards in time, one can view the process as an electron escaping the black hole by moving backwards in time, then changing to an electron and moving forwards and away.

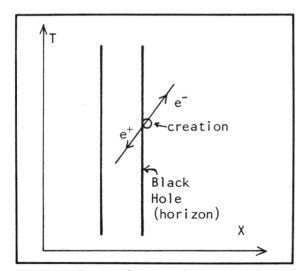

Fig. 16-5. A Feynman diagram of the creation of electron-positron pairs near the horizon of a black hole, one of which enters the hole, reducing its energy, and one of which is seen as energy emitted from the hole.

Radiation (distributed thermally) and particles seem to come spontaneously from the black hole, so it is not truly "black." It shrinks and is said to "evaporate." Small ones, if they were ever formed during the Big Bang, would evaporate very rapidly. A hole the size of a proton would contain about 10^{12} kilograms and would need about 10 billion years to evaporate. It would grow steadily hotter and emit a huge number of gammas as well as electrons and neutrinos, all produced by its effect on virtual particles. A black hole formed from a medium sized star would last about 10^{66} years and be very cool and completely undetectable. All black holes end their careers with a tremendous bang. Toward the end, when a lot of mass remains but the hole is very tiny, energy emission grows stronger very rapidly until there is a titanic explosion of gamma rays from an object much smaller than a proton. Millions of tons of matter are rapidly and totally converted into radiation. The results should be detectable if black holes are dying in space, and searches are being made. The early results are that tiny ones, if any, are very rare. If one could be observed to explode, we would learn much about quarks and GUT, since the energy emitted depends on that physics.

We can be sure that no black holes are nearby, especially not in the Earth, or we would have been gobbled up long ago. Quite a few large ones are expected to be scattered around the galaxy. Some must be members of a pair of stars, so that we can observe their existence by their effects on matter from a more normal companion star. Such is the case with the binary star Cygnus X-1, named as the first strongly emitting x-ray star system and discovered in 1972 by satellite. Observations indicate that gas from a large star is pulled toward the black hole, to orbit it as a thin disk and emit x-rays as it is heated and sucked violently in.

16.7 FUTURE HISTORY OF THE UNIVERSE AND ITS PARTICLES

Although it now seems unlikely, it may turn out the universe is closed and eventually collapses,

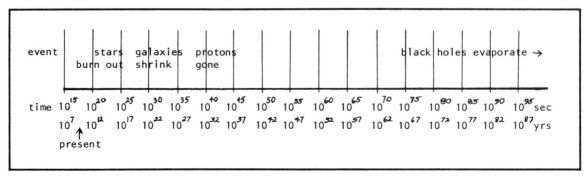

Fig. 16-6. The future and end of the universe on a much compressed logarithmic time scale, as projected from modern particle and cosmological theories and observations.

to form the "big crunch." History would run in reverse until about a million years before the end, when all matter is highly ionized into nuclei and electrons. Then black holes grow rapidly, consuming each other, and the universe disappears into a giant black hole—back to a singularity from whence it came.

Figure 16-6 shows the expected future of the universe, on a logarithmic time scale that extends Fig. 16-4. Recent work in particle physics has made our future much more definite and startling. We have a lot of time left, but eventually almost nothing is going to be remaining. If the universe is flat or open, particles start to play a role in the future after most stars have burned out. The stars need at most about 10^{14} more years (much sooner for massive ones) to use up their fuel, then collapse to white dwarfs and cool to "brown" dwarfs. There will be no more new generations of stars arising from condensation of gas and dust because there will not be enough hydrogen left to start nuclear burning. Most of the elements will be near iron in mass, so that there is a lack of nuclear energy.

Galaxies begin to shrink after 10^{18} years because the random motions of the dead stars eventually fling most stars, dust, and gas out. The center of each galaxy collapses into a giant black hole. In about 10^{30} years protons should have begun to decay in abundance, heating up dead stars. When the dead stars are gone because all their protons have decayed, mostly gamma rays, neutrinos, and black holes remain. The small amount of matter dispersed in space will have a very low density, because the universe has expanded enormously in the meanwhile. Proton decay in intergalactic space results in electrons and positrons which almost never encounter each other. Meanwhile it will take 10^{100} years for the huge black holes to evaporate, leaving photons. The continuing expansion cools the photons to very low energy. The electrons and positrons eventually find each other, form positronium, radiate, then annihilate. If new discoveries are made about particle/field theories, black holes, or the other objects involved, then this dismal but long picture of the future will change. Consideration of entropy shows that the universe will not approach an equilibrium with maximum entropy (disorder). The old theory of a "heat death" at low temperatures is considered invalid.

16.8 COSMOLOGY AND PARTICLE PHYSICS

Cosmology, the study of the universe as a whole, is a vast subject. We can attempt to cover only those aspects which pertain to particle physics. However, an increasing amount of work in astronomy, astrophysics, and cosmology is being integrated with particle physics. Each field is aiding the others.

The collection of all the mass-energy in the universe curves space around it. If there is enough, the space closes, as Einstein first suspected. The two-dimensional analogue of a *closed* universe (spherical, or with positive curvature) is the surface of a sphere, on which there are flat people. A simple geometric property of such a surface, or of a closed universe, is that the angles of any triangle add to more than 180°, depending on the curvature. The closed universe has limited volume, although it may be expanding. Another possibility is an *open* universe (hyperbolic, with negative curvature). The less accurately analogous surface is a saddle shape. On this surface, or in an open universe, the angles of a triangle add to less than 180°.

The special case between closed and open is the *flat* universe, so-named for its analogy to a plane. Here long straight lines remain straight. A better way to express it is that parallel lines can be defined. The open and flat universes have unlimited volume. The flat universe results if the expansion slows down in just the right way that the rate of expansion becomes smaller and smaller as infinite volume is approached. If the universe were much different from flat, it should have a drastically different age than we observe. Only in the closed universe is there the possibility of traveling in a straight line for some 50 billion lightyears and ending up where we started (assuming that other difficulties are ignored).

Other overall properties of the universe that have been pertinent to particle physics are uniformity (homogeneity), expansion, and age. Determination of the uniformity of the overall distribution of matter and radiation (as well as local irregularities) will tell us the exact nature of the Big Bang and what forces, symmetries, and particles must have existed at what stages. Currently the measurement of the expansion rate (the Hubble constant) is uncertain by a factor of two. Better determination of the distances to galaxies and quasars, and of their brightnesses, will improve our knowledge of the expansion rate and of the forces and events that must have determined it. If the expansion has been slowing down, the Hubble constant is not really constant over long periods. Further study of local and extraterrestrial rocks improves knowledge of the age of the universe as reflected in radioisotope abundances.

Dating and study of prehistoric local supernovae are also possible from this information. Globular clusters of stars are another astronomical clue to age.

With the advent of inflationary theories, the possible domain available to our universe has inflated also. Instead of a closed region of about 10^{10} light years in radius, this may be just the region occupied by matter. The domain into which the matter is expanding could be 10^{35} light years because of the huge early inflation. There may be some domain walls far out there—walls in space where GUT is unbroken and protons are unstable.

We have described the particle and field history of the universe but have not faced the issue of a true beginning, however inaccessible. The possibility that the universe began from nothing as one giant quantum fluctuation was proposed by Edward Tryon in 1973. If it were not for the limited speed of light, we might someday see another huge quantum fluctuation within our own domain and a new universe come blasting out at us. This new cosmology has been called "quantum cosmology"!

Some concern should be given to the conservation of energy. It is disturbing that all the mass-energy we see came from nothing. Is there no energy debt to be repaid after all this time? We should realize that the very filling of space with matter represents gravitational energy, which can be defined to be negative using a conventional approach. Then the positive energy we observe can be balanced to zero net energy. The energy budget of the universe could be a big zero now just as it was "before" the beginning.

The best estimates from counting galaxies and gas clouds are that the universe has somewhere between 1/10 and 10 times the amount of mass-energy needed to make it "flat." Expansion will slow but might never reverse. Particle theories of the Big Bang almost demand that the universe be flat. The preponderance of evidence indicates that probably only 1/10 of the mass needed for flatness has been observed. However, the galaxies move as if they are being attracted by more mass, about the total needed for flatness. Hence the "missing mass" problem is vital to both particle physicists and astronomers. For example, some unknown particle in abundance may constitute the rest of the needed mass. The problem might better be called one of "dark matter," since, if we could see it, we could count it. The missing mass must not be radiating. It could be gas far from galaxies, or dim or dead stars, or small black holes, or unknown particles. If it is found, it should also explain why galaxies cluster in the sizes they do, a question that can be traced directly back to the quantum theories of the very early universe.

We live in a sea of high energy neutrinos. Everywhere in the universe, thanks in part to the reactions in stars, there are about 10^8 per cubic meter. Fortunately for us they rarely interact with anything. Recently neutrinos have entered the debate on where the missing mass of the universe is. (Other candidates are the stable particles predicted by supersymmetry, such as gravitinos and photinos.) Neutrinos had been assumed to be stable and massless, without definite experimental proof. Early reports of low mass for them (about 40 eV) have not been born out by further experiments, although this is about the mass needed to provide all the missing mass of the flat universe. On the large scale further studies of galaxies and clustering show that neutrinos with mass provide the wrong explanation of galactic motion.

The number of possible types of neutrinos can be deduced (not very accurately) from the observed abundance of helium, after allowance is made for the small contribution made by stars. At most one more neutrino (and heavy lepton) is allowed beyond the tauon family. This is another curious way that astronomy and particle physics aid each other. Getting a count of neutrino types could tell us how many quarks must exist, too. Some versions of GUT predict masses for neutrinos and instability such that they decay to form each other (neutrino "oscillation"). This has not been confirmed by experiments.

The search for the missing mass recently took a different turn. Cosmic "string," extremely thin long massive defects in space, is proposed. This could be detected by effects on starlight, and the search is on. The ability to put astronomical instruments in space for such subtle studies is now as beneficial to particle physicists as to astronomers and resembles the value of accelerators on the ground.

A way to study matter at Big Bang conditions but in terrestrial laboratories has been found in an unexpected quarter, nuclear physics. Information about quarks can be obtained through the collision of large relativistic nuclei carrying several GeV per nucleon. The Bevatron fed by the HILAC (Heavy Ion Linear Accelerator) at LBL (Color Plate 30) is used to accelerate and collide uranium nuclei against fixed nuclei at these energies, providing about 230 GeV in the lab frame. Effective temperatures in nuclear matter of about 10^{12} K have been obtained, corresponding to the first microsecond of the Big Bang. Physicists may be close to creating a quark-gluon mixture from nuclear matter.

A possible link between cosmology and particle physics has been the Dirac Large Numbers Hypothesis and its relation to the possibility that G, the gravitational constant, changes over time. Recent space radar and astronomical measurements have established that G must change less than one part in 10^{11} per year. Other considerations show that it could not have been much different in the distant past. The link with large numbers comes from the observation Dirac made that the age of the universe is about 10^{40} times the time for light to cross a proton. The electromagnetic force is also 10^{40} times stronger than gravity. And there are about 10^{80} (10^{40} squared) baryons in the universe. The incidence of 10^{40} in physics and cosmology seems more than chance, but speculation is mere numerology in absence of good theoretical reasons. Work continues halfheartedly, with no astounding revelations expected. The relation to changing G is that the rate of atomic process must be slowing down in accord with the increasing age of the universe, to keep the ratio of the rates constant. Conversely, the rate of change in the universe, governed by G, must be speeding up, so that G is getting stronger with time. Measurements showing that G does not change are now so accurate that the theory is probably wrong.

A related unproven principle is *Mach's principle*, from Ernest Mach who philosophized in 1872 against scientific laws and in favor of reliance on experimental facts. He puzzled over the conservation of angular momentum, an invariance with respect to direction. We do see stars move as we rotate, and he wondered if distant matter, which seems to establish absolute direction, had anything to do with local physics. He proposed that inertia is due to an unknown weak influence from all distant matter and further used this as an explanation for the weakness of gravity compared to other forces.

More successful theoretically is the *anthropic principle* as applied to cosmology. It starts with the observation that the evolution of the universe has to be what it is in order to produce the conditions that lead to the origin of life and the eventual appearance of intelligence (human, for example) to observe and learn what we know about the universe. The anthropic principle originated with G. J. Whitrow, Robert Dicke, Brandon Carter, and John Wheeler. There is a subtle connection between the age and the size of the universe. If it were much younger or older we would not exist. If it were much larger or smaller at the present age, galaxies and stars might not have been able to form because the density of matter was inappropriate. The universe also could not be much less uniform and give rise to stars, life, and us. Therefore it seems to be no accident that we are here now to observe our particular universe. It may be that a flat universe is required for intelligence to develop. There is also a very fine balance among the strengths of the fundamental forces that enables the conditions for life. A tiny change in the weak interaction would prohibit long-lived stable stars such as our sun.

The *cosmological principle* states that there is nothing special about where we are located in the universe. Moreover, there is no special location where the universe would look any different than it does to us here (small scale fluctuations aside). This principle originated with Copernicus, when he removed the Earth from the center of the universe. The modern version came from Herman Bondi in 1948. All observational evidence is in support of it. We can tell that atoms and particles everywhere within the range of observation behave exactly the same as they do here. The principle supports the notion from general relativity that there is no "edge" to the universe.

An extended cosmological principle, called "perfect" and stated by Bondi and Thomas Gold, covers the issue of time and age. The universe should look the same at any time. This principle had to be abandoned when the evidence for the Big Bang and expanding universe overwhelmed the earlier "steady state" model of the universe.

The prevailing cosmological principle is now the anthropic one. According to it, we can only be here to observe a universe because in an earlier era its expansion speed was equal to the speed needed for matter to escape against gravity, a unique condition that led to galaxies, stars, planets, and life. The reader is left to wonder at how all the complexity of matter, life, and human intelligence has evolved from some particular GUT with local gauge symmetry and the properties of the quarks, protons, and hydrogen atoms that result. Admittedly GUT and the possibly underlying TUT are also complex. But it is a cause for contemplation that a purely mathematical theory, which could be expressed in perhaps one book (not this one), can have so many implications. Much has been written about how structure evolves from basic laws, and much more remains to be studied as our confidence in GUT and TUT grow.

The anthropic principle also addresses the special strengths and properties of forces needed to arrive at these end results. Of all possible universes, ours seems to be about the only viable one for intelligent life, within extremely narrow limits. This is the subjectivity of quantum mechanics on a huge scale: the observer interacting with the observed to select one of the "many worlds." This is also philosophy or metaphysics rather than physics and thus signals the end of our discussion of the particle view of the universe.

Appendix

Questions and Answers
for each Chapter

A N ASSORTMENT OF QUESTIONS ARE GIVEN TO provoke further thought and practice. Some require verbal answers, and some require simple calculation using relations and facts discussed in the text. The Answers are after the section of Questions. Certain physical constants often needed are listed here to three digit accuracy in SI units:

speed of light: $c = 3.00(10)^8$ m/s
Planck's constant: $h = 6.63(10)^{-34}$ J s
unit of charge: $e = 1.60(10)^{-19}$ C
electron rest mass: $m_e = 9.11(10)^{-31}$ kg
proton rest mass: $m_p = 1.67(10)^{-27}$ kg
Avogadro's number: $N_0 = 6.02(10)^{23}$ mol^{-1}
Boltzmann's constant: $k = 1.38(10)^{-23}$ J/K
gravitational constant: $G = 6.67(10)^{-11}$ N m^2/kg^2
electric force constant: $k = 8.99(10)^9$ N m^2/C^2
permittivity: $\epsilon_0 = 8.85(10)^{-12}$ C^2/V m

Questions

CHAPTER 1.   THE PARTICLE VIEW

Q1-1. Express 5602 meters as a power of ten with both numbers and words.

Q1-2. Calculate, keeping only significant figures: $[2.3(10)^4 + 56]^{-3}$ = ?

Q1-3. List some assumptions about observations made of melting ice that find a change in weight as ice turns to water.

Q1-4. Calculate the densities for Fig. 1-3, to an order of magnitude only (a power of ten).

Q1-5. List some other reasons for measurement problems in Fig. 1-4.

CHAPTER 2.   FORCES, MOTION, AND ENERGY

Q2-1. How are muscle and friction forces traced to the fundamental forces?

Q2-2. Given the mass and charge of the proton, calculate and compare the gravitational and electric forces between two protons.

Q2-3. Look at any photograph in this book and tell how all four fundamental forces appear there.

Q2-4. If the next position in Table 2-2 is 15 m, find v and a.

Q2-5. Calculate the PE of yourself with your center 1 m above the floor. What reference will you use?

Q2-6. If there are 10^{10} galaxies with 10^{10} stars like the sun in each in the universe (an underestimate), what is the total energy produced per second in the universe by its stars, and where does this energy come from?

Q2-7. How can the gravitational force law involving G be used to find the value of g in the simpler gravity law?

CHAPTER 3. ROTATIONS AND INTER-
ACTIONS OF PARTICLES

Q3-1. Calculate the momentum of a 1 ton car going 60 mph (1 mph = 0.447 m/s). How fast must a 0.1 kg baseball be thrown to have the same momentum?

Q3-2. Calculate the angular momentum of the car in Q3-1 as it passes 10 m to the left of a post next to the road. Give the direction of the angular momentum. If the brakes apply a force of 1000 lbs (1 lb = 4.45 N) to the wheels, what is the torque, including direction, with respect to the post?

Q3-3. What happens if the two objects in Q3-1 collide head-on?

Q3-4. If the car in Q3-1 approaches another car going at 30 mph and weighing 3 tons, calculate the velocity of the center of mass of the two cars. If they collide inelastically, what is the speed of the resulting junk while they are not touching the road?

CHAPTER 4. FIELDS

Q4-1. Calculate the electric field strength 1 meter away from 1 coulomb of charge, and the force on another coulomb.

Q4-2. In Fig. 4-2, how could one tell from the field which end is the N pole?

Q4-3. If negative charge is moving east and a magnetic field is pointing down, which way is the force on it?

Q4-4. Calculate the magnetic force on an electron orbiting at 10^5 m/s near the outside of a 1 m cyclotron that is in a field of 1 T. If the field points up, which way does the electron spiral?

Q4-5. If a potential of 20,000 volts is between two linear accelerator plates 0.5 m apart, what is the force on an electron traveling between them, and which plate should be positive?

CHAPTER 5. WAVES

Q5-1. Describe some lower modes for a hanging 2 m string, which has the lower end free.

Q5-2. A 10 m rope vibrates in its fourth harmonic at a rate of 2 Hz (cycles per second). With what speed do waves travel back and forth on the rope, and why don't we see them travel?

Q5-3. Find the units of $1/2kA^2$ for a vibrating system of elastic constant k in N/m and amplitude A in m and interpret them.

Q5-4. Find the period of oscillation of a system with mass 1 kg and spring constant 1 N/m.

CHAPTER 6. LIGHT AND
OTHER RADIATION

Q6-1. How much of the electromagnetic spectrum is visible light?

Q6-2. What would have to be done to Maxwell's equations to make them appear symmetric?

Q6-3. Estimate the energy carried by radio waves in the range used by TV.

Q6-4. Compare the transfer of momentum when light strikes a black and a shiny body.

Q6-5. Estimate the maximum resolution of the eye.

CHAPTER 7. RELATIVITY

Q7-1. What's wrong with a satellite as an inertial frame?

Q7-2. Calculate the relativistic factor, moving mass, total energy, kinetic energy, and rest mass-energy of a proton traveling at 0.9 c.

Q7-3. How much shorter is a meter stick moving at 0.9 c?

Q7-4. Are the meter stick and the hole in the same inertial frame in Fig. 7-4?

Q7-5. How much longer does a particle traveling at 0.9 c live if its rest lifetime is 1 microsecond?

Q7-6. Calculate the energy available in the CM frame when a 1 GeV electron strikes a proton at rest.

CHAPTER 8. THE QUANTIZED MICROWORLD

Q8-1. Find wavelength and frequency for a proton moving at 0.9 c.

Q8-2. Find momentum and energy for a 10^{-10} m x-ray.

Q8-3. If the proton of Q8-1 is located with the x-ray of Q8-2, how accurately can its position be found?

Q8-4. Find the minimum energy an electron can have in a potential well 10^{-10} m big.

Q8-5. What would the world be like if $h = 0$ and there were no quantization?

Q8-6. Estimate the uncertainty in position when the energy is given.

CHAPTER 9. ATOMS

Q9-1. Calculate the second energy level of hydrogen and the classical radius of the second electron orbit.

Q9-2. Calculate the speed in the second orbit for hydrogen and the time for one orbit. Find the frequency of the orbit and compare with the frequency of typical radiation involving that orbit.

Q9-3. For the $n = 4$ orbit, find all possible values of l and m and tell how degenerate this level is for hydrogen.

Q9-4. What is wrong with Fig. 9-3B?

Q9-5. Estimate the time uncertainty for a photon representing a typical pulse of a million cycles of 10^{15} Hz radiation from an atom.

Q9-6. Compare the wavelength of visible light with the size of an atom that might have emitted it.

Q9-7. Estimate the radius of an electron orbit in which it has relativistic speed.

CHAPTER 10. MANY PARTICLES

Q10-1. Calculate the average (rms) speed of a proton at room temperature and at the core of the sun (10^7 K).

Q10-2. Why can hydrogen escape so much more easily from the Earth's atmosphere than oxygen?

Q10-3. What is wrong if the arrangement in Fig. 10-5B is symmetric?

Q10-4. If nuclei are made of fermions, what can be said about the energy states of a nucleus?

CHAPTER 11. NUCLEI

Q11-1. What speed and KE is needed to overcome proton-proton repulsion at 10^{-15} m? What is the corresponding temperature?

Q11-2. Why does hydrogen have no binding energy?

Q11-3. Calculate A for the island of stability at $Z = 114$ and $N = 184$.

Q11-4. Which of these would have to tunnel out of a nucleus and why: neutrino, electron, neutron, proton, alpha?

CHAPTER 12. LEPTONS AND QUANTUM ELECTRODYNAMICS

Q12-1. Why is the electron stable?

Q12-2. Estimate the size of the electron by equating its electric self PE to its rest mass-energy. Can the result be correct?

Q12-3. For how long can 0.5 MeV be borrowed from the vacuum?

Q12-4. Test the conservation of lepton number for beta decay.

Q12-5. Why should a Feynman diagram show less slant for a photon path than for particles?

CHAPTER 13. MODERN ACCELERATORS, HADRONS, AND THE STRONG FORCE

Q13-1. If no farther improvement in magnets is possible, what size synchrotron ring is needed for 20 TeV?

Q13-2. What power is carried by a beam of

10 mA at 10 GeV? How many particles does 10 mA represent if the circumference is 300 m?

Q13-3. Why are muons useful after traveling 1 km at Fermilab?

Q13-4. Estimate how long a relativistic particle can interact with a target in a collision.

Q13-5. Apply the conservation of charge and baryon number to the seven outcomes of a pion-proton collision in the text.

Q13-6. Which way does the magnetic field point in Fig. 13-4?

CHAPTER 14

Q14-1. Invent some new mesons and hadrons from the known quarks.

Q14-2. What would have to be the structure of a baryon with charge −2?

Q14-3. If quarks did not have exactly 1/3 or 2/3 of the charge of an electron, what would happen to atoms?

Q14-4. How many different ways can quark spins be arranged in a proton?

CHAPTER 15.   FIELD THEORIES FOR PARTICLES

Q15-1. Estimate the mass of an intermediate vector boson, given the known range of the weak interaction.

Q15-2. What pion is in Color Plate 5? Explain its colors.

Q15-3. Estimate the lifetime of an X carrier and discuss why they are not observed in present experiments.

Q15-4. How can a particle with a mass of only 1 GeV decay by means of particles of mass 100 GeV or much higher?

CHAPTER 16.   PARTICLES AND GRAVITY, ASTROPHYSICS, AND COSMOLOGY

Q16-1. Calculate the Schwarschild radius of gravitational curvature for the proton and the universe (about 10^{77} protons).

Q16-2. Interpret the Planck energy in chemical terms.

Q16-3. Why do cosmic rays of moderate energy reach the Earth mainly at the poles?

Q16-4. How can cosmic rays conserve charge since we receive mainly positive ions from interstellar space? Does the Earth become charged?

Answers

CHAPTER 1

A1-1. $5.602(10)^3$ m; 5.602 kilometers

A1-2. $8.2(10)^{-14}$

A1-3. Assumed that mass (or weight) was not lost or gained through unseen processes such as evaporation, condensation, accidental spillage, water lost on a thermometer that was inserted and removed, and other perhaps unknown possibilities.

A1-4. See Table 1-2.

A1-5. Heat loss to outside by radiation, conduction, evaporation; heat added by stirring; stratification of liquid into levels of different temperature

CHAPTER 2

A2-1. Both involve the electromagnetic; in muscle, long chains of molecules try to shorten or lengthen through electric forces between atoms; in friction, atoms on the surfaces of the materials tend to attract each other

A2-2. Assume that the protons are held some fixed distance apart, conveniently 1 meter. Then the gravitational force is $1.86(10)^{-64}$ N (attractive), and the electric force is $-2.30(10)^{-28}$ N (repulsive). The electric force is about 10^{36} times stronger, regardless of the separation.

A2-3. Ordinary materials are held together by the electromagnetic. The picture is probably taken on Earth, where gravity acts on all objects shown. The nuclear force enables the existence of the nuclei of all atoms of all materials. There are always a few radioactive nuclei present in all materials, which sometimes emit energy via the weak force. If the scene is lit by sunlight, this light began with nuclear and weak interactions in the sun.

A2-4. $v = 7$ m/s; $a = 3$ m/s²

A2-5. If you assume the floor as reference, then your PE is given by your mass (in kg) times 9.8 times 1 m.

A2-6. At 10^{27} J/s for the sun, total energy output per second (power) is 10^{47} J/s. This comes primarily from nuclear reactions, involving a lowering of nuclear PE.

A2-7. Use the radius and mass of the Earth, about $6.37(10)^6$ m and $5.98(10)^{24}$ kg, in the law to obtain a numerical value of $g = GM/r^2$, good only near the surface of the Earth where $r = 6.37(10)^6$ m.

CHAPTER 3

A3-1. One ton is close to one metric tonne (1000 kg), so the momentum is about 27000 kg m/s. The baseball must travel 270,000 m/s, about 0.1% lightspeed.

A3-2. About 270000 kg m²/s down; about 45000 N m up.

A3-3. They both stop.

A3-4. 17 m/s; the same.

CHAPTER 4

A4-1. Field strength is $9(10)^9$ N/C; force on 1 C is $9(10)^9$ N.

A4-2. The field is symmetric and carries no intrinsic clue to its direction; only the use of a test magnet or charge and the defining of the convention for magnetic polarity can find which is the N pole.

A4-3. North.

A4-4. $1.6(10)^{-14}$ N (note 1 T is nearly the maximum magnetic field strength possible in iron); counterclockwise.

A4-5. The electric field strength is 40,000 V/m or N/C; the electron feels $6.4(10)^{-15}$ N, and the positive plate must be ahead of it.

CHAPTER 5

A5-1. The hanging end swings the most freely of any part of the string so that it is one-fourth of a wavelength away from any nodes. In the

lowest mode, the piece of string carries one-fourth of a wavelength, so the wavelength is 8 meters. In the next mode, the string is 3/4 wavelength long, so the wavelength is 8/3 m. In the next mode the string carries 5/4 of a wavelength, and so forth.

A5-2. Speed is 10 m/s; traveling waves in each direction superimpose to form an apparently standing wave.

A5-3. The units are N m, which should be recognized as joules, units of energy.

A5-4. Period = 6.28 Hz.

CHAPTER 6

A6-1. Since the electromagnetic spectrum is infinitely long, the visible portion occupies a negligible fraction of it, and the portion is very small even if only the useful part of the spectrum is considered.

A6-2. Symmetry is achieved if E and B can be traded without changing the form of the equations. A source for B would have to be found, and either a "current" for E or removal of the current J for B.

A6-3. TV is broadcast at about 10^8 Hz, corresponding to 10^{-25} J per photon.

A6-4. Light waves or photons striking a black surface are absorbed, transferring all momentum to the body. For a shiny surface, the light's momentum is reversed, transferring double momentum to the body.

A6-5. Assuming a midrange visible wavelength of 500 nanometers and an eye aperture (pupil) of 0.5 cm (0.005 m), the ratio of wavelength to aperture 10^{-4} is the same as the ratio of the separation between barely resolved distant objects and the distance to them. Thus at 1 km, the eye could theoretically resolve 0.1 m, although other factors make this unachievable.

CHAPTER 7

A7-1. It is moving in a circular orbit, therefore accelerating inward. (However, we shall see later that it happens to be falling at the same rate as the gravitational field, so no forces are experienced aboard. It is still not a good inertial frame because different parts of the satellite are pulled in different directions by gravity.)

A7-2. Gamma is 2.29, moving mass is $3.38(10)^{-27}$ kg, total energy is $3.44(10)^{-10}$ J, rest mass-energy is $1.51(10)^{-10}$ J, and KE is the difference between these.

A7-3. Contracted by 2.29, to a length of 0.44 m.

A7-4. No, and that is why there is an apparent paradox.

A7-5. 1 microsecond is dilated to 2.29 microseconds.

A7-6. 1.4 GeV is available in the CM frame, compared to 2 GeV in the lab frame, not a bad result, but only because the proton is so much heavier than the electron that they had the same total energies in this case. Note that the calculation had to be done with E and m in the same energy units.

CHAPTER 8

A8-1. First find its momentum and energy (use relativity); then the wavelength is $6.4(10)^{-16}$ m and the frequency is $5.3(10)^{23}$ Hz.

A8-2. $6.6(10)^{-24}$ kg m/s; $2(10)^{-15}$ J.

A8-3. The uncertainty in position is estimated to be 10^{-10} m when the x-rays are shone on the proton, giving it a kick equal to the momentum of the x-rays. This happens to be the wavelength of the x-rays, and greater accuracy could not be obtained under any circumstances.

A8-4. The well size implies an uncertainty in momentum of about 10^{-23} kg m/s. Finding the speed shows it to be nonrelativistic, about 10^7 m/s, giving a KE of at least about 10^{-16} J or 1000 eV.

A8-5. Electrons would not diffract, there would be no inherent uncertainty, and, as we shall learn in the next chapter, atoms could not exist because nothing would keep them from collapsing to zero size. There would be no matter and no world—nothing as we know it.

A8-6. If the energy is given, divide it by c to obtain the momentum, then use the uncertainty principle.

CHAPTER 9

A9-1. The second energy level is $-5.4(10)^{-19}$ J (-3.39 eV) and has classical radius $2.1(10)^{-10}$ meter.

A9-2. The second classical orbit has 2 units of angular momentum, so the speed can be found (using the results of Q9-1) to be $1.1(10)^6$ m/s. The orbital period is found from the circumference and the speed, and the frequency is $8.3(10)^{14}$ Hz. This agrees with the radiation frequency if 3.39 eV is emitted as light.

A9-3. l can have values of 0, 1, 2, 3; m can have values of 0; 0, $+1$, -1; 0, $+1$, -1, $+2$, -2; 0, $+1$, -1, $+2$, -2, $+3$, -3; the $n = 4$ level is 4-fold degenerate.

A9-4. It is a classical picture for a quantized atom. The electron does not orbit like a precessing planet.

A9-5. A burst of a million cycles lasts 10^{-9} s; however, the uncertainty in time is 1 cycle, or 10^{-15} s, as could be estimated by calculating the energy of the photon and using the uncertainty principle.

A9-6. A visible wavelength is about $5(10)^{-7}$ m, 1000 times bigger than a hydrogen atom in its ground state.

A9-7. To have relativistic speed the electron must approach to about 10^{-11} m, which it does in heavier atoms.

CHAPTER 10

A10-1. At room temperature it has about 2700 m/s; in the sun it has about 500,000 m/s.

A10-2. Hydrogen is 16 times lighter than oxygen and so has 4 times more rms speed.

A10-3. Very special initial conditions are needed to cause a multiple collision to result in a symmetric pattern. This is probably even more unlikely than a group of particles coming together to stay, emitting one.

A10-4. Each fermion requires its own state, so a nucleus must have a series of energy levels.

CHAPTER 11

A11-1. The PE between 2 protons at 10^{-15}

m is $2.3(10)^{-13}$ J (1.4 MeV), and this much KE is needed, corresponding to $8.3(10)^6$ m/s. In a gas of protons at 10^{10} K, the average proton would have this speed.

A11-2. Hydrogen has only one nucleon, a proton. (Isotopes of hydrogen do have BE.)

A11-3. $A = 298$.

A11-4. The electron and neutrino are not attracted by nuclear force and do not need to tunnel to escape. The rest are held by nuclear force and must tunnel, despite the electric repulsion felt by the proton and alpha.

CHAPTER 12

A12-1. There is no lighter known charged particle to which it can decay. Alternatively, there is no known violation of conservation of charge.

A12-2. About $3(10)^{-16}$ m, but all evidence is that it has no size.

A12-3. 0.5 MeV is about 10^{-13} J, which can be borrowed for about 10^{-20} second.

A12-4. Neutron and proton have no lepton number; after decay, the electron has an electron lepton number of $+1$ and the antineutrino -1, for a net zero lepton number.

A12-5. The photon moves faster (although Feynman diagrams do not have to be drawn to this degree of accuracy to be useful).

CHAPTER 13

A13-1. Assuming superconducting magnets are used, about 20 km radius.

A13-2. 10 GeV is the equivalent of accelerating the beam through 10^{10} volts, so the power is voltage times current, or 10^8 watts. The beam has speed nearly c, so 1 revolution takes 1 microsecond, and the circulating charge is 10^{-8} C or about 10^{11} charges.

A13-3. Time dilation makes the muons appear to live much longer in the lab frame. To survive for 1 km, they need live only about 10 microseconds, about 10 times longer than in their rest frame.

A13-4. A particle approaches within 10^{-15} m for less than 10^{-23} second.

A13-5. Charge conservation depends on the sign of the pion used, but must be conserved before and after. The proton brings in $B = 1$, so one baryon must result in each case.

A13-6. Downward in the plane of the figure.

CHAPTER 14

A14-1. There are no restrictions except the meson should have a quark and antiquark, automatically giving it charge $+1$, 0, or -1, and the baryon should have three quarks. Example: usb is a baryon with strangeness and bottomness, rather massive, and with charge zero.

A14-2. A baryon with charge -2 must contain (illegally) 6 quarks, or be an antibaryon containing 3 antiquarks.

A14-3. An atom as normally constructed would have a net charge, since its nucleons would each not have the same size charge as the electrons.

A14-4. To obtain net spin 1/2, two spins must be up and one down (not to be confused with u and d quarks). Thus there could be u spin up, u spin up, and d spin down, or there could be u spin up, u spin down, and d spin up.

CHAPTER 15

A15-1. By the uncertainty principle, the uncertainty in momentum for 10^{-17} m is about 10^{-16} in SI units. Since we are in the ultra-relativistic range, the corresponding energy can be found by $E = pc$, giving about 10^{-8} J or 100 GeV.

A15-2. The $u\bar{d}$ pion is positive. It must always have a color and an anticolor—green and magenta, or red and cyan, for example.

A15-3. To obtain the 10^{15} GeV or 10^5 J unification energy, an X can live 10^{-38} second. The chance of one being released is extremely low in ordinary matter, so that observing one during a decay is as unlikely as being able to make one with a superaccelerator.

A15-4. The heavy carrier is borrowed from the vacuum for a very short time and is not always present.

CHAPTER 16

A16-1. Proton: $2.4(10)^{-54}$ m; universe: about 10^{23} m or about 10 million LY. The proton would not make a good black hole, but the universe would make a large one.

A16-2. The Planck energy of about 10^{19} GeV is about 10^9 J, the amount of energy needed to boil about 500 kg of water.

A16-3. Cosmic rays of moderate energy are deflected away by the Earth's magnetic field except near the poles where the field points toward or away from the Earth. High energy cosmic rays are not affected appreciably.

A16-4. On the large scale, as many cosmic rays travel one way in the galaxy as another, conserving charge. Near the Earth are many electrons and ions, so that a positively charged Earth will grab loose electrons, leaving an imbalance of charge in a large region surrounding the Earth where cosmic rays can help provide overall neutrality.

References

BOOKS ARE LISTED FIRST, THEN ARTICLES. *SCI-entific American* is abbreviated *ScAm*. The references include sources for this book, recommended supplementary reading, and many landmark articles by physicists in their particular fields, including Nobel Prize winners. Significant articles on theories and results since proven wrong are included where of historical interest.

* denotes books and articles of a technical nature, relying on mathematics.

Books pertaining to several chapters:

*—. *Novel Experiments in Physics*. AAPT, 1975

Asimov, Isaac. *Understanding Physics*. Walker, 1966.

Baker, Adolph. *Modern Physics and Antiphysics*. Addison-Wesley, 1970.

Bondi, H. *Assumption and Myth in Physical Theory*. Cambridge Univ. Press, 1967.

Born, Max. *The Restless Universe*. Dover, 1951.

Bridgeman, P. W. *The Nature of Physical Theory*. Wiley, 1936.

Charon, Jean. *Cosmology*. McGraw-Hill, 1970.

Cioffari, Bernard. *Experiments in College Physics*, 6th ed. D. C. Heath & Co., 1978.

Davies, J. T. *The Scientific Approach*. Academic Press, 1965.

Davies, P. C. W. *Space and Time in the Modern Universe*. Cambridge Univ. Press, 1977.

*—. *The Physics of Time Asymmetry*. Univ. of California Press, 1977.

Einstein, Albert, & Infeld, Leopold. *The Evolution of Physics*. Simon & Schuster, 1938.

Elton, L. R. B., & Messel, H. *Time and Man*. Pergamon, 1978.

Feynman, Richard. *The Character of Physical Law*. MIT Press, 1965.

*—. *The Feynman Lectures on Physics*. Addison-Wesley, 1965.

Gamow, George. *Biography of Physics*. Harper & Row, 1961.

Gardner, Martin. *The Ambidextrous Universe*. Mentor, 1969.

*Halliday, David, & Resnick, Robert. *Fundamentals of Physics*. Wiley, 1981.

Hawkins, David. *The Language of Nature*. Doubleday, 1964.

Heisenberg, Werner. *Across the Frontiers*. Harper & Row, 1974.

—. *Physics and Beyond: The Revolution in Modern Science*. Harper & Row, 1958.

*Jackson, John. *Classical Electrodynamics*. Wiley, 1962.

*Leighton, Robert. *Principles of Modern Physics*. McGraw-Hill, 1959.

Mason, Stephen. *A History of the Sciences*. Macmillan 1962.

Pickering, Andrew, *Constructing Quarks*. U. Chicago Press 1984.

Postle, Denis. *Fabric of the Universe*. Crown, 1976.

Schlegel, Richard. *Time and the Physical World*. Dover, 1961.

Segre, Emilio. *From Falling Bodies to Radio Waves*. W. H. Freeman, 1984.

—. *From X-Rays to Quarks*. W. H. Freeman, 1980.

*—. *Nuclei and Particles*. Benjamin, 1964.

*Shamos, Morris, ed. *Great Experiments in Physics*. Holt Rinehart & Winston, 1959.

Trefil, James. *From Atoms to Quarks*. Charles Scribner's Sons, 1980.

*Weidner, Richard, and Sells, Robert. *Elementary Modern Physics*. Allyn & Bacon, 1980.

Weinberg, Alvin. *Reflections on Big Science*. M. I. T. Press, 1967.

Weinberg, Steven. *The Discovery of Subatomic Particles*. Freeman, 1983.

Weisskopf, Victor. *Knowledge and Wonder*. Doubleday, 1966.

—. *Physics in the Twentieth Century*. MIT Press, 1972.

Articles pertaining to several chapters:

Dyson, Freeman. "Field Theory". *ScAm* Apr 1953.

—. "Mathematics in the Physical Sciences". *ScAm* Sep 1964.

Feinberg, Gerald, & Goldhaber, Maurice. "The Conservation Laws of Physics". *ScAm* Oct 1963.

Weisskopf, Victor. "Physics in the Twentieth Century". *Science* 22 May 1970.

—. "Three Spectroscopies". *ScAm* May 1968.

—. "Three Steps in the Structure of Matter". *Physics Today* Aug 1970.

Books and articles
pertaining to one particular chapter:

CH 1.   THE PARTICLE VIEW

Books:

Emerton, Norma. *The Scientific Reinterpretation of Form*. Cornel, 1984.

French, A. P., & Hudson, A. M. *Particles and Newtonian Mechanics*. MIT, 1965.

Kuhns, Thomas. *The Structure of Scientific Revolutions*. Univ. of Chicago, 1970.

Nelson, Robert. *SI: The International System of Units*. AAPT, 1981.

Sienko, Michell, & Plane, Robert. *Chemistry*. McGraw-Hill, 1966.

Articles:

Astin, Allen. "Standards of Measurement". *ScAm* Jun 1968.

Feinberg, Gerald. "Ordinary Matter". *ScAm* May 1967.

Hall, Marie. "Robert Boyle". *ScAm* Aug 1967.

Holliday, Leslie. "Early Views on Forces Between Atoms". *ScAm* May 1970.

Weisskopf, Victor. "Is Physics Human". *Physics Today* Jun 1976.

Williams, L. "Humphrey Davy". *ScAm* Jun 60.

CH 2.   FORCES, MOTION, AND ENERGY

Articles:

Cohen, I. B. "Newton's Discovery of Gravity". *ScAm* Mar 81.

Drake, Stillman. "Galileo's Discovery of the Law of Free Fall". *ScAm* May 1973.

—. "Newton's Apple and Galileo's Dialogue". *ScAm* Aug 1980.

—. "The Role of Music in Galileo's Experiments". *ScAm* Jun 1975.

Drake, Stillman, & MacLachlan, James. "Galileo's Discovery of the Parabolic Trajectory". *ScAm* Mar 1975.

McCloskey, Michael. "Intuitive Physics". *ScAm* Apr 1983.

Wilson, Curtis. "How Did Kepler Discover His First Two Laws?" *ScAm* Mar 1972.

CH 4. FIELDS

Book:

Gamow, George. *Gravity*. Doubleday, 1962.

Articles:

Bitter, Francis. "Ultrastrong Magnetic Fields". *ScAm* Jul 1965.

Heilbron, J. L., Seidel, Robert, & Wheaton, Bruce. "Lawrence and his Laboratory". *LBL News Magazine* fall 1981.

Santillana, Giorgio de. "Alessandro Volta". *ScAm* Jan 1965.

CH 5. WAVES

Book:

*Crawford, Frank, Jr. *Waves*. McGraw-Hill, 1968.

CH 6. LIGHT AND OTHER ELECTROMAGNETIC RADIATION

Books:

*Hecht, Eugene, & Zajac, Alfred. *Optics*. Addison-Wesley, 1974.

*Meyer-Arendt, Jurgen. *Introduction to Classical and Modern Optics*. Prentice-Hall, 1972.

Articles:

Connes, Pierre. "How Light is Analyzed". *ScAm* Sep 68.

Feinberg, Gerald. "Light". *ScAm* Sep 1968.

Sharlin, Harold. "From Faraday to the Dynamo". *ScAm* May 61.

Shiers, George. "Induction Coil". *ScAm* May 71.

Weisskopf, Victor. "How Light Interacts with Matter". *ScAm* Sep 1968.

CH 7. RELATIVITY (SPECIAL)

Books:

Bondi, Hermann. *Relativity and Common Sense*. Dover, 1964.

Born, Max. *Einstein's Theory of Relativity*. Dover, 1962.

Calder, Nigel. *Einstein's Universe*. Viking, 1979.

Einstein, Albert. *Relativity*. Crown, 1952.

*French, A. P. *Special Relativity*. Norton, 1968.

Gamow, George, *Mr. Tompkins in Wonderland*. Cambridge Univ. Press, 1944.

Hoffmann, Banesh, & Dukas, Helen. *Albert Einstein: Creator and Rebel*. Viking, 1972.

Taylor, Edwin, & Wheeler, John. *Spacetime Physics*. Freeman, 1966.

Articles:

Feinberg, Gerald. "Particles That Go Faster Than Light". *ScAM* Feb 1970.

*Hafele, J., & Keating, Richard. "Around the World Atomic Clocks". *Science* 14 Jul 1972.

Rothman, Milton. "Things That Go Faster Than Light". *ScAm* Jul 60.

Rowe, Ednor, & Weaver, Joh. "Uses of Synchrotron Radiation". *ScAm* Jun 77.

Shankland, R. S. "The Michelson-Morley Experiment". *ScAm* Nov 64.

CH 8.   THE QUANTIZED MICROWORLD

Books:

Davies, Paul. *Other Worlds: Space, Superspace, and the Quantum Universe*. Simon & Schuster, 1980.

Gamow, George. *Mr. Tompkins Explores the Atom*. Cambridge Univ. Press, 1945.

—. *Thirty Years that Shook Physics*. Doubleday, 1966.

Heisenberg, Werner. *Physics and Beyond: Encounters and Conversations*. Harper & Row, 1971.

Hoffman, Banesh. *The Strange Story of the Quantum*. Dover, 1959.

Jauch, J. M. *Are Quanta Real? A Galilean Dialogue*. Indiana Univ. Press, 1973.

Pagels, Heinz. *The Cosmic Code: Quantum Physics as the Language of Nature*. Simon & Schuster, 1982.

Wheaton, Bruce. *The Tiger and the Shark: Empirical Roots of Wave-Particle Dualism*. Cambridge Univ. Press, 1983.

Articles:

Bernstein, Herbert, & Phillips, Anthony. "Fiber Bundles and Quantum Theory". *ScAm* Jul 1981.

Braginsky, Vladimir, et al. "Quantum Nondemolition Measurements". *Science* 1 Aug 1980.

d'Espagnat, Bernard. "Quantum Theory and Reality". *ScAm* Nov 1979.

Dewitt, Bryce. "Quantum Mechanics and Reality". *Physics Today* Sep 1970.

Dirac, P. A. M. "The Evolution of the Physicist's Picture of Nature". *ScAm* May 1963.

Gale, George. "The Anthropic Principle". *ScAm* Dec 1981.

Gamow, George. "The Exclusion Principle". *ScAm* Jul 1959.

Goldberg, Alfred, and Nieto, Michael. "Mass of the Photon". *ScAm* May 1976.

Hughes, R. I. G. "Quantum Logic". *ScAm* Oct 1981.

Lamb, Willis. "An Operational Interpretation of Nonrelativistic Quantum Mechanics". *Physics Today* Apr 1969.

Lande, Alfred. "New Foundations of Quantum Physics". *Physics Today* Feb 1967.

Pais, A. "Max Born's Statistical Interpretation of Quantum Mechanics". *Science* 17 Dec 1982.

Robinson, Arthur. "Loophole Closed in Quantum Mechanics". *Science* 7 Jan 1983.

—. "Quantum Mechanics Passes Another Test". *Science* 30 Jul 1982.

Rohrlich, Fritz. "Facing Quantum Mechanical Reality". *Science* 23 Sep 1983.

Taylor, Barry, et al. "Fundamental Physical Constants". *ScAm* Oct 1970.

Weinberg, Steven. "Light as a Fundamental Particle". *Physics Today* Jun 1975.

CH 9.   ATOMS

Articles:

Binnig, Gerd, & Rohrer, Heinrich. "The Scanning Tunneling Microscope." Aug. 1985.

Bouchiat, Marie-Anne, & Pottier, Lionel. "Atomic Preference between Left and Right". *ScAm* Jun 1984.

Corben, H. C., & DeBenedetti, S. "The Ultimate Atom". *ScAm* Dec 1954.

Crewe, Albert. "High-resolution Scanning Electron Microscope." *ScAm* Apr 1971.

Dyke, W. "Advances in Field Emission". *ScAm* Jan 1964.

Frisch, O. R. "Molecular Beams". *ScAm* May 1965.

Hansch, Theodor, et al. "The Spectrum of Atomic Hydrogen". *ScAm* Mar 1979.

Hurst, G. S., et al. "Counting the Atoms". *Physics Today* Sep 1980.

Hughes, Vernon. "The Muonium Atom". *ScAm* Apr 1966.

Kleppner, Daniel, et al. "Highly Excited Atoms". *ScAm* May 1981.

Robinson, Arthur. "Laser Light Cools Sodium Atoms to 0.07K". *Science* 10 Dec 1982.

Schawlow, Arthur. "Laser Spectroscopy of Atoms and Molecules". *Science* 13 Oct 1978.

—. "Spectroscopy in a New Light". *Science* 2 Jul 1982.

Stebbings, Ronald. "High Rydberg Atoms: Newcomers to the Atomic Physics Scene". *Science* 13 Aug 1976.

Wahl, Arnold. "Chemistry by Computer". *ScAm* Apr 1970.

Weisskopf, Victor. "Of Atoms, Mountains, and Stars". *Science* 21 Feb 1975.

CH 10. MANY PARTICLES: A STATISTICAL VIEW

Books:

Einstein, Albert. *Investigations on the Theory of the Brownian Movement*. Dover, 1926.

Prigogine, Ilya. *From Being to Becoming*. W. H. Freeman, 1980.

Articles:

Anderson, P. W. "1982 Nobel Prize in Physics". *Science* 19 Nov 1982.

Bertman, Bernard, & Guyer, Robert. "Solid Helium". *ScAm* Aug 1967.

Ehrenberg, W. "Maxwell's Demon". *ScAm* Nov 1967.

Ford, Joseph. "How Random is a Coin Toss?" *Physics Today* Apr 1983.

Lavenda, Bernard. "Brownian Motion". *ScAm* Feb 1985.

Layzer, David. "Arrow of Time". *ScAm* Dec 1975.

Mermin, N., & Lee, David. "Superfluid Helium 3". *ScAm* Dec 1976.

Procaccia, Itamar, & Ross, John. "The 1977 Nobel Prize in Chemistry". *Science* 18 Nov 1977.

Robinson, Arthur. "Metrology: A More Accurate Value for Avogadro's Number". *Science* 20 Sep 1974.

Silvera, Isaac, & Walraven, Jook. "Stabilization of Atomic Hydrogen". *ScAm* Jan 1982.

Wilson, Mitchell. "Count Rumford". *ScAm* Oct 1960.

CH 11. NUCLEI

Book:

Hartcup, Guy, & Allibone, T. *Cockcroft and the Atom*. Hilger, 1984.

Articles:

Andrade, E. N. "Birth of the Nuclear Atom". *ScAm* Nov 1956.

Badash, Lawrence. "How the 'Newer Alchemy' Was Received". *ScAm* Aug 1966.

Baranger, Michael, and Sorenson, Raymond. "The Size and Shape of Atomic Nuclei". *ScAm* Aug 1969.

Cerny, Joseph, & Poskanzer, Arthur. "Exotic Light Nuclei". *ScAm* Jun 1978.

Cranberg, Lawrence. "Fast-Neutron Spectroscopy". *ScAm* Mar 1964.

Ghiorso, Albert, & Seaborg, Glenn. "The Newest Synthetic Elements". *ScAm* Dec 1956.

Greiner, Walter, & Stocker, Horst. "Hot Nuclear Matter". *ScAm* Jan 1985.

Hardy, J. C. "Exotic Nuclei and their Decay". *Science* 1 Mar 1985.

Hedges, Robert, & Gowlett, John. "Radiocarbon Dating by Accelerator Mass Spectrometry". *ScAm* Jan 1986.

Herber, R. H. "Mossbauer Spectroscopy". *ScAm* Oct 1971.

Lightman, Alan. "To Cleave an Atom". *Science 84* Nov 1984.

Marshak, Robert. "Nuclear Force". *ScAm* Mar 1960.

Peierls, R. E. "Atomic Nucleus". *ScAm* Jan 1959.

Platzman, Robert. "What Is Ionizing Radiation?" *ScAm* Sep 1959.

Seaborg, Glenn, & Block, Justin. "The Synthetic Elements: IV". *ScAm* Apr 1969.

Seaborg, Glenn, & Fritsch, A. "The Synthetic Elements: III". *ScAm* Apr 1963.

Shapiro, Gilbert. "Polarized Accelerator Targets". *ScAm* Jul 1966.

*Siegbahn, Kai. "Electron Spectroscopy for Atoms, Molecules, and Condensed Matter". *Science* 9 Jul 1982.

Zafiratos, Chris. "The Texture of the Nuclear Surface". *ScAm* Oct 1972.

Zare, Richard. "Laser Separation of Isotopes". *ScAm* Feb 1977.

CH 12. LEPTONS AND QUANTUM ELECTRODYNAMICS

Books:

Brown, Laurie, & Hoddeson, Lillian, eds. *The Birth of Particle Physics*. Cambridge Univ. Press, 1983.

*Gottfried, Kurt, & Weisskopf, Victor. *Concepts of Particle Physics*. Oxford Univ. Press, 1984.

*Schwinger, Julian, ed. *Quantum Electrodynamics*. Dover, 1958.

Articles:

Corben, H. C. & DeBenedetti, S. "The Ultimate Atom". *ScAm* Dec 1954.

Crane, H. R. "The g Factor of the Electron". *ScAm* Jan 1968.

Danby, G., et al. "Observation of High Energy Neutrino Reactions and the Existence of Two Kinds of Neutrinos". *Phys. Rev. Letters* 1 Jul 1962.

Ekstrom, Philip, & Wineland, David. "The Isolated Electron". *ScAm* Aug 80.

Feynman, Richard. "The Development of the Space-Time View of Quantum Electrodynamics". *Physics Today* Aug 1966.

Fulcher, Lewis, et al. "The Decay of the Vacuum". *ScAm* Dec 1979.

Kusch, Polykarp. "The Electron Dipole Moment". *Physics Today* Feb 1966.

Lederman, Leon. "The Two-Neutrino Experiment". *ScAm* Mar 1963.

Penman, Sheldon. "The Muon". *ScAm* Jul 1961.

Perl, Martin, & Kirk, William. "Heavy Leptons". *ScAm* Mar 1978.

Reines, Frederick. "The Early Days of Experimental Neutrino Physics". *Science* 5 Jan 1979.

—. "Neutrinos from the Atmosphere and Beyond". *ScAm* Feb 1966.

Robinson, Arthur. "Precision Positronium Spectroscopy Tests QED". *Science* 20 Jul 1984.

Tomonaga, Sin-itiro. "Development of Quantum Electrodynamics". *Physics Today* Sep 1966.

Treiman, S. B. "Weak Interactions". *ScAm* Mar 1959.

Wigner, Eugene. "Violations of Symmetry in Physics". *ScAm* Dec 1965.

CH 13. MODERN ACCELERATORS, HADRONS, AND THE STRONG FORCE

Books and pamphlets:

—. *HERA*. Deutsches Elektronen Synchrotron, Mar 1984.

—. *Review of Particle Properties*. *Rev. Mod. Phys.* Apr 1984.

Particle Data Group. *Particle Properties Data Booklet*. CERN, 1984.

Articles:

—. "Accelerators". Lawrence Berkeley Laboratory, 1981.

—. "Stanford Linear Accelerator Center". *Beam Line* Feb 1984.

Amaldi, Ugo. "Proton Interactions at High Energy". *ScAm* Nov 1973.

Barger, Vernon, & Cline, David. "High Energy Scattering". *ScAm* Dec 1967.

Chew, Geoffrey, et al. "Strongly Interacting Particles". *ScAm* Feb 1964.

*Cronin, James. "CP Symmetry Violation: the Search for Its Origin". *Science* 12 Jun 1981.

Dyson, Freeman. "Mathematics in the Physical Sciences". *ScAm* Sep 1964.

Fitch, Val. "The Discovery of Charge Conjugation-Parity Asymmetry". *Science* 29 May 1981.

Fowler, William, & Samios, Nicholas. "The Omega-Minus Experiment". *ScAm* Oct 1964.

Gardner, Martin. "Can Time Go Backward?" *ScAm* Jan 1967.

Ginzton, E. L., & Kirk, William. "The Two-Mile Accelerator". *ScAm* Nov 1961.

Golub, R. et al. "Ultracold Neutrons". *ScAm* Jun 1979.

Hill, R. D. "Resonance Particles". *ScAm* Jan 1963.

Kendall, Henry, & Panofsky, Wolfgang. "The Structure of the Proton and Neutron". *ScAm* Jun 1971.

Litke, Alan, & Wilson, Richard. "Electron-Positron Collisions". *ScAm* Oct 1973.

McMillan, Edwin. "A History of the Synchrotron". *Physics Today* Feb 1984.

Murphy, Frederick, & Yount, David. "Photons as Hadrons". *ScAm* Jul 1971.

O'Neill, Gerard. "Particle Storage Rings". *ScAm* Nov 1966.

—. "The Spark Chamber". *ScAm* Aug 1962.

Oppenheimer, Robert. "Thirty Years of Mesons". *Physics Today* Nov 1966.

Overseth, Oliver. "Experiments in Time Reversal". *ScAm* Oct 1969.

Phelps, Michael, & Mazziotta, John. "Positron Emission Tomography: Human Brain Function and Biochemistry." *Science* 17 May 1985.

Robinson, Arthur. "Fermilab Tests Its Antiproton Factory." *Science* 27 Sep 1985.

—. "Los Alamos Neutron Source Meets First Test." *Science* 21 Jun 1985.

—. "New Cornell Accelerator Stores First Beam". *Science* 18 May 1979.

—. "Proton-Antiproton Collisions at Fermilab." *Science* 1 Nov 1985.

—. "Stanford Pulls off a Novel Accelerator". *Science* 25 Jun 1982.

Telegdi, V. L. "Hypernuclei". *ScAm* Jan 1962.

Ting, Samuel. "The Discovery of the J Particle: a Personal Recollection". *Science* 10 Jun 1977.

Van der Meer, S. "Stochastic Cooling and the Accumulation of Antiprotons". *Science* 15 Nov 1985.

Weigand, Clyde. "Exotic Atoms". *ScAm* Nov 1972.

Weisskopf, Victor. "Three Spectroscopies". *ScAm* May 1968.

Wilson, R. R. "The Batavia Accelerator". *ScAm* Feb 1974.

Yount, David. "The Streamer Chamber". *ScAm* Oct 1967.

CH 14.   THE INSCRUTABLE QUARKS

Articles:

Brown, Gerald, & Rho, Mannque. "The Structure of the Nucleon". *Physics Today* Feb 1983.

Drell, Sidney. "Electron-Positron Annihilation and the New Particles". *ScAm* Jun 1975.

Feynman, Richard. "The Structure of the Proton". *Science* 15 Feb 1974.

Glashow, Sheldon. "The Hunting of the Quark". *New York Times Magazine* 18 Jul 1976.

—. "Quarks with Color and Flavor". *ScAm* Oct 1975.

Jacob, Maurice, & Landshoff, Peter. "Inner Structure of the Proton". *ScAm* Mar 1980.

Krisch, Alan. "The Spin of the Proton". *ScAm* May 1979.

Lederman, Leon. "The Upsilon Particle". *ScAm* Oct 1978.

McHarris, William, & Rasmussen, John. "High-Energy Collisions between Atomic Nuclei". *ScAm* Jan 1984.

Mistry, Nariman, et al. "Particles with Naked Beauty". *ScAm* Jul 1983.

Nambu, Yoichiro. "The Confinement of Quarks". *ScAm* Nov 1976.

Richter, Burton. "From the Psi to Charm: The Experiments of 1975 and 1976". *Science* 17 Jun 1977.

Robinson, Arthur. "CERN Finds Evidence for Top Quark". *Science* 27 Jul 1984.

—. "Evidence for Free Quarks Won't Go Away". *Science* 6 Mar 1981.

—. "Particle Physics: New Evidence from

Germany for Fifth Quark". *Science* 2 Jun 1978.

Schwarz, John. "Dual-Resonance Models of Elementary Particles". *ScAm* Feb 1975.

Schwitters, Roy. "Fundamental Particles with Charm". *ScAm* Oct 1977.

CH 15. FIELD THEORIES FOR PARTICLES

Books and pamphlets:

—. *Fermilab Highlights*. Fermilab, 1983.

—. *High Energy Physics*. U. S. Dept. of Energy, 1979.

—. *Presenting CERN*. CERN, 1983.

Davies, Paul. *Superforce: The Search for a Grand Unified Theory of Nature*. Simon & Schuster, 1984.

Moriyasu, K. *An Elementary Primer for Gauge Theory*. World, 1984.

Southworth, Brian, and Fraser, Gordon. *Achievements with Antimatter*. Reprinted from *CERN Courier*, Nov 1983.

Articles:

—. "Recent Research Highlights at SLAC". SLAC, Apr 1984.

Barish, Barry. "Experiments with Neutrino Beams". *ScAm* Aug 1973.

Bloom, Elliott, & Feldman, Gary. "Quarkonium". *ScAm* May 1982.

Carrigan, Richard, & Trower, W. "Superheavy Magnetic Monopoles". *ScAm* Apr 1982.

Cline, David, et al. "The Detection of Neutral Weak Currents". *ScAm* Dec 1974.

Cline, David, et al. "The Search for Intermediate Vector Bosons". *ScAm* Mar 1982.

Cline, David, et al. "The Search for New Families of Elementary Particles". *ScAm* Jan 1976.

Crease, Robert, & Mann, Charles. "Gambling with the Future of Physics". *New York Times Magazine* 5 Dec 1982.

*Creutz, Michael. "High-Energy Physics". *Physics Today* May 1983.

Diebold, R. "The Desertron: Colliding Beams at 20 TeV". *Science* 7 Oct 1983.

Evans, John, & Steinberg, Richard. "Nucleon Stability: A Geochemical Test Independent of Decay Mode". *Science* 2 Sep 1977.

Ford, Kenneth. "Magnetic Monopoles". *ScAm* Dec 1963.

Georgi, Howard. "A Unified Theory of Elementary Particles and Forces". *ScAm* Apr 1981.

—. "Unified Theory of Elementary Particle Forces". *Physics Today* Sep 1980.

Glashow, Sheldon. "Toward a Unified Theory: Threads in a Tapestry". *Science* 19 Dec 1980.

Goldhaber, M. et al. "Is the Proton Stable?" *Science* 21 Nov 1980.

Harari, Haim. "The Structure of Quarks and Leptons". *ScAm* Apr 1983.

Hooft, Gerard t'. "Gauge Theories of Forces between Elementary Particles". *ScAm* Jun 1980.

*Hung, P. Q., & Quigg, C. "Intermediate Bosons: Weak Interaction Couriers". *Science* 12 Dec 1980.

Ishikawa, Kenzo. "Glueballs". *ScAm* Nov 1982.

Johnson, Kenneth. "The Bag Model of Quark Confinement". *ScAm* Jul 1979.

Lederman, Leon. "The Value of Fundamental Science". *ScAm* Nov 1984.

Losecco, J. M., et al. "The Search for Proton Decay." *ScAm* Jun 1985.

Quigg, Chris. "Elementary Particles and Forces". *ScAm* Apr 1985.

Rebbi, Claudio. "The Lattice Theory of Quark Confinement". *ScAm* Feb 1983.

—. "Solitons". *ScAm* Feb 1979.

Robinson, Arthur. "CERN Sets Intermediate Vector Boson Hunt". *Science* 10 Jul 1981.

—. "CERN Vector Boson Hunt Successful". *Science* 26 Aug 1983.

—. "Particle Theorists in a Quandary". *Science* 17 Sep 1982.

—. "Unexpected Zeta Particle Baffles Physicists". *Science* 31 Aug 1984.

Salam, Abdus. "Gauge Unification of Fundamental Forces". *Science* 14 Nov 1980.

Waldrop, Mitchell. "Blooms in the Desert?" *Science* 11 May 1984.

—. "Do Monopoles Catalyze Proton Decay?" *Science* 15 Oct 1982.

—. "Gambling on the Supercollider". *Science* 9 Sep 1983.

—. "In Search of the Magnetic Monopole". *Science* 11 Jun 1982.

—. "New Ways to Accelerate?" *Science* 17 Jun 1983.

—. "Something Strange from Cygnus X-3." *Science* 14 Jun 1985.

—. "The Supercollider, 1 Year Later". *Science* 8 Aug 1984.

Weinberg, Steven. "Conceptual Foundations of the Unified Theory of Weak and Electromagnetic Interactions". *Science* 12 Dec 1980.

—. "The Decay of the Proton". *ScAm* Jun 1981.

—. "Unified Theories of Elementary Particle Interactions". *ScAm* Jul 1974.

Wilson, Robert. "The Next Generation of Particle Accelerators". *ScAm* Jan 1980.

Science Fiction Novel
(at the frontiers of particle physics):

Preuss, Paul. *Broken Symmetries*. Simon & Schuster, 1983.

CH 16. PARTICLES AND GRAVITY, ASTROPHYSICS, AND COSMOLOGY

Books:

Alfven, Hannes. *Worlds-Antiworlds: Antimatter in Cosmology*. W. H. Freeman, 1966.

Alonso, J. et al. "Acceleration of Uranium at the Bevalac". *Science* 17 Sep 1982.

Bergmann, Peter. *The Riddle of Gravitation*. Scribner's, 1968.

Born, Max. *Einstein's Theory of Relativity*. Dover, 1965.

Einstein, Albert. *Relativity*. Crown, 1952.

Gamow, George. *Gravity*. Doubleday, 1962.

Geroch, Robert. *General Relativity from A to B*. Univ. of Chicago Press, 1978.

Kaufmann, William. *Black Holes and Warped Spacetime*. W. H. Freeman, 1979.

*Misner, Charles; Thorne, Kip; & Wheeler, John. *Gravitation*. W. H. Freeman, 1973.

Sciama, D. W. *The Physical Foundations of General Relativity*. Doubleday, 1969.

Sullivan, Walter. *Black Holes: The Edge of Space, The End of Time*. Warner, 1979.

Weinberg, Steven. *The First Three Minutes*. Basic Books, 1977.

Articles:

Alfven, Hannes. "Antimatter and Cosmology". *ScAm* Apr 1967.

Alvarez, Luis, et al. "The Search for Hidden Chambers in the Pyramids". *Science* 6 Feb 1970.

Bahcall, John. "Neutrinos from the Sun". *ScAm* Jul 1969.

Barrett, Louis. "Acoustic Detection of Cosmic-Ray Air Showers". *Science* 17 Nov 1978.

Barrow, John, & Silk, Joseph. "The Structure of the Early Universe". *ScAm* Apr 1980.

Bekenstein, Jacob. "Black-Hole Thermodynamics". *Physics Today* Jan 1980.

Benedetti, Sergio de. "The Mossbauer Effect". *ScAm* Apr 1960.

Bergmann, Peter. "Unitary Field Theories". *Physics Today* Mar 1979.

Bethe, Hans, & Brown, Gerald. "How a Supernova Explodes." *ScAm* May 1985.

Bethe, Hans, & Salpeter, E. E. "The 1983 Nobel Prize in Physics". *Science* 25 Nov 1983.

Broad, William. "Tracing the Skeins of Matter". *New York Times Magazine* 6 May 1984.

Burbidge, Geoffrey. "The Origin of Cosmic Rays". *ScAm* Aug 1966.

Callahan, J. J. "The Curvature of Space in a Finite Universe". *ScAm* Aug 1976.

de Vaucouleurs, G. "The Case of a Hierarchical Cosmology". *Science* 27 Feb 1970.

Dewitt, Bryce. "Quantum Gravity". *ScAm* Dec 1983.

Dicke, R. H. "The Eotvos Experiment". *ScAm* Dec 1961.

Dicus, Duane, et al. "The Future of the Universe". *ScAm* Mar 1983.

Dyson, Freeman. "Energy in the Universe". *ScAm* Sep 1971.

Flandern, Thomas van. "Is Gravity Getting Weaker?" *ScAm* Feb 1976.

Fowler, William. "The Quest for the Origin of the Elements". *Science* 23 Nov 1984.

Frautschi, Steven. "Entropy in an Expanding Universe". *Science* 13 Aug 1982.

Freedman, Daniel, & van Nieuwenhuizen, Peter. "Hidden Dimensions of Spacetime". *ScAm* Mar 1985.

—. "Supergravity and the Unification of the Laws of Physics". *ScAm* Feb 1978.

Gale, George. "The Anthropic Principle". *ScAm* Dec 1981.

Gamow, George. "Gravity". *ScAm* Mar 1961.

Ginzburg, V. "The Astrophysics of Cosmic Rays". *ScAm* Feb 1969.

Ginzburg, V. L. "Artificial Satellites and Relativity". *ScAm* May 1959.

Gott, J., et al. "Will the Universe Expand Forever?" *ScAm* Mar 1976.

Greenberger, Daniel, & Overhauser, Albert. "The Role of Gravity in Quantum Theory". *ScAm* May 1980.

Greiner, Walter, & Stocker, Horst. "Hot Nuclear Matter". *ScAm* Jan 1985.

Guth, Alan, & Steinhardt, Paul. "The Inflationary Universe". *ScAm* May 1984.

Hartline, Beverly. "In Search of Solar Neutrinos". *Science* 6 Apr 1979.

Harwood, Michael. "The Universe and Dr. Hawking". *New York Times Magazine* 23 Jan 1983.

Hawking, S. W. "The Quantum Mechanics of Black Holes". *ScAm* Jan 1977.

Herbst, William, & Assousa, George. "Supernovas and Star Formation". *ScAm* Aug 1979.

Lake, George. "Windows on a New Cosmology". *Science* 18 May 1984.

Learned, John, & Eichler, David. "The Deep-Sea Neutrino Telescope". *ScAm* Feb 1981.

Linsley, John. "The Highest Energy Cosmic Rays". *ScAm* Jul 1978.

Mackeown, P. Keven, & Weekes, Trevor. "Cosmic Rays from Cygnus X-3." *ScAm* Nov 1985.

Meier, David, & Sunyaev, Rashid. "Primeval Galaxies". *ScAm* Nov 1979.

Morrison, Philip. "Neutrino Astronomy". *ScAm* Aug 1962.

Muller, Richard. "Cosmic Background Radiation and the New Aether Drift". *ScAm* May 1978.

Pasachoff, Jay, & Fowler, William. "Deuterium in the Universe". *ScAm* May 1974.

Peebles, P., & Wilkinson, David. "The Primeval Fireball". *ScAm* Jul 1967.

Penrose, Roger. "Black Holes". *ScAm* May 1972.

Penzias, Arno. "The Origin of the Elements". *Science* 10 Aug 1979.

Pines, David. "Accreting Neutron Stars, Black Holes, and Degenerate Dwarf Stars". *Science* 8 Feb 1980.

Reines, Frederick, & Sellschop, J. "Neutrinos from the Atmosphere and Beyond". *ScAm* Feb 1966.

Robinson, Arthur. "Atomic Physics Tests Lorentz Invariance." *Science* 23 Aug 1985.

—. "New Test of Variable Gravitation Constant". *Science* 23 Dec 1983.

Rossi, Bruno. "High Energy Cosmic Rays". *ScAm* Nov 1959.

Ruffini, Remo, & Wheeler, John. "Introducing the Black Hole". *Physics Today* Jan 1971.

Schaefer, Bradley. "Gamma-Ray Bursters". *ScAm* Feb 1985.

Schramm, David. "The Age of the Elements". *ScAm* Jan 1974.

—. "The Early Universe and High-Energy Physics". *Physics Today* Apr 1983.

Schramm, D. N., & Clayton, R. N. "Did a Supernova Trigger the Formation of the Solar System?" *ScAm* Oct 1978.

Sullivan, Walter. "A Hole in the Sky". *New York Times Magazine* 14 Jul 1974.

—. "Mystery of Cosmic Ray Origin May Be Solved". *New York Times* 20 Jan 1985.

Thorne, Kip. "Gravitational Collapse". *ScAm* Nov 1967.

—. "The Search for Black Holes". *ScAm* Dec 1974.

Turner, Michael, & Schramm, David. "Cosmology and Elementary Particle Physics". *Physics Today* Sep 1979.

Waldrop, Mitchell. "Bubbles upon the River of Time". *Science* 26 Feb 1982.

—. "Inflation and the Arrow of Time". *Science* 25 Mar 1983.

—. "Inflation and the Mysteries of the Cosmos". *Science* 3 Jul 1981.

—. "The Large-Scale Structure of the Universe". *Science* 4 Mar 1983.

—. "Massive Neutrinos: Masters of the Universe?" *Science* 30 Jan 1981.

—. "Matter, Matter, Everywhere..." *Science* 20 Feb 1981.

—. "The New Inflationary Universe". *Science* 28 Jan 1983.

—. "New Light on Dark Matter?" *Science* 1 Jun 1984.

—. "String as a Theory of Everything." *Science* 20 Sep 1985.

—. "Supersymmetry and Supergravity". *Science* 29 Apr 1983.

—. "Why Do Galaxies Exist?" *Science* 24 May 1985.

—. "WIMPs, Cosmions, and Solar Neutrinos." *Science* 6 Sep 1985.

Weber, Joseph. "The Detection of Gravitational Waves". *ScAm* May 1971.

Weisberg, Joel, et al. "Gravitational Waves from an Orbiting Pulsar". *ScAm* Oct 1981.

Wilczek, Frank. "The Cosmic Asymmetry between Matter and Antimatter". *ScAm* Dec 1980.

Will, Clifford. "Gravitation Theory". *ScAm* Nov 1974.

Wilson, R. "The Cosmic Microwave Background Radiation". *Science* 31 Aug 1979.

Science Fiction Novel (combining ultrarelativistic speed and cosmology):

Anderson, Poul. *Tau Zero*. Doubleday, 1970.

Excellent PBS TV programs about particles and cosmology available on videotape:

"What Einstein Never Knew", NOVA, WGBH, 1985.

"The Creation of the Universe", Northstar, 1985.

Index

Index

Edited by Roland S. Phelps

Other Bestsellers From TAB

☐ **COMETS, METEORS AND ASTEROIDS—How They Affect Earth—Gibilisco**

What are comets and meteors? Where do they come from? Have comets affected the Earth's climate, even the evolution of life? What are the chances of a comet actually striking the Earth? And, does Halley's Comet pose a threat to Earth in 1986? The answers to these and many other fascinating questions about comets, meteors, asteroids, and other related space phenomena are here for the taking in this timely and informative selection. Exceptionally well illustrated, it includes a spectacular eight-page section of color photos taken in space. And the book itself is packed with little-known details and fascinating theories covering everything from the origins of the solar system to speculation on what may happen in the future. 224 pp., 148 illus. Plus 8 pages in 4-color.

Paper $12.95   **Book No. 1905**

☐ **TIME GATE: HURTLING BACKWARD THROUGH HISTORY—Pellegrino**

Taking a new approach to time travel, this totally fascinating history of life on Earth transports you backward from today's modern world through the very beginnings of man's existence. Interwoven with stories and anecdotes, and illustrated with exceptional drawings and photographs, this is history as it should always have been written! It will have you spellbound from first page to last! 288 pp., 142 illus. 7″ × 10″.

Paper $16.95   **Book No. 1863**

☐ **333 SCIENCE TRICKS AND EXPERIMENTS—Brown**

Here is a delightful collection of experiments and "tricks" that demonstrate a variety of well-known, and not so well-known, scientific principles and illusions. Find tricks based on inertia, momentum, and sound projects based on biology, water surface tension, gravity and centrifugal force, heat, and light. Every experiment is easy to understand and construct . . . using ordinary household items. 208 pp., 189 illus.

Paper $9.95   **Book No. 1825**

☐ **E = mc²: PICTURE BOOK OF RELATIVITY**

It sounds complicated, but with this exceptional and easy-to-follow handbook, *anyone* child or adult, can grasp the real meaning of Einstein's theories. It's an enlightening, delightfully illustrated look at relativity minus the difficult mathematical equations and confusing scientific jargon. You'll be able to clearly understand how he reached his conclusions and the experiments that proved his theories correct. 128 pp., 138 illus. 7″ × 10″.

Paper $9.95   **Book No. 1580**

☐ **THE PERSONAL ROBOT BOOK**

This state-of-the-art "buyer's guide" fills you in on all the details for buying or building your own *and even how to interface a robot with your personal computer!* Illustrated with dozens of actual photographs, it features details on all the newest models now available on the market. Ideal for the hobbyist who wants to get more involved in robotics without getting in over his head. 192 pp., 105 illus. 7″ × 10″.

Paper $12.95   **Hard $21.95**
Book No. 1896

☐ **ROBOTICS—Cardoza and Vik**

This comprehensive overview traces the historical progression of robotics and the enormous impact robots are making on our society. Includes a look at opportunities and a listing of schools and training programs. Plus, a large comprehensive glossary of robotics terms, a complete bibliography of helpful books, magazines, and information sources, and a listing of the robots now on the market including their manufacturers. 160 pp., 28 illus. 7″ × 10″.

Paper $10.95   **Hard $16.95**
Book No. 1858

☐ **VIOLENT WEATHER: HURRICANES, TORNADOES AND STORMS—Gibilisco**

What causes violent storms at sea? Hurricane force winds? Hail the size of grapefruit? Blinding snowstorms and tornadoes? The answers to all these and many more questions on the causes, effects, and ways to protect life and property from extremes in weather are here in this thoroughly fascinating study of how extremes in weather violence occur. 272 pp., 192 illus. 7″ × 10″.

Paper $13.95   **Book No. 1805**

☐ **HOW TO FORECAST WEATHER—Ramsey**

With the help of this excellent new book, you can learn to make your own accurate weather predictions for tomorrow, next week, even next month! Here are complete plans for setting up a simple, fully operational home weather station. Learn how to use maps and other information supplied by the Weather Service to make your forecasts more accurate for your own geographic area and more! Whether you want to calculate the wind chill factor or know whether it's going to be a good day to take the boat out, try your hand at forecasting the weather yourself—it's easy with this complete how-to guide! 224 pp., 213 illus. 7″ × 10″.

Hard $16.95   **Book No. 1568**

Other Bestsellers From TAB

☐ **BLACK HOLES, QUASARS AND OTHER MYSTERIES OF THE UNIVERSE**

Discover the awesome mysteries and complexities of our universe: black holes, quasars, pulsars, quarks, matter and antimatter, superluminal motions, creation theories, the search for intelligent life on other planets or in other galaxies, the possibilities to travel at speeds faster than light, and more. Includes 16 spectacular color photos of space phenomena! 208 pp., 125 illus. 8 color pages. 7″ × 10″.

Paper $13.50   **Book No. 1525**

☐ **ASTRONOMY AND TELESCOPES A BEGINNER'S HANDBOOK**

If you're fascinated with space phenomenon and the wonders of our universe, here's your chance to discover the solar system, the Milky Way and beyond . . . to follow the path of meteors, asteroids, and comets . . . even to build your own telescope to view the marvels of space! You'll get an amazing view of the brilliance of outer space in 16 pages of full-color photos! 192 pp., 194 illus. 7″ × 10″.

Paper $14.95   **Book No. 1419**

*Prices subject to change without notice.

Look for these and other TAB books at your local bookstore.

TAB BOOKS Inc.
P.O. Box 40
Blue Ridge Summit, PA 17214

Send for FREE TAB catalog describing over 900 current titles in print.

MAY 2 8 1986

LIBRARY